THE INSEAD GLOBAL MANAGEMENT SERIES: PROCESS REENGINEERING ORGANIZATIONAL CHANGE AND PERFORMANCE IMPROVEMENT

The INSEAD Global Management Series

PROCESS REENGINEERING, ORGANIZATIONAL CHANGE AND PERFORMANCE IMPROVEMENT

Soumitra Dutta *and* Jean-François Manzoni

McGraw-Hill Publishing Company

London · New York · St Louis · San Francisco · Auckland · Bogotá · Caracas
Lisbon · Madrid · Mexico · Milan · Montreal · New Delhi · Panama · Paris
San Juan · São Paulo · Singapore · Sydney · Tokyo · Toronto

Published by

McGraw-Hill Publishing Company

Shoppenhangers Road, Maidenhead, Berkshire, SL6 2QL, England
Telephone 01628 23432
Facsimile 01628 770224

British Library Cataloguing in Publication Data
The CIP data of this title is available from the British Library

Library of Congress Cataloging-in-publication Data
The CIP data of this title is available from the Library of Congress, Washington DC, USA

Further information on this and other McGraw-Hill titles is to be found at
http://www.mcgraw-hill.co.uk

Publisher: Alfred Waller
Desk Editor: Alastair Lindsay
Produced by PSP Publishing Services
Cover by: Hybert Design

Typeset by Mackreth Media Services, Hemel Hempstead
Printed and bound in Great Britain at the University Press, Cambridge

CONTENTS

PREFACE

This case and text book series was conceived to meet a need felt by business school faculty for a set of volumes comprising recent international cases which have shown to be effective by the hard test of actual classroom use. Although cases were the motivation, the series' editorial board insisted that there be an accompanying text that would guide the reader, whether professor, student, or just interested party, through the fundamental concepts the case authors wanted to illustrate. There is logic for this insistence.

Learning with cases is learning inductively. The student reads and thinks about one or more specific observation, participates in a class discussion, and then tries to draw some generalizations useful in a different setting. This is the classical method of science; the method of Ptolemy, Maxwell, and Newton. As is true with their work, however, even a large number of observations can prove the truth of any generalization. Many favourable observations under precisely defined and controlled conditions can only raise the researcher's confidence in his or her theory even if it is still unproved.

The business academic using cases, and the student learning from them, must appreciate early on that theirs is a field where large numbers of observations are impossible to make, and that conditions of observation cannot be controlled sufficiently to conduct anything close to experimental work. It might often seem to students that they are pushed to make generalizations from single cases. Worrisome in addition is the fact that the cases are not chosen randomly. They are typically selected to demonstrate a point or to suggest a generalization to the student. Rarely does a professor look for disconfirming evidence.

This is why business cases alone can rarely drive a convincing argument and why in this series the editorial board has sought to give considerable attention to accompanying text. The author's point must be woven of several fabrics: logic, examples (including brief reference to examples typically known to the reader), theories borrowed from formal disciplines such as economics, psychology, or sociology, and a draw on the reader's own experiences and intuition. The case itself is vital but not sufficient alone. It is perhaps the frame on which to weave the analysis.

We trust the reader will find the themes of these volumes well supported, and we thank the various authors for their enthusiasm and hard work.

H. Landis Gabel
Series Editor
10 September 1998

ACKNOWLEDGEMENTS

This book marks an important step in a journey that started about five years ago. Back then, the 'Business Process Reengineering (BPR) revolution' was in full swing! Hammer and Champy had just published their best selling book 'Reengineering the Corporation' and it seemed that every company we visited and every executive we met at or outside INSEAD, was in the middle of a reengineering effort, in some cases several such efforts. In the immense majority of cases, these BPR projects translated into significant downsizing efforts, which carried with them their share of human pain and drama, for those who stayed as well as those who left the organizations.

Back then, we were studying these cases from different points of view: Soumitra was examining the changes from a more technical, 'content of the change' point of view, while Jean-François was more interested in the change process and the impact of the change efforts on the 'people' within these organizations. Beyond our differences, however, we had two important views in common: we believed in the potential of BPR for significant performance improvement, but we experienced uneasiness over the frequency with which BPR efforts led to brutal changes and downsizing within organizations. Somehow, we felt that 'things could not continue this way', and that 'there had to be a better way' to manage these efforts. The basic idea of redesigning one's processes for greater effectiveness and efficiency was right, we felt, but it had to be done in a way that would integrate the human element and would position the BPR effort into a more global change effort.

Starting from this idea that 'there had to be a better way', we decided to join forces to study BSR-based change efforts more systematically, both to test our hunches and to identify some pitfalls and better practices. Over the last five years, this collaboration has given fruit to numerous pedagogical cases, articles and executive development programmes. Along the way, we have been encouraged by the publication of several articles and books emphasising the 'human side' of BPR and the need to anchor BPR projects into growth-oriented strategies.

We decided to prepare this book to help those who, like us, are interested in the use of process redesign as a major pillar of long term performance improvement: management instructors of course, who are responsible for putting together 'courses' on the subject and for educating people who will later manage BPR-like projects, but also managers who will find in the book both recent examples and a number of conceptual frameworks. We hope that the book's combination of chapters and cases will provide readers with a good road map into a complex issue.

Looking back at the work that led to this book, it is clear that we received lots of help along the way. Among our INSEAD colleagues, many thanks go to Deigan Morris for inviting us to join him in creating a new executive programme called 'Achieving Outstanding Performance: Integrative management of people, activities and processes', and for directing the programme since its first delivery in 1995. The programme gave us an invaluable opportunity to test and refine our ideas with executives, as well as excellent leads for research and case writing Arnoud de Meyer and Martine van den Poel, respectively Associate and Assistant Dean for Executive Education, encouraged the development of the AOP programme and helped us make it a success. Thanks are also due to Jean-Louis Barsoux, Arnoud de Meyer, James Teboul and Enver Yücesan who co-authored cases with us and/or gave us helpful comments on the book chapters.

Yves Doz and Landis Gabel, INSEAD's Associate Dean for Research & Development (1993–1995 and 1995–1998, respectively) encouraged our research and provided generous funding for most of the cases included in this book. Claude Michaud and CEDEP also provided funding for the Air France and British Airways cases included in chapters V and VII. The 'The Arthur D Little Fund for the Enrichment of the Learning Experience' also provided support for the second British Airways case and the two Taco Bell cases.

Outside INSEAD, we benefited greatly from the challenge and suggestions of hundreds of participants in executive programmes with whom we have discussed the material presented in the book: participants in the AOP programme, of course, but also in INSEAD's MBA programme and in scores of other executive programmes, both public and company-specific. We are also grateful for the help we received from managers who allowed us to study them and, in many cases, to write cases on their organizations.

We should also thank Alfred Waller and Alastair Lindsay, of McGraw Hill, and Pejay Belland at INSEAD, for helping us keep track of the goal in such a gentle and effective manner.

Last, but by no means least, we want to thank our respective families for their love and support. The material contained in this book has consumed much of our time over the last five years, some of which we might otherwise have devoted to family life.

Looking to the future, we continue to work in the areas discussed in the book. We hope you will find the book helpful and will welcome any comment or suggestion you might have.

Soumitra Dutta
Jean-François Manzoni   *Fontainebleau, August 1998*

To
Lourdes and Sarah,
Anne, Jean-Patrick, Adrien and Matthieu

Process Reengineering, Organizational Change and Performance Improvement

Chapter

1

Introduction

Today's managers are faced with the difficult challenge of continuing to improve over time the performance of the unit or company they are managing. It is no longer sufficient for managers to produce steady organizational performance most of the time, they must do so all the time; there is less and less patience for weak results. This challenge is all the more daunting because the competitive environment that managers are facing is also becoming increasingly demanding. As a result, companies have been bombarded over the last few years with advice on various ways of improving performance: horizontal organization, time-based competition, matrix structures, lean manufacturing, concurrent engineering, Activity-Based-Costing and Management, Business Process Reengineering, to name just a few of the most recent 'management innovations'.

As described in the management literature, each of these 'solutions' holds out the promise of dramatic change. Customers will be thrilled, employees will be happy with, and motivated by hassle-free jobs, and corporate profits will increase. The reality, however, has often proved disappointing. In too many companies, after long periods of work by task forces and significant investment in organizational change programmes, customers were less than delighted, employees became scared and overworked and improvements in the bottom line fell short of expectations.

We have observed three identifiable patterns of failure (see Figure 1.1): Some initiatives simply have no measurable impact; they receive a lot of attention for some period of time, they may even lead to massive plans and long lists of projects, but they die somewhere along the way, too early to have an impact. Such failures tend to leave employees frustrated at having spent so much time on an initiative that never really got off the ground, and more cynical about the chances of success of the *next* change initiative.

Others get off to a promising start, but quickly tail off; the improvement is not sustainable. In some cases, improvements are lost because the novelty and urgency effect wear off, people's attention gets diverted to another initiative and the previous priority starts receiving insufficient attention. In several cases, the improvement proves

Figure 1.1 Results often fall short of objectives

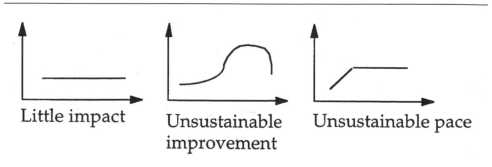

Little impact   Unsustainable   Unsustainable pace
                     improvement

unsustainable because it was achieved through effort and attention beyond the call of duty rather than by changing the way people worked, and it is simply a fact of life that employees cannot function at 110% of capacity for very long. Some companies also eliminated people without eliminating the work they were doing, thus leading to temporary profitability increases, soon eliminated by rehiring of staff. Sometimes the very people that were let go earlier were rehired as part-time consultants, at much higher daily rates.

Finally, some projects did provide performance improvement, but of the 'one-off' variety. That is, performance (measured through quality, customer satisfaction or profitability) increased rapidly and noticeably, but then seemed to hit a plateau. This pattern is not a complete failure, as the improvement itself is maintained over time, but it is not a complete success as the pace of improvement proves unsustainable.

Overall, we have seen few change initiatives that generated on-going improvements over time. The question then, is how can companies approach and manage the change process in a way that triggers sustainable improvement and a positive pace of improvement? That is the subject of this book.

The search for performance improvement

As organizations strive to be more competitive in today's challenging business environment, more of them are taking a radical look at the underlying business processes and systems which make them successful. As mentioned earlier, such efforts typically assume one or more of a number of labels, such as business process reengineering, employee empowerment, total quality and customer focus. The literature describes the core emphasis of these efforts in different ways:

> *The fundamental rethinking and radical redesign of business processes to achieve dramatic improvements in critical, contemporary measures of performance, such as cost, quality, service, and speed.*[1]

> *The analysis and redesign of business and manufacturing processes to eliminate that which adds no value.*[2]

> *A radically new process of organizational change that many companies are using to renew their commitment to customer service.*[3]

Beyond the specificities of each approach, some common aspects can be identified across these performance improvement efforts. For one, there is a shared emphasis on rethinking different aspects of the existing business behaviours:

---

[1] Hammer, M. and Champy, J., *Reengineering the Corporation: A Manifesto for Business Revolution*, Harper Business, 1993.
[2] Parker, K., 'Reengineering the Auto Industry', *Manufacturing Systems*, January 1993.
[3] Janson, R., 'How Reengineering Transforms Organizations to Satisfy Customers', *National Productivity Review*, Winter 1992–93, pp. 45–53.

It isn't about fixing anything, [it] means starting all over, starting from scratch.[4]

Secondly, much of the motivation for this rethinking seems to arise from observations that many current work practices are outdated and no longer suited to today's competitive situation or matched to the capabilities offered by current technology:

The problem appears to be . . . developed systems of production that were remarkably successful in their time but are no longer suited to a changed world.[5]

Also central to many performance improvement efforts is an emphasis on *processes*. The assumption is that a business can be defined as a set of interrelated 'processes' that are logically and continuously evolving to satisfy a set of common customer-oriented objectives. From this point of view, a process can be defined as:

- . . . a sequence of activities that fulfills the needs of an internal or external customer.[6]

- . . . a collection of activities that takes one or more kinds of input and creates an output that is of value to the customer.[7]

In addition, today's performance improvement programmes are themselves becoming a process; performance improvement is no longer looked at as a project only, but increasingly as an ongoing effort of *analysing and radically redefining the key processes* of a company with the ultimate goal of achieving and sustaining significant performance gains.

A framework for change

Change efforts tend to be massive undertakings and involve scores of different decisions and actions. To get a good handle over such complexity and to compare and contrast the efforts of different companies, we need some form of conceptual, or at least an organizing, framework. Many such frameworks designed to analyse businesses and define performance improvement programmes have been developed over the last few years, mostly by authors of management theory and within the major management consultancies. Two representative frameworks are the 7-S Model and the Business Integration Model.

Developed in the late 1970s, the 7-S Model emphasizes that, in order to understand the dynamics of organizational change and develop goals for a performance

---

[4] Hammer, M. and Champy, J., *op. cit.*
[5] Keith, G., *Reengineering History: An Analysis of Business Process Reengineering*, Anglo American Insurance Company Limited, October 1993.
[6] Harrison, D.B. and Pratt, M.D., 'A Methodology for Reengineering Businesses', *Planning Review*, March–April 1993, pp. 6–11.
[7] Hammer, M. and Champy, J., *op. cit.*

improvement programme, one needs to achieve consistency and balance between seven specific dimensions (7Ss).[8] The seven Ss are:

- *Strategy*: a coherent set of actions aimed at gaining a sustainable competitive advantage (and, as such, the approach to allocating resources).

- *Skills*: distinctive capabilities possessed by the organization as a whole as distinct from those of an individual.

- *Shared values*: ideas of what is right and desirable (in corporate and/or individual behaviour) as well as fundamental principles and concepts which are typical of the organization and common to most of its members.

- *Structure*: the organization chart and related concepts that indicate who reports to whom and how tasks are both divided up and integrated (reporting relations and management responsibilities).

- *Systems*: the processes and procedures through which things get done.

- *Staff*: the people in the organization, considered in terms of corporate demographics (not individual personalities), i.e. their skills and abilities.

- *Style*: the way managers collectively behave with respect to use of time, attention and symbolic actions.

Another model worth mentioning is Andersen Consulting's Business Integration Model, which is based on the premise that business performance derives from the alignment of a company's people, processes and technology with its strategy (see Figure 1.2). As a result, the model suggests that a consistent and comprehensive organizational change programme should incorporate, independently and collectively, the following four aspects of the organization:

- *Strategy*: establish a customer-focused strategic vision that will optimize long-term success.

- *People*: organize, motivate and empower people to succeed.

- *Business processes*: redefine and streamline business processes to implement strategic vision and to achieve maximum effectiveness and efficiency of all resources.

- *Technology*: apply appropriate technology to support streamlined process, to provide information and tools to support the entire workforce and to enhance customer/supplier relationships.

A key similarity between the 7-S and the Business Integration Models is their joint emphasis that performance improvement programmes succeed only when they

---

[8] Pascale, R.T. and Athos, A.G., *The Art of Japanese Management*, Simon & Schuster, New York, 1981.

Figure 1.2 Andersen Consulting's Business Integration Model

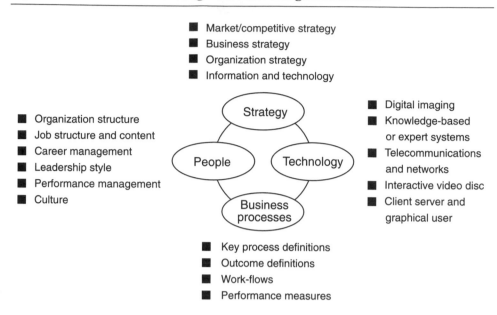

- Market/competitive strategy
- Business strategy
- Organization strategy
- Information and technology

- Organization structure
- Job structure and content
- Career management
- Leadership style
- Performance management
- Culture

- Digital imaging
- Knowledge-based or expert systems
- Telecommunications and networks
- Interactive video disc
- Client server and graphical user

- Key process definitions
- Outcome definitions
- Work-flows
- Performance measures

focus on multiple dimensions of the organization. This observation is one of the foundations of this book and will be a recurrent theme in the following case studies and chapters.

Searching for Performance Improvement

Our own framework builds on the two models presented above. It starts with the observation that every organization is influenced by a number of contextual factors, both external and internal. External elements include the economy, the industry, the competition, the marketplace, customer demands, etc.; internal elements include historical progression of the organization, results of previous change efforts, cultural aspects, etc. This *context* often highlights an organization's need for performance improvement, as well as the constraints it may face in undertaking such change.

The objective(s) of performance improvement programmes may also differ from company to company: cost for some, speed or quality for others. It is also becoming common for organizations to include 'non-traditional' goals, such as increasing creativity, in their performance improvement programmes.

Despite differences in organizational contexts and performance improvement goals, our research and the literature show that successful performance improvement programmes are characterized by a relentless and dedicated focus on rethinking and revitalizing multiple aspects of the organization:

- *Culture and people*, defined as the shared values and experiences and common goals that '. . . a group learns over a period of time as that group solves its problems of survival in an external environment and its problems of internal integration'.[9]

- *Processes*, defined as the sequence of activities that fulfills the needs of an internal or external customer,[10] and more generally the way work is done within the organization and the way the change project is actually realized.

- *Structure and systems*, which define who communicates with whom and how, as well as the degree of individual or collective responsibility and accountability; they can enable or prevent the necessary communication, knowledge transfer and customer contact.

- *Technology*, which plays a critical role in the generation, transfer and management of information.

Table 1.1 depicts in more detail the different aspects of these four sets of dimensions.

Over the last years, we have studied several performance improvement programmes in corporations across the globe. Most of them started with ambitious goals of seeking radical performance improvements. Some programmes were very successful and achieved the desired goals, other were less successful. Several of these are described in the case studies included in this book and discussed in more detail in the following chapters. To develop the framework introduced above, we summarize below some of the key lessons from performance improvement programmes we have witnessed in our research and consulting.

Culture and people

Change in the culture of the company is a common theme in all performance improvement efforts. Some organizations such as Taco Bell (described in Case 2.1) underwent a radical shift in their culture. Such a shift was intimately related to the organization's rethinking of its fundamental business model. Taco Bell moved from considering itself as a manufacturer of Mexican-American fast food to being a retailer of food. The shift in its basic business paradigm (from 'manufacturer' to 'retailer') and its product (from the 'Mexican-American fast food' niche to 'food in general') served to drive through a series of multidimensional changes within the company which radically changed its culture. For example, Taco Bell restaurants went from being points of manufacturing to points of distribution, and its managers evolved from a procedure-driven control mode to a more market-driven, entrepreneurial management style.

While Taco Bell's culture change was driven by a crisis situation, the need for a radical shift in culture and values was less explicit in many other organizations, especially those which were starting from a healthy situation. In fact, companies such as

---

[9] Schein, E.H., 'Organizational Culture', *American Psychologist*, February 1990, pp. 109–119.
[10] Harrison, D.B. and Pratt, M.D., *op. cit.*

Hallmark explicitly reaffirmed some existing cultural values and norms which would *not* change during the performance improvement programme. This reaffirmation was seen as providing a valuable bedrock of stability during the change efforts.

Regardless of the degree of cultural shift, some aspects are common to most organizations. First, there is a need for flexibility and adaptability within the

Table 1.1 Elements of the multiple dimensions

Culture and people	Process	Structure and systems	Technology
Shared values and experiences and common goals of a group of people	How and when actions are implemented	Who communicates with whom and how; responsibility and accountability	How the organization uses technology to support itself
■ level of trust	■ control methods	■ degree of delegation	■ standardization
■ incentive programmes	■ performance objectives	■ career path definition	■ degree of automation
■ amount of information sharing	■ training and education	■ existence and formality of boundaries	■ ease of information sharing
■ degree of formality	■ rules and procedures	■ ties to suppliers and other external firms	■ availability of tools
■ sense of urgency	■ team vs. individual	■ decision-making process	■ maintainability
■ team vs. individual orientation	■ degree of process ownership	■ hierarchy vs. network vs. matrix, etc.	■ speed of development
■ degree of empowerment and autonomy	■ abrupt change vs. pilots	■ degree of empowerment and autonomy	■ effectiveness
■ customer orientation	■ focus on and responsibility for customer	■ formality of rules and communication	■ efficiency
■ proven vs. pioneer mentality	■ task vs. process orientation	■ physical layout and geography	■ technological advancement
■ quality focus	■ problem solving process	■ integration of functions	■ technology readiness
■ ability/willingness to change	■ integration of functions	■ clarity of mission/ objectives	
■ sense of ownership			
■ clarity of mission/objectives			
■ leadership			
■ looking to external sources			

organization, especially for those that are in a relatively good condition at the start of the change programme. In this regard, Hallmark was successful by introducing a theme called 'The Journey', through which all people in the organization were made to understand that reengineering/change was necessary, that it was not a planned process and that it was a never-ending one. They challenged employees to go beyond the rules and dare to innovate.

Secondly, coherent changes to incentive programmes are often needed to modify the organization's culture. Examples include: Nissan, which changed its promotion/pay policies from a seniority-based one to one based on performance to promote innovative ideas; Quantum, which began to link performance assessment to teams instead of to individuals; and Reuters, which began to relate its commission structure to the successful instalment of an order instead of upon order-taking.

Thirdly, the customer is frequently a focal point driving the cultural change process. The customer is seen as the central 'stake in the ground' around which the cultural changes can coalesce. Nissan issued a statement of corporate philosophy concerning the need to be market-driven, to anticipate trends, to understand and be sensitive to customer needs, and to think globally. Some organizations realized that they did not really know their customer's true needs. The conscious discovery of those requirements served to galvanize the culture shift within the organization. For example, Taco Bell discovered that the true need of the customer was simply 'good, hot, clean food' and not other things such as large restaurants, which they had assumed to be of value to the customer.

Fourthly, companies emphasize the need to clarify corporate goals and recognize the need for radical change. After its flotation on the stock market, the Trustees Savings Bank (see Case 1.2) had set the goal to become the fifth largest clearing bank in the United Kingdom. This implied a move away from its core customer base of downmarket retail customers to upmarket retail and corporate customers. This decision accelerated TSB's crisis as its cost/income ratio increased significantly. TSB then reset the corporate mission, 'to become the leading retailer of financial products in the UK', and used this goal to drive through changes.

Clarifying corporate goals also often implies setting a stretch target for the organization. TSB moved its goal from becoming the fifth largest UK clearing bank to being the leading retailer of financial products in the United Kingdom. Taco Bell's corporate goal evolved from being number one in the niche of Mexican-American fast food to being number one in the entire fast-food industry to being number one in the 'share of the stomach'! Stretch goals motivate the cultural shift within the organization and highlight the need for radical performance improvement.

Process

A focus on identifying and improving processes is a pervasive feature of performance improvement programmes. However, the scope and maturity of the business process architectures and the nature of changes within processes varies within organizations.

While the need for organizations to focus on their core processes has been recognized in the literature (see Chapter 2), few organizations have explicitly focused on identifying their core processes and redefining their businesses in terms of these core processes. The definition of these, and an appropriate business process architecture, is a non-trivial, evolutionary procedure requiring a certain level of organizational maturity in process knowledge and awareness.

For example, when Xerox first implemented Leadership Through Quality in the mid-1980s, it focused on process identification and improvement within tasks. As employees became familiar with work processes, Xerox applied it to large-scale, functional processes. Later Xerox moved into cross-functional and cross-organizational processes. This progression in understanding its processes enabled it to define its core business process architecture in the early 1990s, which consisted at a macrolevel of three core processes: the customer interface process, the logistics process and the product delivery process. These core processes defined an integrated process architecture for the organization, which together with their 76 subprocesses served as the template for Xerox's change efforts. While the process architecture has evolved since its inception (see Case 6.1, 'Xerox (France) 2000'), it has continually been used to define new management roles and responsibilities, such as process sponsors and process owners. The core processes are also seen to provide the context for empowering people within the organization through cross-functional process teams and facilitating the cultural change.

Most organizations are not as thorough or systematic as Xerox with the definition of their business process architectures. Typically, they focus on identifying and improving one or two key high-leverage processes. For example, Hallmark chose to improve the process of new product introduction. Improvements in this process through the reduction of non-value-adding steps and the creation of cross-functional work teams helped it to slice from around three years to less than a year the time to market for new products.

The extent of change within processes varies from streamlining (incremental improvements) to the creation of completely new processes (process innovation). Bell Atlantic, for example, streamlined its process for fulfilling a connection request and hooking-up service, thus reducing the time taken from 20 days (involving 13 hand-offs between work groups) down to three days. In contrast, TSB faced the problem of trying to increase sales through branches in the absence of a well-established sales process. Thus, TSB had to design, pilot and test a completely new sales processes within the branches.

Both operational and management processes are addressed in companies. In its shift from a manufacture to a retailer, Taco Bell moved major parts of the kitchen out of each restaurant and consolidated them in a central location to take advantage of economies of scale in the production process. Complementing this redesign of the operational production process, Taco Bell also changed its management process by delayering the organization, changing the roles of restaurant managers to one of general managers and introducing new market-oriented management positions in its hierarchy. TSB made similar changes in its operational and management processes. It moved major parts of its back office processes away from branches to centralized customer service centres. Simultaneously, TSB also restructured the organization significantly, and redefined roles and responsibilities at different levels of the management hierarchy.

Regardless of the starting state of the organization and the nature of its performance improvement efforts, processes are useful for focusing the minds and energies of the organization on the customer. The redesign of appropriate processes can yield significant enhancements in delivered customer value and increase the overall efficiency and effectiveness of the organization.

Structure and systems

Significant changes in structure and systems accompany most performance improvement efforts. A common theme in companies' experiences is an emphasis on cross-functional work teams. Many organizations such as Nissan, Hallmark, Bell Atlantic, Reuters and Capital Holding formed cross-functional teams to help redesign processes. The cross-functional nature of teams helped to increase intraprocess awareness and ownership for the change efforts. The emphasis on cross-functional teams requires a significant change in the nature of skills, education and performance packages. Nissan, Reuters and other organizations all reworked their career systems to be consistent with these new roles and responsibilities. Significant improvements in time, costs and quality also resulted from cross-functional collaboration. As mentioned earlier, at Hallmark the time taken to create and market new products was reduced from between two to three years to less than a year.

Cross-functional teams in many organizations are formed for a specific period of time to focus on a particular change effort. However, some organizations such as Xerox and Quantum emphasized the formation of permanent cross-functional business teams taking end-to-end responsibility for specific business processes (see Chapter 6). These changes were often dramatic. In most examples, they implied a move away from a classical 'command and control' organization consisting of discreet functions (such as sales, marketing, service, manufacturing, personnel and finance) organized on a formal hierarchical basis, to a more cross-functional and participative organization where team orientation and self-managed work groups became commonplace. The formation of such process-based structures (e.g. based on cross-functional business teams taking end-to-end responsibility for processes) also raises important questions regarding the allocation of responsibilities between the new process-based structure and the traditional functional departments.

Complementing the move towards cross-functional teams, is a tendency to move towards flatter organizations with a larger span of control at each level. In Phillips 66, the CEO had nine people reporting to him in the new structure as compared to two previously. In Taco Bell, the span of control for district managers escalated tenfold in the new structure! Delayering and the increase in the span of control requires a corresponding change in the management style.

Job redefinition is an integral part of structural changes in all companies. District managers in Taco Bell could no longer afford to oversee closely (and sometimes interfere in) the functioning of 50 restaurants, while they could when they had five restaurants reporting to them. Their role shifted from one of controlling to coaching restaurant managers, and from enforcing procedures to focusing on market development in their own regions. As a result, their titles were changed from district managers to market managers to reflect the new emphases in their job content. In addition, restaurant managers had to be empowered to make all general management decisions regarding the functioning of their restaurants.

Consistent and effective communication of plans and top management commitment are very important. Personal commitment and communication, including visits by the company president, helped to communicate the right messages throughout Nissan. Hallmark successfully used its 'The Journey' campaign to communicate plans and was also aware of the need to communicate what would *not* change during the change efforts.

Information technology

Due to the rapid improvements in information technology (IT) over the last decade (see Chapter 3), IT forms a core component of the performance improvement programmes of companies. Today, corporations recognize and emphasize the need for business objectives to drive IT requirements and tailor solutions to business goals and desired changes. Simply automating existing tasks with new IT solutions is not seen as adding substantial long-term value.

In many organizations, information systems supporting the existing work processes are fragmented. For example, the process for completing a new service request within Bell Atlantic required transfers across 27 different IT systems. Thus the change efforts required the definition and implementation of new systems to support the new work processes.

Information technology makes possible the change efforts in many organizations. In Phillips 66, the CEO provided the leadership for the creation of a new executive information system to enable himself and his senior management team to manage the expanded span of control. Taco Bell created a new IT system called TACO (Total Automation of Company Operations) to enable the restaurant managers to manage the restaurants effectively in their new jobs. Hallmark developed a new IT system to obtain more accurate marketing information from the retail stores and feed the information in real time both to management and the stores. Frito-Lay pioneered a system based on hand-held computers to drive through a micromarketing strategy. TSB developed an expert system to help overcome the skill deficiencies of front-line branch sales staff and drive through an increase in branch sales effectiveness of over 100%.

The impact of IT in the change efforts can be characterized as organizationally neutral. In some organizations, such as Taco Bell, it helped to empower managers and upgrade jobs. In some other organizations, such as TSB, it helped to automate and move decisions out of bank branches and thus downgrade the jobs of branch managers. Due to this organizational neutrality, it is important to have a strong business leadership and an effective partnership between the business and IT divisions to ensure that IT does have the desired business impact. Most successful implementations of IT such as in Frito-Lay and Phillips 66 were characterized by such business leadership and organizational partnership between the IT and business divisions.

Looking forward

Commonalities in different performance improvement efforts are summarized in Table 1.2. We have observed that companies starting from a crisis situation (such as Taco Bell, TSB and Philips 66) tend to exhibit a stronger reorientation towards the customer with increased redefinition of jobs and an emphasis on cross-functional teams and empowerment. This perhaps reflects the urgent need to make radical changes in such organizations. In contrast, successful companies embarking on performance improvement projects (such as Hallmark or Singapore Airlines) seem to emphasize the need for adaptability to a greater extent. The focus on processes, pilot projects, top

Table 1.2 Summary of commonalities

Culture and people	Process	Structure and systems	Technology
■ Recognition of the need for flexibility and adaptability	■ Identify core processes	■ Cross-functional teams	■ Recognition of the need for business objectives to drive IT requirements
■ Coherent changes to incentive programmes	■ Define business process architecture	■ Process-based structures	
	■ Seek improvements in high leverage processes	■ Relation between functional and process-based structures	■ Specific solutions according to goals and changes implemented
■ Customer as focal point of cultural change		■ Flatter hierarchies	
■ Clarification of corporate goals	■ Process streamlining and process innovation	■ Redefinition of jobs	■ Integration of fragmented IT systems
		■ Empowerment and delegation	
■ Recognition of the importance of change	■ Pilot project approach	■ Communication of plans	■ Organizationally neutral
	■ Business and management processes	■ Top management commitment	■ Business leadership
	■ Focus on customer		■ Effective partnership between business staff and IT specialists

management commitment and effective communication of plans seems to be equally high in all companies.

As stated earlier, it is our observation that successful performance improvement programmes are characterized by simultaneous coordinated changes in multiple dimensions of the organization. Success rests on the ability to create conditions that motivate individuals to behave in a desired manner. Such conditions can only be created by working simultaneously on all dimensions shown in Figure 1.3.

Implementing a performance improvement programme within a large corporation is no easy task, especially when one realizes that the changes have to be effected without disrupting business. It is not possible to 'close' a corporation for one year, 'change' and then 'reopen'! This is the acid test for managers intent on successfully leading and managing performance improvement efforts. Real-life heroes of performance improvement efforts liken their efforts to 'changing the wheels of a car while driving down the freeway at 90mph'!

Each performance improvement programme is rich, complex and unique. That is the reason we have chosen to include in this book several detailed case studies from a number of corporations across the globe and across sectors. To do justice to the richness of the change stories, we have frequently included several case studies on the same company—each case shedding light on one or more aspects of the change programme. We hope that, collectively, they will give the reader a better feel for the 'blood, sweat and tears' that lie behind each of these company experiences.

For ease of presentation, we have structured the remainder of this book into sections that mirror the four boxes of Figure 1.3. Each chapter contains an introductory section and a set of cases. The purpose of these sections is to provide sufficient contextual and theoretical information related to the topic of the chapter. The cases are

Figure 1.3 Behaviour is shaped by multiple aspects

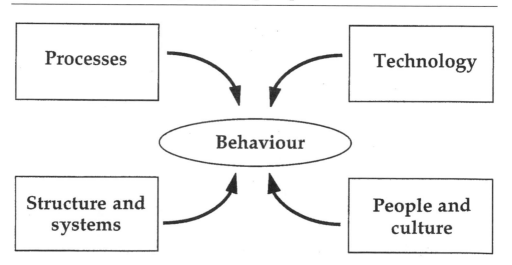

linked to the introductory sections, and serve to provide real-life illustrations of the concepts discussed.

This first chapter concludes with cases on two companies that have often been mentioned above:

Taco Bell (A): Reengineering the Business (1983–95)
Transforming TSB Group (A): Industry and Company Background
Transforming TSB Group (B): The First Wave (1989–91)

Chapter 2 sheds more light on the process redesign dimension of performance improvement programmes and focuses particularly on the notion of core process. This theme is exemplified by three cases:

Rank Xerox France (A): The logistics process
Rank Xerox France (B): Designing the integrated supply chain
Transforming TSB Group (D): The second wave (1992–95)—Focusing on Quality and Processes

Chapter 3 focuses on the role of IT in enabling new ways of working and improving performance. It is illustrated with three case studies:

Knowledge Management at Arthur Andersen (Denmark): Building assets in Real Time and in Virtual Space
Aligning IT with the Business: Banco Comercial Português and Continental Bank
Vandelay Industries, Inc.

Chapter 4 discusses the role of systems, particularly the systems involved with measuring, evaluating and reward employee performance, and their relationship with organizational structure. It is illustrated by four cases:

Friends Provident: Reengineering Customer Services
A Measure of Delight (A): The Pursuit of Quality at AT&T Universal Card Services
Mobil USM&R (A): Linking the Balanced Scorecard
Mobil USM&R (B): New England Sales and Distribution

Chapter 5 focuses on resistance to change, modifying the culture of an organization and management of the change process itself, both at the project level and at the large-scale performance improvement programme level. It is illustrated by four cases:

Transforming TSB Group (C): Managing change teams
Resistance to Change at Air France: Bernard Attali's experience
Pulling Out of Its Dive: Air France under Christian Blanc
The Making of 'The World's Favourite Airline': British Airways (1980–93)

Chapter 6 discusses how organizations can enhance process orientation within their organizations by changing their structures and systems. Cases linked to this topic are:

Rank Xerox (France) 2000: A Race Without a Finish Line
Quantum Corporation: Business and Product Teams
Revolution at Oticon A/S: The Spaghetti Organization

Finally, Chapter 7 discusses the difficulty of maintaining success over long periods of time. Even when the change process is well managed, staying at the top and maintaining peak performance may be more difficult than getting there in the first place! It is only fitting that the cases illustrating this issue pertain to two of the success stories presented earlier in the book:

Taco Bell (B): Reengineering the Business (1995–97)
Remaining The World's Favourite Airline: British Airways (1993–97)

Case 1.1

This case was written by Soumitra Dutta, Professor at INSEAD and Niklas Moe, MBA, student at the Haas School of Business at the University of California at Berkeley. It is intended to be used as a basis for class discussion, rather than to illustrate either effective or ineffective handling of an administrative situation.*

Taco Bell: Reengineering the Business (A) (1983–95)

INTRODUCTION

Staying in tune with the world around us . . . and accepting and mastering change within our own organizations . . . enables us to not merely survive . . . but thrive in today's competitive marketplace.[1]

John Martin, the CEO and chairman of Taco Bell, and arguably one of the most influential figures in the fast-food industry, could take justifiable pride in his achievements at Taco Bell over the last decade. His value-oriented reengineering of Taco Bell had marked an era in fast-food. He had revolutionized the fast-food business by selling tacos profitably for as little as 59 cents. Since joining the firm in 1983, Martin had led the chain through a complete transformation from a dull regional player to a major national force and trendsetter. Under Martin, Taco Bell had grown from a chain of 1500 restaurants and $500 million in annual sales to more than 20 000 feeding locations worldwide, and volume approaching $4.7 billion in 1995 (see Exhibit CS1.1.1). Martin's achievements had not been unnoticed. His boss, PepsiCo's CEO and chairman, Calloway, once declared Martin to be a *bona fide* 'oracle' among food service leaders. Martin also received the first ever *Nation's Restaurant News* Innovator Award for his contribution to the food industry in 1994.

> *For us, the process of reengineering has been like a voyage of discovery—a voyage we have been on for more than a decade, and one that we realize will continue as long as Taco Bell is in the business of serving customers.[2]*

Martin knew that the reengineering process would never stop, but he worried about whether Taco Bell was still in tune with the world around it. After several years of runaway growth and

---

*Copyright © 1997 INSEAD, Fontainebleau, France.
[1] Mastering Change, Business Forum, University of California at Irvine, 18 December 1992.
[2] Hammer, M. and Champy, J. *Reengineering the Corporation*, Harper Business, 1993, p. 171.

increasing same-store sales (see Exhibits CS1.1.2 and CS1.1.3), Taco Bell was experiencing stagnant sales and decreased same-store sales in 1995. And prospects in the immediate future did not look very bright either.

CATCHING UP (MID 1980s)

It may be somewhat surprising, but back in the early '80s, Taco Bell was badly broken. You see, it was very much like many other companies in the fast-food business. We were a top-down, 'command-and-control' organization with multiple layers of management—each concerned primarily with keeping tabs on the layers below them.

The company was control-driven, control-obsessed, really—with operational binders for everything, including handbooks to explain other handbooks. We strove for bigger, better and more complex in just about everything we did.

If it was simple, we made it hard. If it was complex, we made it more so. And the irony was that while labouring in the name of improvement, all we were doing was making things slower and more costly for the customer.[3]

Kenneth Stevens, Taco Bell's President, was describing the drive for change within Taco Bell in the early 1980s. Taco Bell's financial situation was also critical. In real terms, its cumulative growth since PepsiCo's acquisition in 1978, had been a negative 16% compared to the industry's positive 6%.

Taco Bell's production systems were designed to minimize the occurrence of errors, but at the same time, the design eliminated all opportunity for crew members to be creative or give feedback. Operational responsibility was placed at the restaurant manager (RM) level. The responsibility of the RM was to meet company standards regarding restaurant and crew cleanliness, timely delivery of food, food quality, cost of food, and cost of labour. As a result of the inconsistent quality of food delivered and the inconsistent preparation process, a significant part of the RM's daily job was focused on product preparation. The average monthly salary of a RM was around $2000 with a bonus of around 10%. RM turnover was around 50%.

The RMs reported to district managers (DM). Typically a DM oversaw six RMs and acted as a policeman in that the DM pointed out problems in restaurants and audited the financial results, leaving little time for the coaching or development of RMs. The average monthly salary for a DM was around $3000 with a bonus of around 15%.

The production process was labour intensive, low risk and used a low level of technology. For example, a manual process was used for communicating orders between the cashier and the kitchen. The kitchen was the primary part of each restaurant, with a typical design layout being 70% kitchen and 30% dining area.

John Martin described the challenge facing Taco Bell:

Our biggest problem was that we didn't know what we were. We thought maybe we were in the Mexican food business. We thought maybe we were in the fast-food

[3] 'Reengineering for your customers: How Taco-Bell did it', *Restaurants USA*, June–July 1994, pp. 36–38.

business . . . The reality was, we are in the fast-food business, and by not understanding who we were, who our potential customer was, we were just slightly missing the mark.[4]

In 1983, John Martin initiated a series of changes within the Taco Bell organization. He believed these changes were crucial for Taco Bell to remain competitive with other major fast-food chains. Among other changes, he:

- replaced the old 1600 ft^2 mission-style restaurants with new, more modern and larger units
- introduced new signage and a new logo
- gave the crew more comfortable, modern looking uniforms
- emphasized additional training of restaurant crew and managers
- increased seating capacity in the restaurants
- introduced new technology in the restaurants. Electronic point-of-sale (cash registers) were tied to television monitors on the assembly line indicating what had been ordered
- introduced new products such as nachos and taco salad
- increased the pace of growth. Prior to John Martin the growth rate had been fewer than 100 units a year. During the period 1983 through 1988 the growth increased to average about 250 stores a year.

DELIVERING VALUE (1989–91)

At the end of 1988, Taco Bell was still a fairly small player in the fast-food business. John Martin realized that in order to fulfil the company's growth vision, Taco Bell had to come up with something different from the bloated prices and swollen menus that had dimmed fast-food's traditional attributes of affordability and speed.

Kenneth Stevens described the soul-searching within Taco Bell in the late 1980s:

In the ever-increasing effort to micro-manage virtually every aspect of the business, we forgot to ask a very basic question: 'What in the heck do our customers think about all of this?' Did they care that we managed to turn the relatively simple business of fast-food into rocket science, all under the presumption that this was good for them? No.

What they cared about was that a family of four was now paying upwards of $20 to eat at a typical fast-food restaurant. And they weren't happy about it. After a lot of market research, we finally woke up and began listening to what our customers wanted. And what they wanted was real value: better quality and more quantity, all at a lower price. And that's where our initial reengineering efforts began. It began with reengineering for value.[5]

---

[4] John Martin as quoted in 'Taco Bell Corp.', Harvard Business School Case Study, No. 9-692-058.
[5] 'Reengineering for your customers: How Taco Bell did it', *Restaurants USA*, June–July 1994, pp. 36–38.

Stevens continued:

> *We started by taking a hard look at what customers got for every dollar they slid across the counter. And what they were getting at the time was essentially 29 cents' worth of food and packaging. The other 71% went to things the customer didn't care about: operations, labour, rent, marketing and advertising. So we went to work on the remaining 71 cents and began reengineering literally every facet of the business that would give our customers more in return for the dollar they gave us.*
>
> *This philosophy was exactly the opposite of what everyone else in the industry was doing. The industry's goal was to reduce the cost percentage of food and packaging that they returned to the customer. Our goal was to increase it—to give the customer more.*

John Martin commented that the reactions of industry observers to Taco Bells' focus on value were quite critical:

> *When we introduced a taco for 49 cents, . . . a lot of people in our business thought we were crazy. When we moved again in 1990, with a 59, 79 and 99 cent menu the cries got even louder. People said we were ruining the business. That we were being short-sighted. That we were taking a huge risk that would leave us in the fast-food graveyard. What they didn't understand was that our move to value wasn't a short term proposition—it was a long term commitment—it was imaginative.*
>
> *It wasn't even a tremendous risk because we knew we were right because our customers told us so—we never lost sight of what the customer wanted. We listened and kept their needs firmly in mind. And we put ourselves into their shoes.*

Normally in the fast-food industry, quality and price were seen as contradictory. John Martin believed that if Taco Bell were to deliver value, it would have to provide both price and quality to customers. However, this would need a radical shift in the way it did business. Taco Bell had to change its fundamental organizational principles and management outlook if it were to succeed. Martin elaborated:

> *The (monetary part of the) value concept depended on a strategic reallocation of overhead, not mere discounting—what you have to do is to get in and change the way you're doing business.*[6]

Cost-cutting approaches in the fast-food industry traditionally focused on lowering cost of goods sold, i.e. through cutting ingredient costs, quality and direct labour. Taco Bell believed that this approach was not consistent with the notion of customer value. Besides, Taco Bell's margins were stretched even before the value concept was introduced and improved sales volume alone would not be the answer. Instead, Taco Bell thought outside the box and decided to create a fundamental transformation process to place resources in the hands of managers closest to the customer, eliminate several middle management layers within the organization, and introduce

---

[6] Martin, R., 'John Martin', *Nation's Restaurant News*, January 1995, p. 137.

technological innovations both at the food preparation level and in cross-functional information systems.

A RETAILING COMPANY

Instead of being a manufacturing company, we've turned things around to becoming more of a customer-focused retailing company.[7]

Kenneth Stevens described the mindset driving the dramatic changes at Taco Bell. Central to this was the concept of K-Minus. Stevens continued:

> The floor space in a typical Taco Bell restaurant was comprised of 70% kitchen and 30% customer seating area. Like everyone else in the fast-food business, we had complicated our operations to the point where our internal needs were pushing the customer right out the door.

> But we've turned it around. Our newer restaurants average only 30% kitchen and 70% customer area. And the kitchen area is getting smaller all the time. Thanks to new processes and new technology, instead of being in the back slicing, dicing and chopping, our crew people are now out in front serving customers.

K-Minus required a simplification of the food preparation process. Taco Bell solved the problem by having the ingredients delivered precooked to the restaurants—the kitchen was used as a heating and assembly area. This also increased the consistency of food preparation across restaurants.

ORGANIZATIONAL CHANGES

Consistent with the strategy to empower managers close to customers, Taco Bell expected a new behaviour from these managers. To signal the shift in tasks and expectations strongly, the restaurant manager position was renamed restaurant general manager (RGM). The role of the RGM was broadened—the RGM needed to understand functional statements, make operative decisions, take initiative, and interact intelligently with customers.

John Martin believed that restaurants could operate by themselves with help from technological innovations, active communication of the new expectations, redesign of the compensation package, and adequate training. Taco Bell changed how they recruited for the RGM position by focusing on people with general management potential and also technological awareness and curiosity about what technology could accomplish.[8] Once hired, the new RGM was trained and educated in management and production skills. They also worked several months at a Taco Bell restaurant in various positions enabling them to get first hand experience in what it took to run a fast-food restaurant.

In order to attract and keep quality RGMs, Taco Bell overhauled its compensation programme. Previously the RM was paid a salary plus a regular bonus equal to the industry standard at the

---

[7] 'Reengineering for your customers: How Taco-Bell did it', *Restaurants USA*, June–July 1994, pp. 36–38.
[8] Karlgaard, R., 'Susan Cramm & John Martin', *Forbes ASAP*, 29 August 1994, p. 67.

appraisal

time. The new system was intended to be similar to that of a franchisee's and to create commitment, ownership, and irreplaceable value[9] among the RGMs.

The role and recruiting process for the district manager changed as well. Taco Bell started to recruit people from outside the fast-food industry—it was looking for sales and product managers with leadership and management experience from large corporations. These traits were needed because the position's span of control increased to cover twenty instead of the previous six restaurants. To successfully manage twenty restaurants, it was no longer possible to manage in the traditional way. Instead the role assumed a more encouraging, positive coaching style. To signal the shift in responsibilities, the position was renamed market manager (MM).

Because there were fewer managers per restaurant, Taco Bell needed to create ways to ensure that restaurants met the quality requirements. Taco Bell did this by setting up a 1-800 number (toll-free) for customers to call with complaints, sending out mystery shoppers, and by regularly conducting extensive marketing surveys. The results from the surveys were included in the criteria used to determine the MM's bonus.

In order to allow MMs to 'grow' within their jobs, Taco Bell created a wide range of salary levels ranging from around $50k to $100k The MM's salary was based on performance, seniority and the complexity of the market.

INFORMATION TECHNOLOGY

Taco Bell also implemented the TACO (Total Automation of Company Operations) project, which was a computer network linking every restaurant with headquarters. TACO enabled the RGMs to receive information on food and labour costs, perishable food and inventory. The RGMs now had the information needed to run the restaurant as smoothly and effectively as possible. Taco Bell's restaurant managers had more information and operational control than any other restaurant managers in the entire fast-food industry. TACO also tracked sales and forwarded the information to a central computer giving the company immediate financial information at any given time.

Kenneth Stevens described the results of the reengineering efforts to create value for customers:

> Thanks to our 'kitchenless kitchens', we've essentially doubled the capacity in our restaurants while reducing the costs to consumers by some 25%. We also reduced customer waiting time by 70%. Add to this better overall quality, better employee morale, the industry's lowest turnover rate, and fewer accidents and injuries. And most important of all, our customer satisfaction levels are at an all-time high.[10]

THRIVING IN THE 1990s (1991–94)

> I believe that there are three separate but equal leadership qualities that are going to differentiate the thrivers from the survivors in the years to come. The first quality is imagination: Constantly looking and thinking outside the box. Changing the game whenever you can. Quality number two is technology . . . it's an enabler . . . a powerful

---

[9] The new average base salary was around $33 000 and target incentive bonus was approximately $12 000.
[10] 'Reengineering for your customers: How Taco-Bell did it', *Restaurants USA*, June–July 1994, pp. 36–38.

tool, that when used correctly . . . maximizes the results of everything we do. And finally, the third leadership quality is empowerment. Giving people the chance to take on greater responsibility, to become more self-sufficient and to perform at higher levels than ever before.[11]

Despite Taco Bell's tremendous success in the late 1980s, John Martin believed there was more potential for the chain and the brand. He continued:

To give you an idea of what I mean, just consider this: there are almost one billion meal occasions in the US alone every day. But, by operating in a conventional restaurant manner, we only reach a tiny proportion of those billion opportunities . . . as convenient as we've tried to become, it still takes a huge effort for our customers to get what it is we have to offer. They still have to come to us . . . By thinking outside the box, we're taking our business beyond the $78 billion fast-food arena to a much larger playing field. Instead of making people come to us, we're developing new concepts which will let us take what we have to them. We are changing the game once again.

Taco Bell started to redefine its goal in terms of maximizing the share of the stomach. The new vision of growth included creating points of access (PoA) at any place where people congregated. Several new ways to reach customers were introduced. For example, carts and kiosks were installed in high school and college cafeterias, airports, malls, convenience stores and gas stations. Taco Bell participated in a programme to provide school lunches in over 2000 schools. John Martin added:

There are 90 000 public schools. And then we take our products to hospitals, movie theaters, etc.[12]

His vision was a worldwide retail system consisting of 200 000 PoA by the year 2000.

The pricing revolution initiated by Taco Bell affected the entire fast-food industry. Taco Bell's competitors started, one by one, to copy the value concept. McDonald's began to cut prices in a big way—introducing a value menu with hamburgers being sold for as little as 49 cents. The company also started to change the design of its restaurant to save costs and tested mini-mart restaurants located in petrol stations. Also, an increasing number of McDonald's franchisees started to use technology to improve their drive-through service. The most advanced restaurants equipped crew members with wireless battery-powered headsets that allowed crew members to hear an order arriving over the intercom. The employees did not even have to wait for a customer to finish speaking.[13]

In order to react to the competitors' moves as well as fulfil the expanded vision, Taco Bell had to continue to innovate to reduce costs and increase customer value. John Martin thought that nothing less than a total reinvention of Taco Bell would enable it to be ready for the next wave of price cuts.

---

[11] Mastering Change, Business Forum – University of California at Irvine, 18 December 1992, p. 1.

[12] Farkas, D., 'The very secure saddle of John Martin', *Restaurant Hospitality*, May 1993.

[13] Berg, E., 'An American wrestles with a troubled future', *New York Times*, 12 May 1991, p. 1.

Case 1.1

> *Taco Bell's entire operation has to be turned inside out to squeeze costs—our competitors have figured out value pricing, so I think it is time for Taco Bell to make its next big move. Just when they all think that we have stopped, maybe we are going to fool them and take another cut like they have never seen before.*[14]

John Martin saw the Taco Bell brand as a brand that had high awareness across many categories and that altered the competitive landscape:

> *By thinking innovatively, we'll become a superbrand like Disney by maximizing the equity of a brand name that people trust.*[15]

He therefore started to build the chain into a brandname identity. Taco Bell started to consider new products to reach more hungry people than just its previous 18–24 year-old target market. Taco Bell developed a new line of snack foods that were distributed in some 3000 supermarkets. The introduction was initiated by marketing research indicating that the Taco Bell brand had higher brand awareness than Doritos. John Martin and other Taco Bell executives viewed each of the 150 000 US supermarkets as a potential PoA.

Taco Bell also acquired the burger chain Hot-n-Now and the upmarket Chevys Mexican Restaurants in the early 1990s and started managing three multiple brands. John Martin added:

> *Buying Hot-n-Now allows Taco Bell to benefit from that chain's seven year experience (in drive-through units)—over the next 10 years, (Taco Bell) hopes to open 2000 to 3000 such drive-through units that could produce operating profits 3% to 5% higher than at traditional Taco Bells.*[16]

He also expected Hot-n-Now to provide Taco Bell with significant growth potential by increasing the 77 store chain to 5000 stores within a decade.

ORGANIZATIONAL CHANGES

John Martin commented on Taco Bell's strategy for growth:

> *As we began looking for ways to accelerate growth, it quickly became apparent that we needed to institutionalize a 'passion for change' throughout the company. We had to give people more responsibility, more self-sufficiency and greater freedom to make an impact. We wanted to build a culture where literally everyone working inside it was an agent for change.*[17]

Kenneth Stevens added his view:

> *Even though we had reengineered for value back in the late 1980s, we were still comprised of a number of divisions and departments all pancaked on top of one*

---

[14] Crain, R., 'Taco Bell again poised to chisel away costs', *Advertising Age*, 12 September 1994, p. 26.
[15] Mastering Change, Business Forum – University of California at Irvine, 18 December 1992, p. 1.
[16] Mc Carthy, M., 'Pepsi spilling into burgers, drive-up units', *The Wall Street Journal*, 20 December 1990, Sec. B, p. 1.
[17] Mastering Change, Business Forum – University of California at Irvine, 18 December 1992, p. 1.

another in a mass of functional silos. This structure, which is still the standard in most large corporations today, keeps people running into walls instead of enabling them to manage an innovative, highly customer-focused business.

In short, while we wanted and needed to be tremendously entrepreneurial for the '90s, the structure of our organization wouldn't allow us to be that way. We needed to create an organization that ran on self-sufficiency and on empowerment. And so, we began that process of creating a flatter, more responsive and more nimble organization.[18]

In order to increase the span of control and to reduce its traditional field management structure in both absolute and relative terms, Taco Bell developed team managed units (TMUs) as a natural evolution of Taco Bell's empowerment strategy. The TMUs were teams of employees with enough training to run the restaurant without a full-time manager supervising them. The teams were responsible for identifying problems and had the authority and know-how to solve them. Taco Bell's philosophy was that the crews interacting with the customers were best able to provide them with value. By the end of 1993, TMUs were operating in 90% of company-owned restaurants.

The TMU programme also enhanced the role of the RGM by reducing the time needed to be spent on daily operational responsibilities and by giving the latter more time to coach and encourage staff, interact with customers, and develop community relations. An important part of the RGM's job now was to find ways to develop the business into new channels.

Kenneth Stevens commented on the changing roles of the MMs:

The bottom line is this: give people the opportunity, give them the tools and the training, let them take ownership, stop micro-managing—and they will astound you with their achievements. In the past, the market manager was basically an extension of the restaurant manager but much less hands-on . . . Our market managers were each responsible for overseeing five restaurants—which is pretty standard for the fast-food industry. Today, that same person manages, on average, 50 points of access. They're responsible for multi-million-dollar portfolios. And a number of them are managing not only traditional restaurants but our food service programmes and our other new concepts as well. We asked these managers to stop being extensions of the restaurant managers and instead become complete business managers.[19]

As Taco Bell continued to expand its multiple brands and increase the span of control, a challenge was the company's ability to quickly transmit information throughout the company in ways that added value. Taco Bell's response was to create a 'learning organization'. Kenneth Stevens described it as follows:

Our goal is to become a learning organization—an organization where information can be captured, distributed and shared throughout our entire company rapidly.

---

[18] 'Reengineering for your customers: How Taco-Bell did it', *Restaurants USA*, June–July 1994, pp. 36–38.
[19] 'Reengineering for your customers: How Taco-Bell did it', *Restaurants USA*, June–July 1994, pp. 36–38.

John Martin felt that the learning organization concept would give each employee a sense of ownership and belonging to the company. Employees were encouraged to seek change, think and act creatively and independently and communicate their ideas to the rest of the organization. Sharing ideas would increase the collective organizational knowledge and give Taco Bell a very flexible and adaptive organizational design. In order to foster this company culture, the company introduced the new technological innovations described below that allowed each employee to access operational information with a user-friendly interface. This resulted in new and innovative ideas coming from individual stores. In order to create incentives to share thoughts and resources with each other, MMs in one zone voluntarily decided to pool their entire bonuses to be shared equally.

When asked if the reengineering process was hard to sell internally at Taco Bell, John Martin said:

> When people can clearly understand that they're doing this for the customer, that's a lot different from saying, I'm doing it for shareholders or so that Martin's office can be nicer.[20]

TECHNOLOGICAL CHANGES

John Martin commented on the role of technology in Taco Bell's strategy:

> At Taco Bell we view technology as an enabler, a tool that makes everything we do better for both the customer and the people who serve those customers. Technology is helping move us from being a manufacturer of food to a retailer of food. It's getting our people away from the tedious work in the back of the house to the front where they can better serve our customers . . . technology is helping us to move ahead by driving more value back to our customers.[21]

The company was changing to a culture based on trust and empowerment. Taco Bell believed that employees now had the right mindset to continue the company's growth and that the biggest challenge was to get even more technology in place to support a learning organization.

TACO had given the stores quick access to detailed data and automated a number of existing processes. When TACO was introduced in 1989, information was delivered to the GMs rather than to the crew. At the crew level, there was still a command and control process. Next, TACO II was implemented to supersede the automation of processes and change the way people made decisions. TACO II was directed at TMUs and provided them with access to operational information and relevant 'best practices'. In order to facilitate acceptance and wide usage, Taco Bell focused on providing the TMUs with usable user-friendly information rather than just data. Susan Cramm, Taco Bell's Vice President of Information Technology, explained:

> I would rather have acceptance and wide spread use of technology than own state of the art systems. The key difference between our (Taco Bell's) information sources and our competitors' has do to with the level of acceptance and usage within the culture.[22]

---

[20] Karlgaard, R., 'Susan Cramm & John Martin', *Forbes ASAP*, 29 August 1994, p. 67.
[21] Mastering Change, Business Forum, University of California at Irvine, 18 December 1992, p. 1.
[22] Karlgaard, R., 'Susan Cramm & John Martin', *Forbes ASAP*, 29 August 1994, p. 67.

In addition to TACO II, Taco Bell developed an intellectual network that was intended to be an on-line communications system that allowed every Taco Bell employee to disseminate information, ask questions, get answers and perform his or her job better.

Taco Bell developed a 'restaurant of the future' testing site where operational innovations were continually at test. Some new innovation introductions were the Flex-Station—an automatic taco assembler, and Customer Activated Terminal (CAT)—a graphical touch screen ordering terminal that enabled customers to order their meals easily via a terminal instead of a cashier. Susan Cramm commented on the CAT terminals:

> Our customer ordering entry system is really putting the customer in charge of their own ordering experience. It is very interesting when you see how inefficient the traditional ordering process is. The system provides an opportunity to offer frequent dining incentives based on purchasing patterns. We can customize the experience for them.[23]

FACING UP TO THE CHALLENGE (1995)

At the end of 1994 there was some concern about Taco Bell's financial performance. Growth had slowed down significantly in 1994, with operating income increasing only 7%. The results were affected by the costs of increasing the number of nontraditional 'points of access' from 9000 in 1993 to almost 25 000 locations. In the fourth quarter of 1994, for the first time since the value concept was introduced in 1988, same-store sales decreased.[24]

The first quarter of 1995 did not show any improvements and same-store sales continued to drop. Industry analysts argued that the value concept had lost its edge. John Martin conceded that 'value alone is no longer a competitive advantage'.[25] Some analysts also pointed to changing consumer priorities as the baby boomers aged (see Exhibit CS1.1.4).

To compound the problems, growth in the recently acquired chains had been much slower than anticipated. For example, Chevys only had around 60 units; this stood in stark contrast to the prediction which John Martin had made while making the acquisition of Chevys: 300 units and $1 billion in sales by 1988.

As Taco Bell's sales stagnated, the pressure for change intensified. A source close to the company commented: 'They're soul-searching to identify the next big thing, but nobody's figured the focus yet.'

---

[23] Howard, T., 'Taco-Bell: Still on the innovation's cutting edge', *Nation's Restaurant News*, 19 September 1994, p. 120.

[24] Martin, R., 'Accounting changes, closures hit PepsiCo's profits', *Nation's Restaurant News*, 19 February 1996.

[25] Martin, R., 'Taco Bell's "Lights" menu seeks fix for setback', *Nation's Restaurant News*, 20 February 1995.

Exhibits

Exhibit CS1.1.1 Taco Bell Summary (1995)

Taco Bell: Taco Bell Corporation, a subsidiary of PepsiCo Inc., is the largest quick-service Mexican-style restaurant chain in the world, with approximately 4600 locations in 50 states and a growing international market. At this time, nearly 200 restaurants are operating in Canada, Guam, Japan, Grand Cayman, Dominican Republic, Honduras, Chile, Egypt, Oman, Poland, Costa Rica, Guatemala, Puerto Rico, Saudi Arabia, Aruba, Qatar, Ecuador, Jamaica, Peru and Hawaii. Approximately 30% of the units are owned and operated by independent franchisees.

Organization: Taco Bell's corporate headquarters is located in Irvine, California. There are six zone offices located throughout the country to serve both company and franchise restaurants.

History: Originated by Glen Bell, Taco Bell became a reality on 21 March 1962. The first Taco Bell restaurant was built in Downey, California. The first franchise was sold in 1964. Taco Bell went public in 1969 and was acquired by PepsiCo in 1978.

Systemwide growth: Systemwide sales have grown 100% over the last five years. At the end of 1995, system sales had grown to $4.7 billion.

New concepts: Taco Bell has greatly increased its number of nontraditional units in 1995. These units include mobile carts, kiosks, school lunch programmes and Express units in such locations as sports stadiums, universities, airports and movie theatres.

Full service dining: In 1993, Taco Bell purchased the 'fresh-mex' casual dining chain, Chevys, based in San Francisco. With over 70 restaurants in California, Arizona, Colorado, Nevada, Missouri, Illinois, Florida, Oregon and Washington, Chevys promises to be the next growth chapter in Taco Bell's long line of success.

Employment: Taco Bell employs over 100 000 people in company-operated and franchised units, and corporate offices across the United States.

Marketing statistics: The company serves more than 55 million people weekly in restaurants across the country. Nearly 123 million people see a Taco Bell commercial once a week—about half the US population. Over 4.5 million tacos are served in Taco Bell restaurants each day.

Raw materials: Taco Bell is the largest restaurant consumer of whole Iceberg lettuce in the world. In one year, franchise and company restaurants consume 4 billion corn and flour tortillas, 53 million pounds of fresh tomatoes, 114 million pounds of lettuce, 60 million pounds of pinto beans, 294 million pounds of ground beef and 73 million pounds of cheddar cheese.

Source: Information provided by Public Affairs Department, Taco Bell Corp.

Exhibit CS1.1.2 Financial highlights 1983 to 1995

	1983	1984	1985	1986	1987	1988	1989	1990	1991	1992	1993	1994	1995
Total system sales ($ billion)	0.7	0.9	1.1	1.3	1.5	1.6	2.1	2.4	2.8	3.3	3.9	4.4	4.6
Operating profit ($ billion)	0.043	0.059	0.068	0.078	0.085	0.076	0.113	0.15	0.181	0.215	0.253	0.273	0.105
Same-store sales ($000)	439	539	550	560	579	589	686	771	814	866	925	953	925
No. of Taco Bell locations not including carts or retail						3125	3349	3670	4152	4921	5684	6490	

Source: PepsiCo Annual Reports.

Exhibit CS1.1.3 1995 Restaurant unit activity—Taco Bell—US

	Company	Franchised	Licensed	Total
Beginning of year	3232	1523	929	5684
New builds and Acquisitions	190	98	668	956
Refranchising and Licensing	(214)	169	45	—
Closures	(75)	(11)	(64)	(150)
End of Year	3133	1779	1578	6490

Source: PepsiCo Annual Reports.

Case 1.1

Exhibit CS1.1.4 Consumers' priorities as % of age group

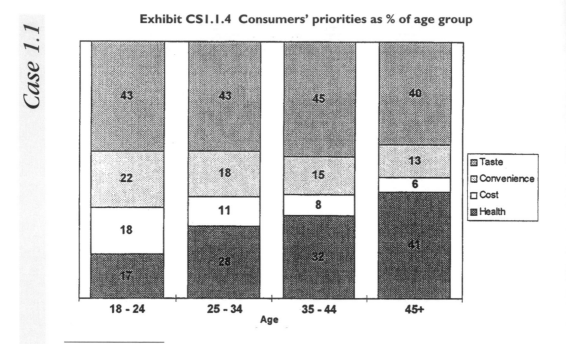

Source: Taco Bell Corporation as reported in *Nation's Restaurant News*, **29**, (8), 20 February, 1995.

Case 1.2

This case was written by Francesca Gee, Research Associate, Professor Soumitra Dutta and Associate Professor Jean-François Manzoni, at INSEAD. It is intended to be used as a basis for class discussion rather than to illustrate either effective or ineffective handling of an administrative situation.*

Transforming TSB Group (A): Industry and Company Background

> 'The sheer magnitude of what had to be done made it easier in a way', mused Peter Ellwood as he recalled his arrival at the TSB Group in the spring of 1989. 'The company was in a difficult situation financially. That made it easier for us to bring about big change, it made it easier for us to take bigger risks.'

The TSB Group, privatized two and a half years earlier, had enjoyed fast growth. But it was facing daunting challenges in a rapidly evolving marketplace, and Peter Ellwood had been appointed chief executive of the core retail banking division with a clear mandate: to implement far-reaching change. Eight weeks after moving into his new office in the City of London, he was about to announce sweeping changes to the bank's staff and shareholders.

THE BRITISH BANKING INDUSTRY

London had a long tradition as a leading financial centre—a result of the UK's past international leadership, as well as of its liberal approach to financial regulation. Its financial district, the City was one of the oldest trading centres in the world. The industrialization of Britain in the eighteenth and nineteenth centuries had turned it into Europe's largest financial marketplace. The abolition of exchange controls in 1979 also contributed to the market's internationalization. Financial services played a major part in the UK's economy: in 1985 banks and insurance companies generated more than 7% of GDP, compared with 4.7% in the United States and 5.7% in France.

British banks had a long history of avoiding political controversy, working informally with the government through the Bank of England. The banking sector had for a long time been characterized by a high degree of concentration in the hands of a social élite—a network of families. It had been competitive in the nineteenth century, but by World War One amalgamation had produced a loose, unstable cartel that fixed interest rates on deposits. In 1918, a five-bank cartel became effective. Economies of scale encouraged concentration, as did the

---

Bank of England's belief that larger units would better withstand periodic financial crises. This cosy arrangement ended in May 1971 when the central bank abolished all lending restrictions and freed the banks to compete with one another for deposits.

The advent of competition in retail banking was associated with service innovations such as credit cards or personal loans instead of overdrafts, the rapid spread of branch networks, and a new emphasis on advertising and marketing (Barclays was the first major London bank to open a marketing department in 1968). Change gathered pace when Margaret Thatcher became prime minister in 1979. She ushered in a period of financial deregulation that culminated with the so-called 'Big Bang' of October 1986, and many commercial banks embarked on expensive diversification ventures, acquiring stock exchange member firms. The government accompanied the rapid deregulation with stricter supervision through the 1986 Financial Services Act. The UK banking sector is described in Exhibits CS1.2.1 and CS1.2.2.

The main players

In the late 1980s, UK retail banking was dominated by the 'Big Four': Barclays Bank, National Westminster Bank, Midland Bank, and Lloyds Bank. Another four medium-sized banks stood out: Standard Chartered Bank, TSB Group and two Scottish-based banks: Royal Bank of Scotland and Bank of Scotland. Together these eight banks employed 75% of all banking industry personnel and occupied 93% of all bank branches. They were the 'high street' banks, which served the general public; they were also known as 'clearing banks' because they were members of the cheque clearing system.

The Main Issues

In September 1988 the *Financial Times* described the late 1980s as 'a golden era of UK banking':

> *'Seldom have the clearing banks enjoyed such prosperity—at least in so far as their traditional high street markets are concerned. The country's almost insatiable demand for credit has brought the banks record levels of business—and profits.'*

While business had been brisk for most of the decade, the question in most bankers' minds was how much longer would the good times last? Making money had become easier for banks. Increasing numbers of wealthier customers required services such as higher rate deposit accounts, share dealing schemes or unit trusts.[1] In this lucrative, but competitive, environment the battle was for market share, with advertising and promotional campaigns becoming steadily more important as customers became increasingly discerning and sophisticated in satisfying their financial needs.

New international rules on capital adequacy imposed more stringent standards for capital strength. The unification of Germany and the 1992 Single European Market challenged London's dominance as a financial centre. The 1990s also promised major upheavals with the evolution of new competitors and further refinements of technology, particularly in the manner in which banks delivered their products.

---

[1] Unit trusts (the UK equivalent of mutual funds) invested the combined contributions from many people in various securities and paid them dividends in proportion to their holdings.

An even more serious challenge to the banks came from building societies—organizations that made loans secured by mortgages to members who wished to build or buy a house. Building societies played an important role in the centralization of savings. At the end of 1987, they held 13.2% of the system's total assets, 15.5% of deposits and granted 75% of the financing for the purchase of homes.

High street banks had been able to resist the competitive onslaught from building societies as long as the loan market was expanding fast, particularly in mortgages. But 1988 saw the start of a sharp rise in UK interest rates which placed pressure on intermediation margins (the margins added on by banks to their own money costs), particularly in personal lending and mortgages, while consumer demand fell. The marketplace was increasingly crowded, with growing competition from other high street retailers and niche operators:

- Completely new players (for instance, HFC Bank, which claimed to cater especially for the family) or formerly peripheral institutions (TSB or the Post Office's Girobank) had begun to offer a full product-range from current accounts to gold cards and mortgages.

- Building societies, which enjoyed a ready-made customer base and enormous goodwill, were using their new powers to establish themselves as fully fledged loan institutions.

- Banks from other European countries, some of which had already entered the mortgage market, were attracted by the profitability and openness of UK retail banking. Its margins were among the highest in the EC (except for mortgages, which were the EC's cheapest). However, these margins were due to fall.

- Large retailers, especially suburban hypermarkets, were introducing automated cash dispensers (ATMs). The advent of EFTPOS[2] shopping, using a plastic card in an electronic terminal, forced the banks to negotiate seriously with the retailers, who made their own plastic cards. Distribution giant Marks and Spencer, which never accepted credit cards issued by banks, had acquired a banking licence and entered the personal loan business.

The clearing banks reacted to the increased competition by diversifying into insurance, estate agencies and stock broking. They also competed to lend to cash-poor developing countries and to company start-ups, despite evidence that many small businesses collapsed within five years. This imprudent lending often resulted in losses. The Third World loan crisis, for instance, cost UK banks £3bn. High Street customers were faced with increased charges as a result, and investment in their domestic retail activities had to be trimmed.

The search for new customers was also becoming more important. Despite the proliferation of institutions that offered current accounts, very few adults changed their bank accounts each year. Some banks succeeded in growing customer bases by finding vacant niches, but the main search was for young and student customers. Competition was made even tougher by a sharp reduction in the number of teenagers. Banks, therefore, felt that they had to catch customers as early as possible—at the piggy-bank stage.

---

[2] Electronic fund transfer at the point of sale.

Operating costs, already higher in the United Kingdom than on the continent, were rising further because of heavy investment in new technology. Electronic innovations, which made it possible to provide mass financial services on an unprecedented scale, played an ever-increasing role as banks developed new forms of cashless payment and automated their back offices. Technological progress, which fostered interbank cooperation for clearing and settlement purposes, also heightened competition as electronic funds and information transfer systems became the principal vehicles for delivering services to customers. Despite the high costs, banks saw investment in technology as an important component of their overall strategy.

THE TSB GROUP

The Trustee Savings Banks: early history

The TSB Group's origins date back to 1810, when a Scottish clergyman set up the first parish savings bank to encourage the poor to save. From the beginning the so-called Trustee Savings Banks—they were managed by boards of trustees—were tightly regulated: all savings were to be invested in public debt and trustees were not to derive any benefit from their office. They were also an immediate success. By 1815 there were 465 savings banks operating in Britain. For the next 150 years or so, the network flourished as an association of loosely connected institutions. Branch managers enjoyed so much autonomy that two branches in the same region could offer different interest rates on the same type of savings account. As late as 1981 a newly-recruited manager found the structure unfamiliar and frustrating: 'I found that what one thought of as subsidiaries were regarded as laws unto themselves. It was very much a rule-by-committee environment.'[3]

The 1960s saw the first attempt at modernization as the Trustees Savings Banks introduced cheque accounts, unit trusts and set up a life assurance arm. Deregulation, which began in the 1970s, enabled them to offer a full banking and financial services. In 1976 the government authorized the Trustee Savings Banks to grant credit, personal loans and overdrafts, mortgages, and the Trustcard credit card business grew quickly, along with general insurance and pensions. The banks also turned their attention to business customers, initially to small- and medium-sized companies. Diversification began in earnest in 1981 with the acquisition of a consumer-finance group and business development in hire purchase and commercial lending.

Consolidation began in 1975 as the 73 Trustee Savings Banks merged into 20 to form a rather loose confederation supervised by a central board. In 1983 their number was reduced to four: TSB England & Wales, TSB Scotland, TSB Northern Ireland and TSB Channel Islands; a Group central executive was formed, bringing the various operating units together under one management. The 1985 Trustee Savings Bank Act turned TSB into a group of commercial companies. At the time it had 6 million customers with 13 million accounts and more than £8bn in deposits.

The TSB Group was floated in September 1986 after considerable political wrangling. The Scottish lobby was particularly wary of TSB Scotland losing its Scottish identity and a number of concessions had to be made. TSB Group plc, the Group holding company, was incorporated as a Scottish company, whose annual shareholder meeting would be held in Scotland; more

---

[3] Moss, M., and Rusell, I. *An Invaluable Treasure: A History of the TSB*, Weidenfeld and Nicolson, 1994, p. 295.

importantly, the four banks were promised a large measure of independence. The flotation was a success, raising nearly £1.5 billion in equity through a public offering of ordinary shares and TSB acquired 3.1 million shareholders, many of them customers.

The Privatized TSB: A strategy of diversification and fast growth

Buoyed by its enormous capital base, the newly privatized TSB Group set out on an ambitious acquisition and diversification programme that was meant to transform it into a broadly based financial services concern. Its strategy involved the creation, through a mixture of organic growth and acquisitions, of a centralized mortgage lending arm, a private bank for wealthy customers in Luxembourg and a sprawling estate agency business, among others. TSB's main acquisitions are listed in Exhibit CS1.2.3.

The Group's most costly acquisitions were the Target life insurance company and Hill Samuel Group plc, a leading City merchant bank, which together cost over £1 billion. Financial analysts questioned the wisdom of the Group's attempts to become a diversified financial services conglomerate. One later commented:

> *In terms of failure few companies could rival TSB's diversification strategy of the late 1980s . . . The diversification was an attempt to raise the return on equity by leveraging up on extensive surplus capital. But this only increased the credit risk within the Group's portfolio that subsequently crystallized during the recession.*

The *Financial Times* was equally critical, describing TSB as 'the privatization that failed' and noting 'Its performance since the 1986 flotation has been a disgrace. Its shares have underperformed [the market] by a third.'[4]

Aside from diversifying the business, the Group's managers also tried to change the customer base. Because of its origins as a savings bank, TSB had far fewer upper middle-class customers (known in socio-economic terms as groups A and B) than its clearing bank competitors. In the words of a senior executive, TSB was 'a bank for the man in the street, the people's bank'. Most of its customers came from the lower middle classes and the working classes (groups C1, C2, D and E). The bank's managers, however, wanted to attract wealthier customers as well as more corporate accounts—areas in which the bank lacked skills.

The Group was reorganized in 1988. Three business divisions were created under a new bank holding company headed by TSB deputy managing director Don McCrickard. Hill Samuel closed down or sold some of its operations, while others (investment management, life assurance and unit trusts) were merged into relevant TSB operations. In exchange, Hill Samuel took over all Group treasury, corporate banking and development capital activities. In 1988 the TSB Group's corporate lending soared by 66% to £3.3 billion.

A NEW CHAIRMAN: SIR NICHOLAS GOODISON

In 1988, the TSB Group appointed a new chairman, Sir Nicholas Goodison, who took up the position in January 1989. Sir Nicholas, described in the press as 'a skilled City diplomat', had

---

[4] *Banking World*, April 1989.

become, in 1976, the youngest chairman ever of the London Stock Exchange. 'As Stock Exchange chairman for 13 years, Goodison masterminded London's transition from a parochial monopoly with fixed commissions to a competitive international market place.'[5] He was also a board member (and later deputy chairman) of British Steel and held a number of prominent appointments in the world of the visual arts.

The *Financial Times* commented: 'It is hard not to see the appointment of Sir Nicholas Goodison to the chairmanship of TSB as yet another missed opportunity . . . What [the TSB] surely needs is someone experienced in the mass marketing of financial products.'

Sir Nicholas, however, took control firmly and pushed hard and fast for restructuring and integration. He quickly spotted some urgent problems:

- The Group had too many boards and too many directors: the main board alone had 32. 'It wasn't a board of directors, it was a debating chamber,' said Sir Nicholas. 'You need a compact board with a necessary range of skills. [This] was really people representing different parts of the organization.' TSB also had eight regional advisory boards with a total of 74 non-executive directors—the old trustees—whose chairmen sat on the main board. Several wholly owned business units also had their own boards and non-executive directors. The structure confused reporting lines.

- When TSB was floated, an undertaking had been given to the four TSB banks that they would remain 'independently managed within the Group'. As a result, reporting lines were unclear. 'It was a sort of federal structure. There wasn't even an entity called TSB bank. There was the TSB Group, the four TSB banks, Hill Samuel, the Hill Samuel holding company. The legal structure was an extraordinary mess', said Sir Nicholas.

- While TSB was the 43rd largest company in the United Kingdom, it lacked what Sir Nicholas described as 'suitable' top managers. One reason was, he said, that 'too many people see the TSB as a sort of down-market organization'.[6] Large scale management changes at the top were, therefore, necessary.

Sir Nicholas tackled the boards first in what he called 'a very symbolic action':

> I decided to start at the top with the board and the advisory boards. I went to all [directors] in the board meetings asking that rather difficult question, 'What value do you add to the organization?' The theory was that they were generating business and keeping an eye on their region. But the truth was that they weren't bringing in much business, and keeping an eye on the business is better done on the executive line. You don't want a lot of non-executive directors at every level in the organization doing that.

In May 1989 he presented a plan for dissolving the advisory boards and the non-executive boards of subsidiaries.

---

[5] *Sunday Telegraph*, 8 May 1994.
[6] *Financial Times*, 14 April 1989.

It was quite an emotional thing for them, they were being cut off from half a lifetime of service as trustees. The advisory boards all said that they were valuable, that they represented the community, that we would be crazy to abolish them because we would be abolishing TSB's traditions and that they were valuable business contributors.

However, over 100 directors across the Group left; the main board sank to a dozen directors. Sir Nicholas said that the changes were intended to 'establish clear accountability for line management and maximum delegation throughout the Group. We are engaged in a process of restructuring and rationalization, starting at the top.'[7] The Group's head office staff was also substantially reduced. He described the changes as being effected with 'great goodwill and understanding'.

He also introduced a new Group structure based on recommendations from consultants. In the new structure, shown in Exhibit CS1.2.4, a holding company which handled strategy, financial control, budgets, and planning oversaw three operational divisions. For the first time, a unified banking division named TSB Bank grouped TSB's previously separate banking operations: TSB Scotland, TSB Northern Ireland, TSB Channel Islands, the much larger TSB England & Wales and Hill Samuel's banking activities. All reported to Don McCrickard, the division's chief executive. A second division looked after life assurance, unit trusts and pensions. The Group's commercial interests, which included a vehicle rental business, an employee-benefits consultancy and a shipping services group, were rolled up into a third division.

A NEW CHIEF FOR TSB – PETER ELLWOOD

On 27 May 1989, Peter Ellwood was appointed chief executive of the newly redesigned TSB retail banking division. Don McCrickard explained:

I went outside the organization for a financial retailer, someone sensitive to the needs of consumers and someone capable of pushing through fundamental change necessary at the branch level.[8]

At 46, Peter Ellwood was the youngest chief executive in British banking. He came from Barclays, where he was known as 'Action Man'. He had joined Barclays as a cashier at the age of 18, rising through the ranks to become the head of its central retail services division (Barclaycard). *The Economist* described him as 'a highly-regarded former chief executive of Barclaycard and an experienced marketer of financial services.'

The extent of the problem

Peter Ellwood spent his first eight weeks at TSB assessing the retail banking business. In November 1988, the Group's board had asked consultants to perform a detailed review of prospects for the retail banking division. Task forces composed of bank staff from different functions, from senior managers to front-desk workers, and external consultants carried out a detailed study to assess the bank's strengths and weaknesses, clarify objectives and identify opportunities for improving performance. They found that TSB's history had left it with duplicated jobs and functions, too much reliance on low-interest deposits, an ageing customer base and a poor image. The division also failed to capitalize on strengths which distinguished it from most of its competitors. The TSB had the following advantages:

---

[7] *Financial Times,* 14 April 1989.
[8] *Financial Times,* 22 November 1989.

- A very profitable life insurance and unit trust unit, whose salesforce cross-sold insurance and investment services to current account customers in TSB branches.

- The biggest customer base of any UK retail bank (together with Barclays) with 7 million account holders.

- A clear lead (estimated at about two years) in on-line, real-time technology and remote delivery systems. TSB had more ATMs than any other UK bank except NatWest, and its telephone banking system, launched in 1987, had over 200 000 subscribers. Exhibit CS1.2.5 shows the number of ATMs in the UK banking system.

- Few risky loans to corporate, developing country or property developers.

- A dedicated staff.

But TSB's history of progressive amalgamation of smaller savings banks had left it with an unwieldy regional infrastructure managing the branch network and costly duplication in head office functions such as technology, marketing and finance. TSB England & Wales, for example, consisted of six regions, each supported in part by a head office with up to 100 staff, covering all major functions (which were already present at divisional head office functions) as well as minor ones, such as architects who designed branches. TSB Scotland, TSB Northern Ireland and TSB Channel Islands also had their own head offices. Branches were generously staffed (Exhibit CS1.2.6 shows network costs). Peter Ellwood explained:

> The traditional branch manager in TSB ran his branch as a micro business unit, with responsibility for administration and paperwork, lending with defined discretionary limits, staff training and development, handling customer queries, service levels, branch profit and loss, and sales. Distributing these responsibilities throughout our network increased TSB's cost base and diluted the time a manager could spend with his customers.

The extent of TSB's troubles was not immediately obvious because the income stream was still growing, albeit at a rate lower than cost growth. The bank was greatly over-capitalized, and earnings on surplus capital inflated profits. The bank paid less interest than its competitors on deposits. 'We were making enormous profit from paying people 1 or 2% on their savings balances when everyone else was paying them 6, 7 or 8%,' explained Peter Ellwood.

As a result, customer account balances had declined from £4279 million in 1985 to £3267 million in 1989—an annual compound rate of 6.5%. Analyses conducted at the time showed that if that trend continued, the balances would fall to £2668 million by 1992—a £40 million loss in income. Meanwhile, the bank was not lending enough. It had an assets/deposits ratio of 63%—it lent only 63% of the deposits it received—compared with an average of 78% for the Big Four. Exhibit CS1.2.7 shows a five-year financial summary for the group.

TSB also suffered from a lack of market penetration, particularly in the southeast. While it had a large branch network, its branches were smaller than those of its competitors, making economies of scale difficult to achieve. In addition, many of its branches were in poor locations. It also lacked innovative, competitive products: this reduced cross-selling opportunities and made it harder to attract new customers.

TSB's cost-income ratio was rising. By 1989, it stood at around 70%, compared with an industry norm in the low 60% (Exhibit CS1.2.8 compares TSB's cost-income ratio with those of the Big Four). For building societies, the average ratio was around 50%. In 1986, it cost the bank 63p to generate £1 of income, but by 1989 it cost 71p. This was because between 1986 and 1989, costs had gone up 14% while income had grown by only 10%. The review concluded that unless significant action was taken quickly, costs would continue to increase faster than income, dragging down profits.

Peter Ellwood's first moves

Peter Ellwood invited a dozen of the bank's top managers to attend a three-day retreat at a secluded hotel at the end of June 1989. They set out to answer some basic questions: 'What kind of organization do we want to be? What face do we want to present to our customers? Do we want to be like a traditional bank? Or do we want to be a world-class retailer of financial products?' One of the participants explained:

> The retreat was a leadership exercise, binding these people together. Peter's excellent at that.

But the goal was also to gain a common understanding of the problems facing the bank, to set out a strategy to overcome them and to adopt an action plan and a communication programme. Drawing partly on a preparatory report by the consulting firm, they developed a blueprint for the organization. Major change projects and, equally importantly, a mission statement, were adopted at the retreat. 'The major tasks and the major building blocks were outlined at that session. We had the basic blueprints for the projects. Not in great detail, but we identified what the priorities were,' said the participant. The group also agreed on a mission statement:

> To be the UK's leading financial retailer through understanding and meeting customers' needs and by being more professional and innovative than our competitors.

Peter Ellwood explained later why the mission statement deliberately did not include the word 'bank':

> I saw a bank as traditional, bureaucratic, slow moving, averse to change, not very creative, unimaginative, dull. I saw a retailer, the best retailer, as very customer-oriented, close to changing consumer behaviour, fleet of foot, entrepreneurial, good at really understanding what the customer needs and then meeting that need. I came to the conclusion that that's the sort of organization that would succeed in the UK financial services market.

He concluded:

> We want to embody all that is best in retailing practice. We don't want to be stuffy, pompous or traditional in the sense of nineteenth century bankers. We can be, and will be, the best UK financial retailer. Not the biggest, but the best.

Exhibits

Exhibit CS1.2.1 The British banking system (1986)

	No. of banks	No. of branches	No. of employees
Banks	633	14300	403000
Building societies	151	6952	69200
Total	784	21252	472000

Source: Bank of England.

Exhibit CS1.2.2 Market share: Deposits (%)

	1974	1980	1983	1985
Banks	32.0	29.8	26.6	25.8
Building societies	17.4	19.7	17.9	17.8
Insurance companies	26.5	25.4	26.2	25.7
Pension funds	15.5	21.0	25.4	26.7
Investment trusts	7.3	3.3	3.1	3.1
Finance houses	1.3	0.8	0.8	0.8

Source: Rose, H. 'Change in Financial Intermediation in the United Kingdom', *Oxford Review of Economic Policy*, **2**(4), 1986.

Exhibit CS1.2.3 TSB Group: Acquisitions (1986–90)

Name of business	Activity	Date of acquisition	Price (£m)
Slater Hogg & Howison and other estate agencies	Estate agencies	1986–87	75
Boston Financial	Factoring company	1987	
Target Group plc	Life insurance, unit trusts, pensions	August 1987	229
Hill Samuel Group plc	Merchant bank	November 1987	777
Atlanta Capital Mgt. Co.		April 1990	
The Johnson Companies		June 1990	

Exhibit CS1.2.4 TSB Group (April 1989)

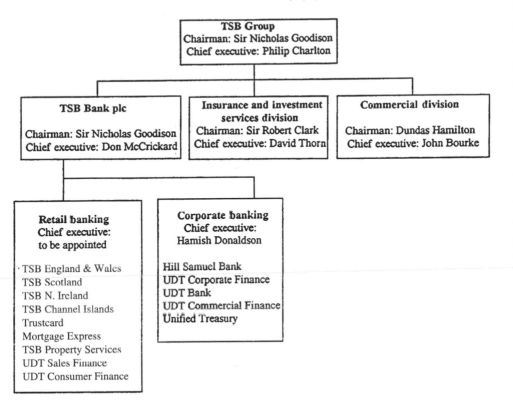

Exhibit CS1.2.5 High Street banks' ATMs in service (30 June 1989)

Barclays	1972
Lloyds	2103
Midland	1773
NatWest	2517
Royal Bank of Scotland	622
TSB	2192

Source: Banking World, November 1989

Case 1.2

Exhibit CS1.2.6 TSB Retail Banking Division: Network costs (June 1989)

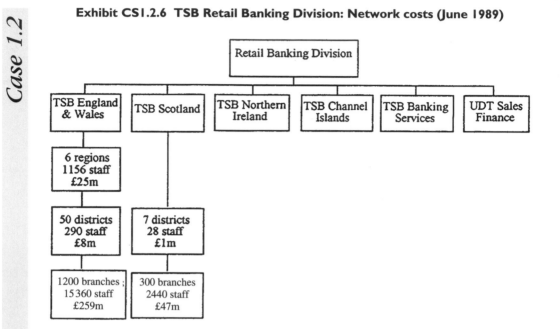

Exhibit CS1.2.7 TSB Group plc, five year financial summary

	1988	1987	1986	1985	1984
Group operating profit before tax	420	340	238	201	171
Total income (£ m)	1442	1062	859	721	626
Total costs (£ m)	977	667	563	471	411
Total assets (£ m)	22516	17194	13303	11579	10207
Value of long-term life insurance and pension businesses (£ m)	237	169	103	80	59
Total advances (£ m)	11189	6800	5137	4651	3406
Total current and deposit accounts (£ m)	18953	14148	11128	10258	9070
Shareholders' funds (£ m)	1866	2269	1620	902	770
Return on average total assets after tax (%)	1.4	1.5	1.3	1.2	1.1
Staff numbers (average)	42243	32050	29936	26665	24865

Source: TSB's Annual Report and Accounts, 1988. Figures for 1987 and prior years have been adjusted to reflect the change of accounting policy for the value of long-term life insurance and pension businesses.

Exhibit   CS1.2.8 Cost-income ratio (%)

	1989
Barclays	64.0
Lloyds	63.8
Midland	72.4
NatWest	66.6
TSB	72.4

Source: TSB Network News, March 1991.

Case 1.2

Case 1.3

This case was written by Francesca Gee, Research Associate, under the supervision of Soumitra Dutta and Jean-François Manzoni, Professors at INSEAD. It is intended to be used as a basis for class discussion rather than to illustrate either effective or ineffective handling of an administrative situation.*

Transforming the TSB Group (B): The First Wave (1989–91)

TSB BANK: STAGE I OF THE TRANSFORMATION (1989–92)

Peter Ellwood envisaged nothing short of shock therapy as he contemplated the changes needed at TSB. When he was appointed chief executive of TSB's retail banking division in May 1989, he was determined that the 175 year-old former association of savings banks should be shaken out of its complacency:

> It wasn't a choice, I had no alternative. If I'd sat there and thought, 'Well, I'm the CEO now, isn't it great?' and watched it, I would have watched it sink . . . I was totally convinced that we would succeed, I knew we couldn't afford to fail. And I knew that if I wasn't convinced, how could I convince other people? I didn't even contemplate that it could not succeed.

The institution which the *Financial Times* called, somewhat patronisingly, 'dear old TSB', was spawned by the protracted amalgamation of dozens of Trustees Savings Banks. In 1986, the TSB Group (which then included four separate banks: TSB England & Wales, TSB Scotland, TSB Northern Ireland and TSB Channel Islands) was floated on the London Stock Exchange. It acquired 3.1 million shareholders and £1.5 billion in capital which it quickly spent on costly acquisitions. By 1989, when Sir Nicholas Goodison, the former chairman of the London Stock Exchange, became chairman of the TSB Group, it was clear that strategically and structurally much remained to be done.

EARLY CHANGES

One of Sir Nicholas's first actions was to dissolve the many regional advisory boards and the non-executive boards of subsidiaries and to endorse a plan which created a new TSB retail

---

banking division, looking after the four regional banks—England & Wales, Scotland, Northern Ireland, Channel Islands—and related activities, such as UDT—a hire purchase company; Mortgage Express—a centralized mortgage lending business; Trustcard—the credit card business and TSB Property Services—an estate agency chain.

All the businesses had been run in a fairly autonomous way and all had their own head offices with their own support functions.

Peter Ellwood remembers his first day:

> *I was introduced to a new divisional team brought in to be the nucleus of the new divisional head office. There was a finance director, an information technology director, a legal director—all of whom had been working elsewhere in the Group, and a new marketing director recruited from outside. However, each of the businesses in the new division had their own support functions and, in addition, above us were a number of Group functions such as finance and personnel. It did not take long to realize that we had an unworkable, unwieldy structure. Simply improving the new divisional head office, which sat on top of businesses which had their own head offices and were happily reporting higher up the line to 'Group', was not going to work.*

Within two months, Ellwood had identified the key players whom he assessed could help him bring about the massive change he realized was needed. They met in July 1989, created the vision and the mission statement for the business, designed a new structure and decided that the change programme would best be brought about by means of a series of major projects.

Structurally, the various head offices were amalgamated and seven regions, including Scotland, were brought down to three, and 54 district offices were reduced to 22.

> *The clash between the old and the new structures made symphony orchestra cymbals sound like cotton wool. Just imagine how it would feel if you were the chief executive of TSB Northern Ireland, with your own board, and had historically reported to the TSB group chief executive, to be told that in future you would report into the western region of the TSB retail bank. It was an interesting process to manage.*

The new structure, shown in Exhibit CS1.3.1, featured six profit centres (including three new banking regions), which were responsible for bottom line profits and bought services from five support functions.

The new chief executive had a clear vision for the business, as he recalled in 1993:

> *Three years ago, if you had asked a member of TSB staff 'Where is the business going?' they may have made some comment about becoming the fifth UK clearer. I can assure you that no-one at TSB now is interested in being the fifth anything. Our mission is 'To be the UK's leading financial retailer through understanding and meeting customer needs and by being more professional and innovative than our competitors'. By 'leading financial retailer' we do not mean the biggest but we do mean the best.*

Ellwood was determined to cut costs, which would require a leaner, fitter structure. He was also

Case 1.3

determined to increase revenue. The main changes would be implemented by five project teams which were set up in the summer of 1989. The largest projects were:

- The head office review aimed at centralizing and streamlining head office functions.

- The network redesign project to improve the 'efficiency and effectiveness' of the branch network.

In addition, there were projects reviewing commercial banking, best practice and the bank's range of products and services.

THE HEAD OFFICE REVIEW

TSB's history as a loose federation of independent savings banks had left it with a large head office in London as well as several smaller head offices in each of the four former banks. The six regions of TSB England & Wales also had miniature head offices with more than 100 staff handling functions such as payroll, personnel and lending—even architects to design branches.

External consultants were retained to work alongside TSB staff to consolidate and streamline the various head offices. The team had three months to prepare a blueprint for a new head office structure that included a detailed implementation plan. The objective was to save £30–40 million annually by creating larger spans of control and fewer layers of management, the centralization and rationalization of support functions and the relocation of some activities.

The team reviewed head office costs in detail and examined whether tasks undertaken by the regions (for instance, mortgage processing) should be centralized. They found many weaknesses within head office functions, particularly in marketing, finance and personnel. Control over costs and staffing levels was weak, and everywhere fragmentation had resulted in duplication and non-standardization. Their conclusion was that centralizing the administrative tasks carried out by the old banking regions would save money and enable regional management to focus on developing the business and offering support to the branches. They also recommended combining most of the four regional banks' head office functions.

Peter Ellwood explained:

> We started off with a blank sheet of paper and asked what shape should head office be and how many layers should it have? Let's forget about what we have now, let's work out what is the optimum design.

The team designed an 'ideal' structure which was compared with the existing structure.

> We had hundreds of extra people, too many layers, one-on-one relationships and people who didn't know whether they were responsible for this or that.

In late 1989, the proposals were accepted by the TSB Bank board and the team set about creating the new organization.

The centralization of head office functions, such as IT, marketing, premises and finance, was completed in May 1990. In the streamlined structure, each regional director had three senior reports: a director of sales, in charge of regional planning, sales support and the area directors; a director of resources, who looked after personnel, training and development, office services and finance; and a director of credit management, responsible for securities, control and monitoring, and lending.

The new structure resulted in 19% of head office staff becoming redundant with 845 of 4500 posts eliminated by October 1990. Non-unionized senior managers with personal contracts were the first to go. Exhibit CS1.3.2 shows how many positions were cut at different head office locations. An additional 650 jobs were cut at the regional and district level in the course of 1990. Head office personnel costs were eventually cut by 25%.

Other changes included moving TSB's clearing centre for cheques from the City to a cheaper site in London in early 1992 which was expected to produce annual savings of £4 million. Additionally, seven mortgage processing plants were amalgamated into one.

The bank's head office was also moved to cheaper, less crowded premises. TSB chose Birmingham, a large city 100 miles north of London. It was more central than London in terms of the TSB network, whose historical roots and much of its customer base were in Scotland and northern England. TSB opted for a gradual relocation, moving one department at a time to minimize risk and make the transfer of skills easier. The then TSB finance director Mervyn Pedelty, chairman of the head office review's steering committee explained, 'We spread the relocation over 18 months and it was remarkably painless in terms of organizational disruption'.

Only managers and senior staff were encouraged to move. A relocation office was set up to produce videos and other documents, provide financial advice and arrange visits to Birmingham. TSB also offered generous financial relocation packages and gave staff time off to visit Birmingham and to move. The move was completed in the spring of 1992. By then, 600 head office staff were working at the newly refurbished Victoria House, 250 had moved from London and 350 were hired locally, many from other TSB departments. The relocation was expected to save up to £10 million annually in due course.

NETWORK REDESIGN

The Network Redesign Project had two major objectives:

- To reduce the cost of running the branch network by cutting the number of regional offices, streamlining branch management and increasing the productivity of administrative work by moving it to dedicated customer service centres.

- To increase income by using the space that was now available in branches for interview rooms, by creating 1250 more salespeople (known as customer service executives or CSEs) and by supporting the selling effort with a rule-based sales system called Super Service.

Areas and clusters

After TSB became a publicly traded company in 1986, it conducted international benchmarking—comparing its business practices with those of other companies—with help from external consultants. Its main benchmark was Wells Fargo, an American retail bank, which had improved its sales by removing administration from branches. The consultants suggested this approach to TSB, and in 1989 Peter Ellwood's new management team decided to adopt it.

Until 1989, TSB's banking network had consisted of over 1600 branches divided into 54 districts. In the new structure, as shown in Exhibit CS1.3.3, the districts were replaced by 159 sales clusters divided into 22 areas whose directors reported to the three regional directors. Area offices performed many of the activities of the old regions and districts, handling sales, marketing support, personnel, training and development functions with an average of 11 staff. Peter Ellwood explained:

> We made sure that the organization structure was composed of units with clear responsibilities, so that each area director had a more manageable span of control, being responsible for some eight clusters of branches. The old structure had a district manager responsible for 27 branches. The span of control was seriously adrift.

Clusters became self-contained profit centres with clearly defined sales objectives. Each cluster was made up of a main branch and an average of eight linked branches. Main branches were big, full-service branches located at busy city centre sites. They were led by a senior branch manager who supervised the cluster's linked branches, and they often included a commercial banking section. Linked branches had fewer staff who handled a smaller range of activities and their managers were paid less.

Many of TSB's 1600 branches were small, which made economies of scale difficult to achieve, and were in poor locations: some had not moved premises since the 19th century. The 200 least profitable branches were earmarked for closure although this was a risky proposition because of the danger of annoying customers, or losing them altogether, explained network redesign project manager, Finian O'Boyle:[1]

> Many branches were too small and gave a poor image of the organization. We closed a branch if it was very shabby, in a poor location, and if we had a brand new refurbished branch nearby.

Customer service centres

A major objective of clustering was to remove as much processing work as possible from branches, which could then be refitted to accommodate more salespeople. Customer Service Centres (CSCs) were set up to perform this processing. Initially 81 CSCs were set up, each serving two sales clusters on average. Finian O'Boyle explained:

> We modeled having between one and 150 CSCs and looked at the cost. With one CSC, the cost of technology was high because we would need a new mainframe. With a great number of CSCs by comparison we got no economies of scale.

Four factors determined the number of CSCs: the availability of premises, the geographical density of branches, staff mobility and technological constraints. Wherever possible, CSCs were

---

[1] The consultants' analysis showed that branch closures would become unprofitable if more than 10% of customers were lost.

housed at existing premises such as old regional offices and, while they were originally to function as physically separate units, some were grouped together to save on premises costs.

The network redesign team expected to achieve economies of scale of 10% by transferring 25% of the branch workload to the CSCs. Significant savings were also expected from redesigning and standardizing tasks and procedures, as shown in Exhibit 1.3.4, and procedures were made uniform across all CSCs.

The larger the CSCs, the greater the risk of their becoming 'electronic sweatshops', where large numbers of staff performed highly specialized and repetitive tasks. TSB intended staff to rotate between CSCs and branches, and between different tasks within the CSC for the sake of cross-fertilization and career development. While CSC staff worked in shifts in a factory-like environment, they also needed a strong sense of service and they had to understand the network's systems and procedures and to be familiar with the operations and activities of branches whose mail and phone calls they handled. A CSC manager explained:

> We answer telephone calls for our 10 biggest branches. The customer dials the branch's number but we intercept the call. If we can answer the query, we do. If we can't or if the customer is asking for a specific person, we log the call, and several times a day, we fax a list to the branch so they can call back the customer. We try to relieve branches of interruptions so they are free for face to face service.

The same amount of branch work was able to be performed by fewer staff at CSCs once processes were streamlined and standardized. In addition, as CSCs merged and became larger, they needed fewer staff to cover for illness and holidays. The number of staff in the branch network fell by 1500 to 17 500 in 1990 purely as a result of the changes in the branches and the creation of CSCs.

Product factories

Several data processing centres or 'product factories' were also set up between 1990 and 1992 to centralize product-specific back office activities at a single location. Mortgage administration and processing (Homeloans) was centralized in Scotland. Its structure was redesigned to mirror that of the branch network which it served, as shown in Exhibit CS1.3.5. Mortgage processing had previously been undertaken at seven centres, and significant annual savings were expected from the move.

Meanwhile, the credit card business moved from being a stand-alone strategic business unit to become an operational and customer service centre with card marketing becoming an integral part of the retail banking marketing department. This was done to enable marketing to look after the total customer proposition in terms of product offering. The same approach was adopted with the life and pensions operation and debt recovery was also centralized.

Technology

Technology played a key part in the Network Redesign Project. TSB boasted that:

> . . . the branch systems delivered as part of the project are at least five years ahead of the other high street banks . . . Major competitors like Barclays and National

Westminster freely admit to being at least five years behind us . . . The branch point-of-sale systems . . . have been viewed with envy by many foreign banks who regard the TSB as being amongst the world's best at bringing together advanced technology and enthusiastic staff who are ready to adapt to new systems.[2]

TSB's technological lead dated back to 1969, when a handful of local Trustees Savings Banks decided to invest in an on-line real time (OLRT) computer system. This allowed transactions to be entered directly onto customers' accounts held on a mainframe computer through terminals at the counters. Other British banks batch-processed transactions in a central unit, normally at night. The Trustee Savings Banks were tightly supervised by the government which prohibited them from letting customers run overdrafts and 'off-line' solutions failed to provide the flexibility that would allow transactions to be made at several branches without accounts being overdrawn.[3] TSB's first OLRT system went live in 1972.

In 1988, the bank installed an updated Unisys computer system and 7500 terminals to improve branches' access to customer information. As part of Network Redesign, TSB started focusing on the management of customer information and cross-selling. From 1991, different systems were consolidated to form the customer information database (CID), an off-line database updated monthly from the transaction systems of TSB's various operating units (such as bank, life insurance, general insurance). CID was mainly used for marketing purposes, and branches could access it through an interface located at the main branch of each cluster.

The branch establishment database

The branch establishment database (BED) had helped to set annual branch staff levels and to calculate the grades of branch staff and managers since the 1970s. A major challenge was to standardize their practices, so a database was created to keep scores of all measurable activities. Des Glover, network redesign project director, recalled:

We were constantly trying to standardize grades, structures, job descriptions and staffing levels. Everything had to be negotiated with the union. So we designed procedures for everything, we timed activities and measured the number of transactions. We had transaction volume data for the previous year, we multiplied these by the measured activities, added some allowances, and thus we obtained the required branch staffing level.

The database played an important role in the Network Redesign Project; it helped determine which tasks should be moved to CSCs as well as new staffing levels for CSCs and branches. For example, automating the referred payments function, which was the single most time-consuming activity for branches, enabled TSB to cut 520 positions (see Exhibit CS1.3.6).

THE IMPACT ON ROLES AND RESPONSIBILITIES

The combination of technology and the new network structure enabled TSB to reduce significantly the number of staff and managers in its branches as well as managerial salary costs.

---

[2] *Network News* (Network Redesign newsletter), April 1991.
[3] Moss, M. and Russell, I., *An Invaluable Treasure: A History of the TSB*, 1994.

Exhibit CS1.3.7 shows the impact of clustering on branch staffing, while Exhibit CS1.3.8 shows that over 600 network management positions (out of 2600) were expected to vanish as a result of clustering and that the number of managers with the highest salaries would fall by nearly 1000.[4]

A senior personnel executive explained:

> *Without mincing words, we have taken away a lot of responsibility from the job of the branch manager. You can remove the need for decision-making through technology. Automation and computerized systems, such as credit scoring techniques, allow the person in charge of the branch to press a button and get an answer without necessarily using his judgment. Therefore, we don't need to pay for that judgment any more.*

TSB attacked the three levers determining branch managers' salaries: the number of staff they were responsible for, their level of autonomy and their level of technical expertise.

Branch managers supervised fewer staff because part of the work that used to be done at branches was centralized at CSCs. Many assistant branch manager and supervisor positions disappeared, and branch managers were no longer expected to deal with commercial business which was ring-fenced to control risk and bring focused expertise.

The creation of sales clusters, meanwhile, reduced the autonomy of nearly all branch managers, since a senior branch manager was now responsible for the whole cluster. Supervision was now closer with a senior manager overseeing nine branches on average while in the previous structure a district manager had supervised some 27 branch managers.

The branch manager's responsibility for making decisions on lending decreased with the introduction of stricter credit guidelines, automation and centralization. Nor did branch managers have to help supervisors chase up small bad debts any longer —an activity which was centralized at the debt recovery 'product factory'. In the new structure they were able to focus much more on sales. They were expected to spend more time building and developing relationships with customers so as to increase sales and profit, and less time managing, although they still had to train young members of staff. They were given precise business objectives and had to meet stretching business growth targets.

The Banking, Insurance and Finance Union (BIFU) was well established at TSB, and the bank had to negotiate with union officials. There were some interesting meetings with union officials, but generally they accepted that if costs were going up at 14% per annum and income was going up at 10%, then the future of the whole organization looked bleak. Tough action was needed if the costs were to be reduced and the income increased. Eventually, BIFU also accepted that salaries should be determined by the level of responsibility in making decisions rather than by the number of years staff had spent in the business. The Hay system of job evaluation was used to define the new network positions, a personnel manager explained:

---

[4] Total redundancy costs were estimated at some £35 million (£53 million including pension rights) in a worst-case scenario.

Case 1.3

The Hay system was first used in TSB England & Wales prior to 1989. Peter Ellwood decided to extend it and make it the dominant job evaluation system. The job evaluation scheme was a joint process with the trade union. Job descriptions were prepared by the incumbent or by personnel; then they were evaluated by a small panel of four to six people, half management, half BIFU. The panel agreed upon a score by consensus and that formed the basis of remuneration.

Refurbishment of branches

In 1987-88, TSB started refitting and redecorating branches to a new open plan layout. The refurbishment programme was accelerated as part of the Network Redesign Project as the creation of CSCs liberated more space in the branches. The 'new design' branches featured more congenial interview facilities at the front and a fast cash transaction area at the rear with the traditional counter removed. A senior manager in charge of premises explained in a 1988 interview:

> *Our new branch design bears more resemblance to other successful high street retailers than the traditional image banks used to display. Our branches should be seen as our display shelves.*

Peter Ellwood explained in a December 1989 article:

> *Banks must emulate retailers in the efficient and user friendly delivery of services. . . . Today, every large store devotes as much space as it can to sales. One of the ways this has been achieved is by reducing the size of the stock room, freeing valuable sales floor space. We had to reverse the space ratio which historically saw two-thirds of branch space devoted to staff activities and one-third devoted to customers.*

The search for innovation in the sales process

TSB had been a pioneer in cross-selling—the selling of additional products and services to existing customers—which it had practised since its life assurance arm was set up in 1967. Branch staff referred selected customers (known as 'warm leads') to insurance salespeople who were based at TSB branches, but who would visit customers at home. As a result TSB's insurance salesmen enjoyed far higher productivity than any of their competitors.

In the 1980s, it became clear that a less costly sales process was needed to cross-sell non-regulated banking products to all customers. Creating an in-branch salesforce was the cheapest option. Staff were instructed to approach customers while they waited and to offer them an interview. Customers did not object and a significant number bought new products. But the process was far from efficient. Staff struggled to build the skills to sell more than two or three basic products with confidence. The logistics, too, were unsatisfactory such that whenever a salesperson needed further information, he or she had to log into other systems or search frantically through a filing cabinet while the customer waited. Said Des Glover:

> *We put in time lapse photography and video cameras in a number of branches. For about a month we filmed customer flows and interviews. The videos showed customers becoming fidgety while staff went to get things.*

Case 1.3

Des Glover also pointed to a contradiction:

> We were driving people out of the branches with automatic tellers and touch-tone telephone banking, yet profits depended on cross-selling, which could only happen at the branch. We ended up always approaching the same people. Someone even came in once with a placard that said: 'No, I don't want a pension!'

As part of the new emphasis on selling, 1250 customer service executives (CSEs) were appointed. Their role was to sell the bank's products and services, both individually and as part of the branch team, and they were given targets for sales and customer interviews. They were expected to maintain and enhance customer relations *and* increase profitability. CSE's were not authorized to sell 'regulated' insurance products which were sold separately by the 850 strong insurance salesforce.

A 'Super Service' rule-based system was developed to help them sell a new range of products. The system prompted the CSE throughout an interview, which was conducted with the computer screen turned towards the customer who could therefore see all the financial information TSB had on him or her. A full Super Service interview took 45 minutes, although many interviewers used only one segment of the system. Super Service analysed the customer's needs and identified the most appropriate products from TSB's range. It also incorporated a quotation process, completed sales documentation and printed out the authorization.

AN IDEAL STRUCTURE?

TSB's transformation of its branch network had been an impressive effort which dramatically reduced costs and improved productivity. (Exhibit 1.3.9 presents the evolution of staff levels in the retail banking and insurance operations). By early 1992, however, senior management was wondering what was the most appropriate structure for the bank's branch network. Was it indeed a combination of clusters of branches and CSCs? What was the most effective role for the bank manager? How could we get the right balance between sales and service? What was the optimum distribution channel strategy for meeting customer needs?

Another important issue was that of staff morale. As the number of staff in the network fell, management felt that it had got rid of the lowest performers. But what exactly would be the skills required to serve financial services customers in the 1990s and beyond? Had the bank been right in automating so many aspects of decision-making? What impact would that have on jobs? Should clerical staff alternate between working at a branch and a CSC, or should TSB, on the contrary, try to reap the benefits of ever-increasing specialization?

There were plenty of challenges left for the future, Peter Ellwood concluded:

> All financial institutions use roughly the same words: they all talk about looking after the customer and providing good products because that's basically what financial retailing is about. But the battle will not be won on words, it will be won through execution of the strategy. Since 1989 we have invested a lot in our branch network, in our products and in our staff. We need to invest more and we need to spend more on training.

Case 1.3

We are not sitting here saying 'Great, we've done it', we are saying, 'We've done the easy bit'. Now we have to raise the level of service to the point where our customers genuinely delight in coming to speak to us, and we've got to make it so that all our staff feel that what they do is really important. That's our next goal.

Exhibits

Exhibit CS1.3.1 TSB Retail Bank: Management structure (from July 1989)

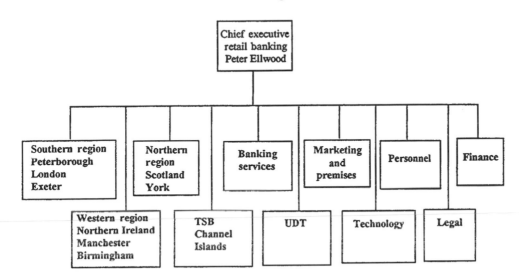

Note: Banking services included credit card services, Mortgage Express, TSB Direct (telephone banking), property services and TSB Private Banking based in Luxembourg. UDT was a consumer finance subsidiary.

Exhibit CS1.3.2 Central head office functions eliminated (October 1990)

	Head office	Regions	Banking services	Total jobs cut	% change	Staff costs (£m)		
						Oct 89	Oct 90	%
Marketing and premises	(52)	86	40	74	−10	17	15.5	−9
Technology	161	37	41	239	−12	42	35.0	−17
Personnel	12	(7)	11	16	−13	6	5.5	−10
Finance	30	340	82	452	−34	26	17.5	−32
Legal	4	3	4	11	−26	2	1.5	−25
Other	110	(33)	(24)	53	−22	8	5.5	−29
Total	265	426	154	845	−19	101	80.5	−20

Case 1.3

Exhibit CS1.3.3 TSB's retail banking network management structure

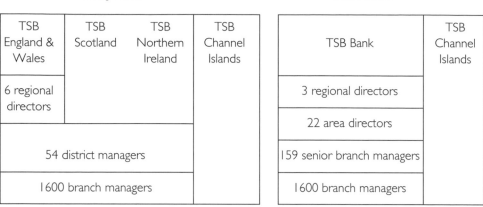

Early 1989

TSB England & Wales	TSB Scotland	TSB Northern Ireland	TSB Channel Islands
6 regional directors			
54 district managers			
1600 branch managers			

Late 1989

TSB Bank	TSB Channel Islands
3 regional directors	
22 area directors	
159 senior branch managers	
1600 branch managers	

Note: The new structure was implemented in stages throughout 1989. TSB Channel Islands was a subsidiary. TSB Northern Ireland, which had a 56-branch network, was sold in May 1991.

Exhibit CS1.3.4 Expected sources of savings (for 100 CSCs, June 1989)

Clerical staff	
Peak and pool management	19%
Basic task efficiency	55.4%
Deskilling of tasks	3.3%
Unspecified work	1.1%
Administrative supervisors	
Restructuring for CSC	29.6%
Premises	
Incremental costs for 75% of CSCs	(25.7%)
Computers	
Back office equipment in CSCs	(22.7%)
Other costs	
Variable costs	14.7%
Branch clerical staff	
Reduction from peak and pool management	25.3%
Total savings	100%

Source: LEK, *TSB Group, Retail Banking Strategy,* consulting report by the LEK Partnership.

Exhibit CS1.3.5 Homeloans structure (October 1990)

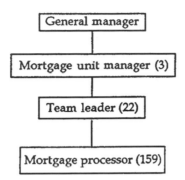

Note: Each mortgage unit manager was responsible for a region—Northern, Southern, Eastern. Each team leader was responsible for an area. The long term objective was that each mortgage processor would eventually cover a cluster.

Exhibit CS1.3.6 Automation at work: Referred payments

Before the introduction of automation, every morning the branches printed a list of accounts that were nearly overdrawn. In each case, a decision had to be made the same day, before 4 p.m., whether to bounce the cheque or charge the customer for unauthorized overdraft. (After 4 p.m., it was too late to return the cheque, which had to be returned the same day.)

The new system preprocessed the information and presented the lending officer with all the information needed to make each decision on a single screen: the customer's profile, what accounts that person had, the monthly average balance, monthly salary or wages, and details of the problem transaction. The officer pressed a key to make the decision, which remained temporary until 3:55 p.m. in case the customer made a payment into his or her account during the day. If the balance changed, the system prompted the officer to make a new decision.

Automation also took care of the next steps. The system produced lists of cheques to be returned, of customer accounts debited, of bank accounts credited, as well as standard letters in which the customer's name and address and details of the charges had been inserted.

The automation process took place over time in four stages:

1. All in branch (the system printed the letters at the branch)

2. Decisions made at the branch, letters printed at the customer service center

3. Decisions prompted (the system advised the lending officer in line with the bank's rules and asked for a justification of any decision that didn't follow these rules)

4. Decisions made at a remote location by a specialized lender

Case 1.3

Exhibit CS1.3.7 Expected impact of clustering on branch staffing, Birmingham region (June 1989)

	A branches		B branches		C branches		D branches	
	Before	After	Before	After	Before	After	Before	After
BM3	3	3	1	0	1	0	0	0
BM1/2	1	1	1	1	0	1	1	0
AM	1	1	1	1	1	0	0	1
Clerical	42	33	18	13	12	9	5	3
Total	47	38	21	15	14	10	6	4

Note: In order of decreasing seniority, branch manager grades ranged from BM3 to BM1; assistant managers were grades AM2 or AM1.

In 1989, the Birmingham region had   31 A (or hub) branches, the future main branches

32 B (large) branches   (>3500 accounts)

81 C (medium) branches   (2000–3500 accounts)

71 D (small) branches   (<2000 accounts)

Source: LEK, *TSB Group, Retail Banking Strategy*, consulting report by the LEK Partnership.

Exhibit CS1.3.8 Expected changes in branch management before and after clustering (June 1989)

Grade	Before	After
BM3	1327	340
BM2	307	321
BM1	260	613
AM2	347	193
AM1	359	513
Total	2600	1980

Source: LEK, *TSB Group, Retail Banking Strategy*, consulting report by the LEK Partnership.

Exhibit CS1.3.9 Total staff numbers in retail banking and insurance

1989	1990	1991	1992
27 922	23 980	23 444	22 943

Case 1.3

Core Processes and Performance Improvement

Chapter

2

A different view of organizations

Try a simple experiment. Ask a manager to draw a picture of his or her organization. It is quite likely that the result will be a picture with boxes and connecting lines which looks something like that shown in Figure 2.1. Why? Simply because such structure charts have been used for decades by organizations. They have been plastered on walls, they have been repeated faithfully in annual reports and employees have witnessed various permutations of the boxes as their firms have experimented with reorganizations in one form or the other.

Over the years, the ubiquitous structure charts have become strong symbolic manifestations of organizations. After all, they are a communication device. And they have some great strengths—such as simplicity. They show how firms are structured; employees can position themselves within a box or a department, and managers can see their position in the hierarchy and develop an understanding of their reporting responsibility.

Despite their strengths, traditional structure charts have several weaknesses. Perhaps the most glaring omission is the lack of a customer or market in the representation. These structure charts have been designed primarily for 'internal' use, to show how the company is organized. Their target audience is the organizations' employees. Customers cannot view such a chart and figure out how they relate to the organization. This would not matter, except for the fact that organizations exist to create value for customers. While structure charts show how firms are organized, precious little is shown about the way they actually function. More specifically, the charts do not show how the different departments work together, in a process, to create value for customers.

Figure 2.1 A typical representation of an organization

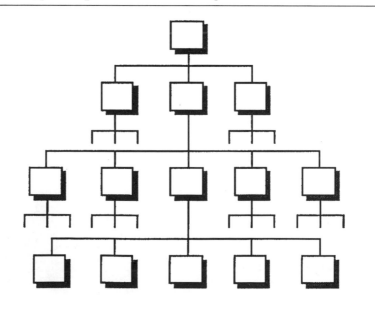

As a communication and orientation device, structure charts have been unusually successful. Organizations have designed hiring, incentives and career path plans in alignment with the departmental boxes. In fact, their success has contributed in part to the now widely acknowledged fragmentation in processes, which have often created problems in terms of delivered end-quality of products, increased costs and extended time delays in the creation of value for customers (see Figure 2.2). Didier Groz, the director of quality for Xerox France comments on how functional silos within Xerox France created organizational problems:

> *The company's classical functional structure was not the way the customer perceived the company. The customer did not wish to go from one function to another. Due to the functional silos, all cross-functional decision-making was pushed up the hierarchy within the organization. This reduced our ability to respond quickly and to be closer to the customer. For example, in France we had eight regional organizations responsible for sales. Service reported directly to the managing director through a separate functional hierarchy. Sales staff had a mindset of 'sell and forget' They had little incentive to be concerned about the relationship with our customer after sales. In the same manner, service staff collected valuable information from the customer during their site visits, but rarely passed them on to the sales staff. In case of any conflicts between sales and service, decision-making had to be pushed up to the managing director.*

Competitive pressures, increased customer demands and shortening business cycles in the global marketplace have required organizations to give serious thought to their organizational structure and systems. Many senior managers have felt the need for change, not simple incremental tweaking but *dramatic* change. Paul Allaire, the CEO of Xerox, expressed his view of the need for change in the early 1990s as follows:

Figure 2.2 Fragmentation in organizational processes

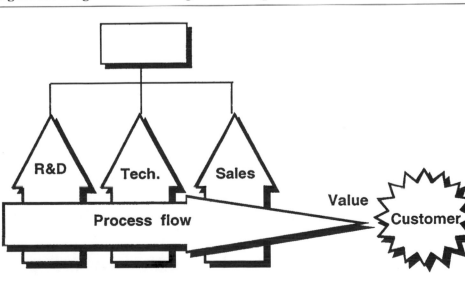

If we are to accelerate the rate of change, then we must change our basic approach to managing the company. For decades, we've run the organization as a large functional machine, which is governed by decisions made at the centre. We've created a system that is complex and which prevents people from taking responsibility . . . We need to create discontinuous change; incremental changes will not get us to our vision.

Associated with this transformation is a different view of an organization—a *process* view that emphasizes how an organization actually 'does what it is required to do' across departments and functions. The focus in the process view is on trying to communicate how an organization 'works together' to create value for its customers, as opposed to how it is structured. Led by a few pioneer firms such as Texas Instruments and Xerox, the process-oriented view shifts the traditional structural chart on its side and brings the customer into sharper focus for all to see.

Core process maps

Core process maps are used to represent a process view of organizations. Figure 2.3 depicts the core process map for Texas Instruments, one of the first companies to publicize its efforts in core process mapping in the early 1990s. Note that the customer is explicitly represented on the core process map. Further, the customer also has certain processes (the processes labelled 'Concept', 'Development' and 'Manufacturing' in Figure 2.3), and these processes interact with the processes of Texas Instruments.[1] This representation highlights the relationships between what Texas Instruments does and what its customers do.

Figures 2.4 and 2.5 depict the core process maps of, respectively, a leading international brewery and Xerox. From Figures 2.3 through 2.5, it is clear that a core process map usually features five to eight core processes and tends to be much simpler than the organizational structure charts for the same organizations. Interestingly, core process maps for small- and medium-sized corporations also typically contain the same number of processes; the complexity of core process maps is relatively independent of the size of organizations. This is not surprising when one realizes that at a certain level, all organizations do similar things (see Figure 2.6): create new products/services, satisfy customers and manage resources in order to achieve the organizational goals.

However, Figures 2.3 through 2.5 also show that organizations choose different core processes. The definition of these is relative and to a certain degree arbitrary. There are several reasons for this. A process that is *core* for one organization may be noncore for another. For example, billing is considered a core process by MCI, but not by most other telecom operators in the world. The billing process strengths of MCI allowed it to launch innovative new marketing programmes which formed a fundamental part of its strategy. For example, the 'Friends and Family' programme allowed individuals to get discounts for calling family and friends. The programme was

---

[1] Texas Intstruments develops components that are used by its customers to create products.

Figure 2.3 Core process map for Texas Instruments

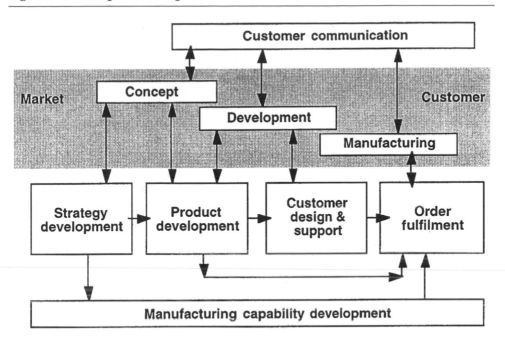

- *Strategy development process*: converts market requirements into a business strategy.

- *Product development process*: uses output of the above process to produce new product designs.

- *Customer design and support process*: produces general product designs for particular customers.

- *Manufacturing capability development*: takes a strategy as input and produces a factory as output.

- *Customer communications*: inputs are customer queries and questions; outputs are increased interest in Texas Instrument products and consolidated responses to customers.

- *Order fulfilment*: converts an order request, a product design and a factory into a product that is delivered to a customer.

key for MCI as it changed the concept of customer from a single individual to a network of individuals, thus driving up the total volume of calls significantly. This would never have been possible without a strong billing process. Except in the United Kingdom, almost no other European telecom operator can match this feature more than five years after its launch in the United States.

Furthermore, there is no unique vocabulary to describe core processes. The core processes represented in Figures 2.3 through 2.5 are all different. These differences are

Figure 2.4 Core process map for a leading international brewery

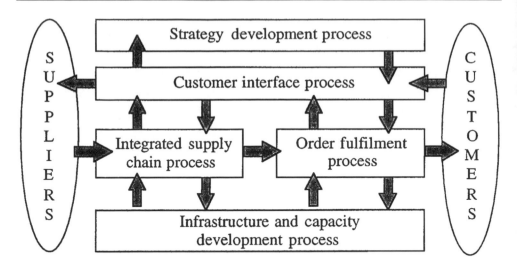

- *Strategy development process:* this process encompasses the decision-making processes responsible for all aspects of corporate direction. The principal goal of this process is to integrate the activities of the organization to best meet customer needs within the framework of longer term strategic intent.

- *Customer interface process:* at the front line of the organization, this process aims to understand and, if possible, exceed customer expectations. This goal is achieved through proactive management of customer perceptions and emphasis of the importance of maximization of customer satisfaction.

- *Integrated supply chain process:* at the tactical level, this process integrates market information and operational planning to optimize the organizational resource allocation. It is a holistic coordination of information and physical flows from the suppliers through to manufacturing and including primary distribution, to minimize total system cost at a specified service level.

- *Order fulfilment process:* the essential purpose of this process is to accurately deliver customer requirements in the most cost-effective and efficient manner. It incorporates a secondary distribution process based on the presales order system and the expansion of empty bottle stocks at the retailer level.

- *Infrastructure and capacity development process:* this process seeks to coordinate development of the physical assets of the company with other core processes so as to create a responsive and cost effective platform for organizational functioning both today and in the future.

Figure 2.5 Core process map for Xerox

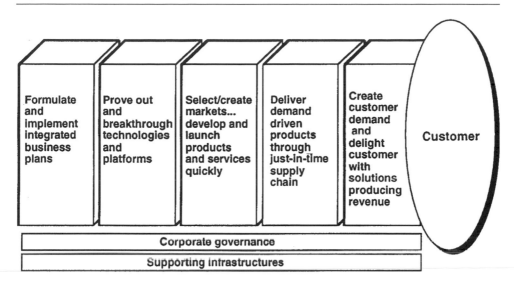

Figure 2.6 Generic core processes for an organization

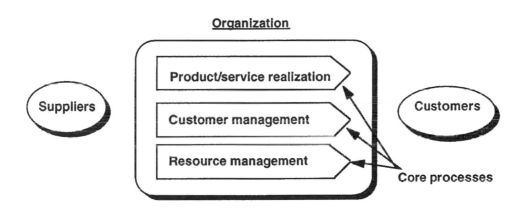

deeper than simply the use of different labels. There is no universally accepted rule about how to set either the boundaries of processes or the granularity—that is, the level of abstraction—of their definitions. In the absence of a consensus on the labels and definitions of processes, organizations have little choice but to resort to using 'extended labels', such as those used by Xerox, which describe the main emphases of the core processes. The importance of the labels and definitions of processes cannot be overemphasized because core process maps, analogous to structure charts, are essentially communication tools. The greater the ease with which they get the message across, the greater their strength.

It is quite common for an individual's view of an organization's core processes to be biased by his or her position within the firm. As a result, two groups could independently come up with different core process maps for the same organization. Top management involvement in the process is vital because the strategic goals of the organization have to be represented in the core process map. Typically, the process of defining these maps involves people from all levels across the organization and takes several months.

There are two generic approaches to defining core processes, as indicated in Figure 2.7. The *internal* perspective is driven by an observation of what the organization does now. The *external* perspective is determined by customer/market demands and defines what the organization *should be* doing to satisfy its customers/market. The two views—internal and external—need to be reconciled before any meaningful action can be taken on core processes.

The experience of the Trustees Savings Bank (TSB) is quite illustrative of the challenges in defining core processes. Its efforts in process mapping began in 1992 and consumed a good part of the next two years. During the first few months, senior management from all corners of the organization were involved in a series of workshops designed to think through the processes. But process mapping was a new concept to the banking sector in the early 1990s, and the lack of an established methodology forced TSB to proceed largely by trial and error.

Frustrated by the slow progress, TSB set up a small department of trained business process analysts to model processes. These analysts spent a good six months plotting 74 processes in detail using the IDEF methodology.[2] The 74 processes were

Figure 2.7 Approaches to defining core processes for an organization

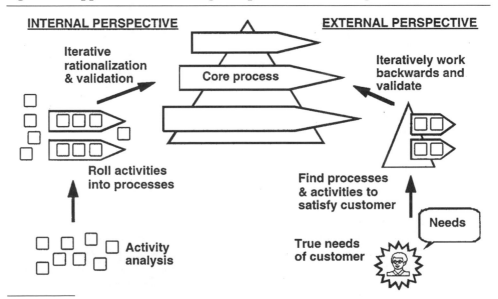

<hr />

[2] IDEF is a formal process mapping methodology that maps a process as interconnected blocks with inputs, outputs, controls and support mechanisms. See Exhibit 2.3.5 of Case 2.3, TSB(D).

grouped into three categories: direction-setting, which included strategic planning, budgeting, top management control; support, which included support processes such as human resource management and IT; and operations, which dealt with the heart of the business including initial contact, transactions, operational centres and the branch network.

After having spent nearly a year looking internally at its processes, and despite having generated volumes of documents with formal process descriptions, TSB faced the challenge of making the processes meaningful and relevant for its management. More workshops were held and little progress was made until a 'new' concept emerged, that of focusing on customer needs. Archie Kane, the director of operations for TSB commented:

> *It was pretty radical for us. We had to get ourselves out of supply-side thinking, which is very difficult for banks. We had a whole range of people from traditionally separate businesses like the credit card business, the pensions business, the branch network, and we had to get them to think in terms of customer-facing processes.*

The next few months were spent questioning the true needs of customers and reconceptualizing the 'internal' processes of TSB in relation to customer needs. In this way, five customer-facing processes were defined: homebuying services, payment services, insurance services, savings and investment services and lending services. The identification of these customer-facing processes also lead to the subsequent realization that TSB did not manage its business that way. Archie Kane commented thus:

> *What does the customer need? If he wants to buy a home, there is a homebuying process, which will involve a bundle of separate products handled in different parts of the organization, but it is one process; if anything goes wrong, in any of those places, the customer will be unhappy.'*

Such insights, which came from understanding and mapping core processes, laid the basis for many of the performance improvement efforts that TSB undertook in the following years.

Leveraging core processes

The story of TSB is not unique. Several organizations make considerable efforts to determine core processes because it gives them a different understanding of their organization, an understanding of how they should organize and what they should change to best meet their customers' needs. In the late 1980s and early 1990s, many organizations tried to leverage core processes in an *ad hoc* manner, and often by trial and error. Looking retrospectively at the process improvement efforts of several organizations, we find a pattern emerging. The pattern, which is depicted in Figure 2.8 consists of a few distinct phases.

Figure 2.8 A methodology for leveraging core processes for performance improvement

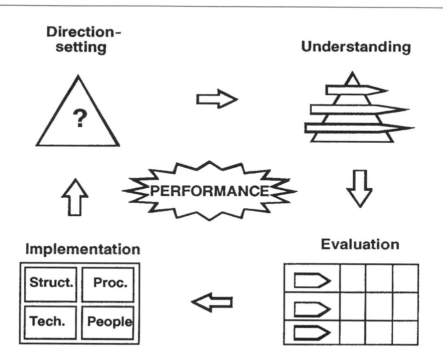

Direction-setting

The performance improvement cycle typically begins with an initial phase of questioning and direction-setting. Several important questions typically guide this phase:

- What are the true needs of our customers?

- What kind of an organization do we want to be?

- What are the goals of our organization?

- What are the guiding values of our organization?

The above questions look deceptively simple; answering them can be surprisingly difficult and often requires a gut-wrenching reexamination of the organization's strategic direction at the highest level. For example, prior to its well-publicized reengineering efforts of the late 1980s, Taco Bell undertook a thorough questioning of the basic purpose and mission of the entire organization. John Martin, the CEO of Taco Bell described the fundamental problem it faced at that time:

> *Our biggest problem was that we didn't know what we were. We thought maybe we were in the Mexican food business. We thought maybe we were*

in the fast-food business . . . The reality was, we are in the fast-food
business, and by not understanding who we were, who our potential
customer was, we were just slightly missing the mark.[3]

TSB also went through a similar process of questioning at the start of its
performance improvement programme. Soon after taking over as the head of TSB's
retail banking division, Peter Ellwood invited a dozen of the bank's top managers to
attend a retreat where they set out to answer some basic questions about the nature
of organization they wanted TSB to become. The result was the adoption of
a mission statement which emphasized a shift away from a 'bank' to a 'financial
retailer':

To be the UK's leading financial retailer through understanding and
meeting customers' needs and by being more professional and innovative
than our competitors.

The word 'bank' was deliberately omitted from the mission statement. Peter
Ellwood explained the reasons for this:

I saw a bank as traditional, bureaucratic, slow moving, averse to change,
not very creative, unimaginative, dull. I saw a retailer, the best retailer, as
very customer-oriented, close to changing customer behaviour, fleet of foot,
entrepreneurial, good at really understanding what the customer needs
and then meeting that need.

Like Taco Bell, this initial phase of questioning was critical for guiding much of
the subsequent performance improvement efforts within TSB. While many organizations
may not necessarily change their fundamental positions as dramatically as Taco Bell
and TSB did, this initial phase of questioning and direction-setting is very valuable for
reaffirming the strategic thrust of the organization. In an ideal world, each organization
should review its positioning and direction on a regular basis. In practice, however,
such phases of reevaluation tend to occur more often in response to intense
competition and market crises. This was certainly the case at both Taco Bell and TSB,
as well as in many other organizations we studied.

Understanding

This phase is concerned with creating an understanding of the core processes of the
organization. As explained earlier, core processes represent key activities and outline how these interlink with each other to create customer value. An understanding
of the organization's strategic direction is very critical for identifying its core
processes. A process is core because a firm *chooses* to make it core. A process
cannot be core if it does not form a critical support for the strategic direction of the
organization.

---

[3] John Martin as quoted in 'Taco Bell Corp.', *Harvard Business School Case Study*, No. 9-692-058.

Drawing comprehensive core process maps can be a time-consuming chore. Though they may look relatively simple (see Figures 2.3 through 2.5), each process can be further decomposed into subprocesses. For example, Table 2.1 depicts the major subprocesses of the international brewery company represented in Figure 2.4. Each of these subprocesses has its own associated set of inputs and outputs and can be further

Table 2.1 Subprocesses for the leading brewery company described in Figure 2.4

Core process	Major subprocesses
Strategy development process	■ Secondary distribution and sales strategy ■ Marketing strategy ■ Product offering development ■ Operations strategy
Customer interface process	■ Market research ■ Market strategy planning ■ Advertising ■ Secondary distribution planning ■ Promotion campaign processing ■ Customer satisfaction services ■ New product development ■ Integrated customer demand planning
Integrated supply chain process	■ Customer demand integration ■ Integrated operations planning ■ Procurement processing ■ Production and packaging ■ Bidirectional primary distribution
Order fulfilment process	■ Presales order taking ■ Depot inventory processing ■ Seeding* ■ Routing decision-making ■ Presales and seeding delivery ■ Commercial transaction processing
Infrastructure and capacity development process	■ Long-term aggregate demand forecasting ■ Long-term commercial zone and seeding forecasting ■ Network optimization of supplier–plants–depots ■ Network optimization of depots–retailers routing ■ Technical and financial project management

*This is the practice of keeping empty bottle stocks at the retailer level.

decomposed into subsubprocesses. Figure 2.9 depicts a two-level decomposition of one of the core processes at Xerox: 'Deliver demand-driven products through just in time supply chain'.

While this decomposition of processes into subprocesses can be carried out down to several levels, it is typically not very useful to proceed beyond two levels. As the number of decomposed levels increases, the complexity of the core process maps

Figure 2.9 A two-level decomposition for the 'supply chain' core process of Xerox

Manage Supply

Sub-sub-processes
■ Plan, design and implement the supply network
■ Plan inventory strategies and levels
■ Develop and qualify suppliers
■ Integrate and supply network capacities
■ Implement short and long-term integrated production plan
■ Plan, design and implement the service spares network

Produce Product to Meet Customer Demand

Sub-sub-processes
■ Design and provide product capabilities/capacities
■ Schedule productions operations (staff/machines/materials)
■ Generate pull signals for shipment from supply network
■ Move materials through manufacturing operations
■ Set up manufacturing and assembly
■ Perform value-added manufacturing steps
■ Stage final product integration for customer shipment
■ Maintain product infrastructure (machines/processes/human skills)

Manage Logistics

Sub-sub-processes
■ Managing and executing logistics operations
■ Receiving and handling materials
■ Operating warehouses and intermediate materials staging facilities and mechanisms
■ Perform packing and shipping
■ Manage transportation across the integrated supply network

increases and the usefulness of the tool to aid understanding for performance improvement decreases. Further, representations of the inter-relationships of each subprocess with other organizational processes can result in very complex charts with a maze of arrows and links.

Some pioneer organizations have chosen to chart our their entire set of organizational processes in detail. For example, Xerox has published a complete book on its process definitions and inter-relationships for use by its employees. Such a route is feasible for organizations with high levels of prior investment in process thinking. Xerox has invested significantly in educating its employees about processes since the start of its quality movement in the early 1980s. Its employees can hence relate fairly easily to processes and process-related performance improvement programmes.

An alternative approach, adopted by firms such as Alcatel, is to identify a few key processes and focus on improving them. This is more useful for organizations that have either not made significant investments in process-related thinking or do not have the luxury of waiting for the year or more that it usually takes to produce a comprehensive core process map.

Evaluation

This third phase is concerned with (a) the identification of performance gaps for each core process and (b) the setting of appropriate performance goals for the organization. For each core process, resource utilization and output effectiveness have to be carefully examined. This phase is key for changing the organizational context before launching specific performance improvement programmes.

There are two generic sources for identifying performance gaps: external and internal. External sources include customers and competitors. Because customers are both ultimate recipients and judges of organizational performance, they are probably the best source for the identification of performance gaps. Even when the firm thinks it knows its customers well, it can still be surprised by careful research on this subject, as was the case with Taco Bell.

When Taco Bell undertook a thorough reexamination of its customers' needs, it found that customers most valued the quality and quantity of the food they received across the counter. Other features, such as large, fancy premises, mattered much less to customers than Taco Bell managers had thought. While with hindsight this revelation does not seem earth-shattering, it implied a major performance gap for Taco Bell as they had built their business, much like their competitors, on the premise of reducing the cost of food given to customers. As it turned out, they had been actively working on reducing the value delivered to customers! This realization guided much of Taco Bell's reengineering efforts in the following years and revolutionized the fast-food industry. Kenneth Stevens, Taco Bell's president described it as follows:

> *We started by taking a hard look at what customers got for every dollar they slid across the counter. And what they were getting at the time was*

essentially 29 cents' worth of food and packaging. The other 71 cents went to things the customer didn't care about: operations, labour, rent, marketing and advertising. So we went to work on the remaining 71 cents and began reengineering literally every facet of the business that would give our customers more in return for the dollar they gave us. This philosophy was exactly the opposite of what everyone else in the industry was doing. The industry's goal was to reduce the cost percentage of food and packaging that they returned to the customer. Our goal was to increase it—to give the customer more.[4]

Benchmarking against other firms can also make underperformance visible and help set performance goals. During its 'catch-up' phase in the mid-1980s, for example, Taco Bell actively benchmarked against its leading competitors in the fast-food industry, McDonald's in particular. Sometimes, it is easier to benchmark against competitors in different geographical regions. TSB, for example, used consulting firms to help them benchmark against leading-edge practices in Californian retail banks such as Wells Fargo.

While benchmarking performance indicators is commonly done against competitors, benchmarking *processes* is very powerful as one can easily cross industry barriers and benchmark against best-in-class firms. Certain organizations are renowned for their strengths in particular processes, such as Federal Express in the order tracking and management process, LL Bean in the inventory management process, Milliken in customer management processes and MCI in the billing process. A bank can benchmark against Milliken for customer management processes and a manufacturing company can benchmark inventory management with LL Bean. TSB regularly benchmarked itself against leading retailers such as Marks and Spencers as it moved ahead towards its goal of becoming the United Kingdom's leading financial retailer.

More accessible than external mechanisms, an organization's *employees* can be valuable sources of information about performance gaps. Front-line employees have valuable knowledge of problems and complaints at points of contact with customers and partner organizations. Employees on the shop-floor have first-hand knowledge of problems with manufacturing processes. Managers have data on performance differentials across different parts of the organization. The challenge is to set up mechanisms to obtain and communicate valuable performance related feedback from these 'internal' sources. Cross-departmental meetings and quality improvement teams are frequently used mechanisms for such feedback. 'Brown paper exercises' are also useful to allow employees oppressed by organizational bureaucracies to express their frustrations and views on organizational deficiencies.[5]

---

[4] Reengineering for Your Customers: How Taco-Bell Did It', *Restaurants USA*, June–July 1994, pp. 36–38.
[5] Brown paper exercises involve collaborative mapping of a process and the identification of its limitations by different employees across an organization. Typically, a process is outlined on a large sheet of brown paper (hence the name) and this sheet is circulated across relevant departments. Employees are encouraged to post coloured Post-It notes on the brown sheet outlining major (red colour) to minor (yellow) problems with the process. At the end of the process, the Post-It notes are tabulated and the performance gaps identified for further action.

Combining inputs from internal and external sources, the evaluation phase typically results in a summary of key goals and performance gaps for the organization's core processes. Table 2.2 depicts such a summary for the leading international brewery company whose process map is presented in Figure 2.4.

The determination of performance gaps provides the basis for identifying high leverage processes, i.e. processes which, upon appropriate change, will lead to the maximum improvement in organizational performance. This focus is important because organizational constraints make it difficult for a firm to try to achieve significant change in all processes simultaneously. While performance gaps point the direction towards high leverage processes, the actual choice of the degree of change desired in different processes is dependent upon a number of other factors: strategic importance of the process, timing of benefits from the process change, risk associated with the change and amount of resources needed for the change. Table 2.3 provides a simplified summary of the identification of high leverage processes by the brewery company considered in Figures 2.4 and 2.9 and Table 2.2. The high leverage processes are indicated in italics in Table 2.3. For practical implementation, a similar analysis would typically be done for the major subprocesses of the high leverage processes.

Implementation

The last, but perhaps the most important phase, in the process improvement 'pattern' of Figure 2.8 is that of *implementation*. Successful implementation is a critical differentiator between successful and unsuccessful performance improvement programmes. That is because, in many cases, the fundamental concepts for process redesign are easily available from competitors or best-in-class organizations. TSB obtained most of the building blocks of its performance improvement programme, such as the concept of clustering bank branches, by benchmarking against Californian banks. Other UK and European banks also had access to this information. TSB differentiated itself by deciding to implement the new concepts a few years earlier than its competitors, and by managing the implementation successfully; several other European banks which later tried to implement the same concepts as TSB failed to get the implementation right, and other banks are still working on it!

Generally, effort spent upfront in understanding core processes and evaluating high leverage processes, pays off in facilitating successful implementation programmes. As discussed in Chapter 1, most performance improvement programmes require changes along multiple fronts: processes, structures and systems, and technology and human resource management. It is important to exploit synergies across these fronts and to ensure that changes in one dimension complement those in other aspects.

For example, a major part of the change programme within Taco Bell was a shift away from a manufacturer to a customer-focused, retailer of food. This decision implied major changes within multiple aspects in the organization. The operational processes of food preparation were simplified and kitchen space in restaurants was minimized. Managerial processes were modified to allow restaurant managers to grow into new, expanded 'general management-oriented' job definitions. Advanced technological

Table 2.2 Key goals and performance gaps for the international brewery of Figure 2.4

Core process	Goal of core process	Goal measurement	Goal evaluation	Needs (L,M,H)	Current performance (L,M,H)	Performance gaps
Integrated supply chain	Minimize total supply chain costs	System unit cost	Internal	H	L	Large
	Assure technical quality	6 sigma	External / Internal	H M	M	Medium
	Increase SKU service fill rate (primary distribution)	% of orders fulfilled from depot stock per SKU	External / Internal	H L	L	Large
Customer interface process	Increase net revenues	• Net contribution margin per $ of transaction • Net contribution margin of product mix	Internal	H	M	Small
	Increase penetration of retailers	% vertical and horizontal coverage	Internal	H	M	Small
	Achieve increasing level of customer satisfaction	• Purchase intention • Perceived quality survey	External	H	M	Small
Order fulfilment process	Increase service fill rate —retailers (secondary distribution)	% of orders fulfilled from retailer stock to end customer per SKU	External	H	L	Large
	Minimize secondary distribution costs	Transportation and inventory holding unit/cost	Internal / External	H	L	Large
Infrastructure and product development process	Minimize time to market	Weeks	Internal	M	M	None
	Minimize supply chain infrastructure development costs	Current vs. budget	Internal	M	M	None
Strategy development process	Maximize EVA	ROIC	Internal	H	M	Small
	Increase shareholder value	ROE	Internal	H	M	Small

Key: L = low; M = medium; H = high; internal = company defined; external = customer defined

77

Table 2.3 High leverage processes for the international brewery of Figure 2.4

Core process	Perform-ance gap (S,M,L)	Strategic import-ance 1–10	Economic benefits (L,M,H)	Timing of benefits (St, Mt,Lt)	Risk of change (L,M,H)	Resources needed (L,M,H)	Strategic leverage (L,M,H)
Integrated supply chain process	H	10	H	Mt	M	M	H
Customer interface process	L	10	M	Mt	M	L	M
Order fulfilment process	L	10	H	St-Mt	M-H	L	H
Infrastructure and product develop-ment process	L	8	L	Lt	L	H	L
Strategy develop-ment process	L	9	L	Lt	M	L	L

Key: (S,M,L) = (small, medium, large); (L,M,H) = (low, medium, high); (St,Mt,Lt) = (short term, medium term, long term)

support was implemented in restaurants to support restaurant managers in order to create time for customer care. Hiring and incentive packages were modified to attract and retain the right kind of employees. A critical success factor in this project was that each individual element of change complemented and reinforced the other. All the pieces came together in one coherent whole to create the new Taco Bell.

Implementation also raises several important managerial issues, especially when change is undertaken across the organization. One important decision is whether to entrust change projects to line managers or to dedicated teams. While line management can handle small-scale change projects in addition to their line responsibilities, dedicated project teams are often favoured for large-scale change projects. Another problem is the level of appropriate project management skills available; in too many organizations shortage of such skills hampers implementation progress. We will come back to these issues in Chapter 5.

Another frequent cause of implementation failure is the lack of success at managing the IT side of the change project. This will be the subject of the next chapter.

Case 2.1

This case was written by Nikos Christodoulou, Research Associate, and Soumitra Dutta and Enver Yücesan, Associate Professors at INSEAD. The case is intended to be used as a basis for class discussion, rather than to illustrate either effective or ineffective handling of an administrative situation. Financial support from the INSEAD Alumni Fund European Case Programme is gratefully acknowledged.*

Rank Xerox France (A): The Logistics Process

INTRODUCTION

> Rank Xerox will, over the next few years, bring to the Market place an even greater product array. It will include the traditional reprographic products supplemented by Fax and Electronic Transmission devices, Industrial Reprographics, Document Systems and Centralized and Decentralized electronic printing equipment.
>
> This wide array of technologies will blend with ever increasing demands for service from our customers. Requirements will be made still more complex by the distribution channels and account type classification employed by Rank Xerox to tailor service at an affordable price to our customers' individual needs. In this scenario the competitive pressures will increase and the drive to maintain profit margins through reduced support will continue unabated.
>
> Logistics Development in the '90s
Internal Rank Xerox Document, November 1988

Rank Xerox had spent most of the 1980s painfully regaining, inch by inch, the market share it had so dramatically lost to new entrants (primarily Japanese competitors) from its dominant position in the market during the 1970s. Central to this comeback was drastic cost-cutting with the introduction of just-in-time (JIT) in manufacturing, and an obsessive dedication to quality which had gradually changed the culture and management of the company.

As Rank Xerox looked forward to designing its strategy for the 1990s, there seemed to be a never-ending barrage of challenges. Competition was intense and showed no signs of slowing

*Copyright © 1996 INSEAD, Fontainebleau, France.

down. In fact, competitors were preparing to take the challenge to the core of Rank Xerox's most profitable segment—high-end reprographic equipment. Intense price competition had held Rank Xerox's profit margins down to 4% (from 17% at the start of the 1980s). Shrinking lead times in the market were constantly pushing the boundaries of what the Rank Xerox organization could support. Customers were also raising their expectations of the quality of equipment and services delivered by Rank Xerox .

Central to Rank Xerox's efforts to meet the challenges of the 1990s was an ambitious logistics strategy which could provide to the customer (a) an up-front commitment on delivery dates, installation, training and other customer services, and (b) to meet these commitments consistently on a timely basis with the highest degree of quality to achieve high levels of customer satisfaction and market share, (c) while enabling the Rank Xerox organization to operate productively to maintain its competitive edge. However, satisfying these aims for the logistics strategy was not easy. Claude Joigneault, the logistics director for Rank Xerox France in 1993, elaborated further:

> In 1988, the total planned logistics cost for Rank Xerox was around 12% of revenue. Using cost of quality techniques, we were able to establish that the cost of non-conformance within the logistics network of the organization amounted to around 30%. Thus, there was significant room for improvement.

> However, there were considerable organizational challenges in moving ahead. The logistics process was broken—fragmented into independent parts each of which operated independently with little concern for either the end customer or the effectiveness of the whole process. For example, the manufacturing units had no links to the level of service offered to end customers. They were measured only by their ability to deliver a fixed number of products every month to each country unit. Thus, we often faced situations when we had either too much inventory or were unable to keep up with the demand and satisfy customers. We tried to adjust within Rank Xerox France by managing our warehousing space, but this increased the level of stocks—which then was 73 days of supply, a figure much above benchmark standards. This made us uncompetitive.

> Due to the diversity of countries and markets within Europe, the logistics network for Rank Xerox is more complex than that for Xerox in the USA. Differences caused by the international nature of markets and relative independence of the country operating units further complicated our problems. For example, it was not possible to move stock easily across countries to balance demands. This meant that we often did not have the right product demanded by a customer at the right time. This made us unattractive to our customers.

In view of these concerns, it was evident that change had to happen. An effective and efficient logistics process was vital for enhancing the competitive position of Rank Xerox.

RANK XEROX

Rank Xerox is one of Europe's leading high technology companies providing a wide range of products, systems and services to handle customers' document management requirements. Rank

Xerox was formed in 1956 as a joint venture between the Rank Organization of the United Kingdom and Xerox Corporation of the United States to manufacture and market copiers in Europe and Africa. Subsequent major joint ventures of Rank Xerox included the creation of Fuji Xerox in 1962 to develop and market products in Japan, and Modi Xerox in 1983 to manufacture and market products in India. Members of the group share research and development costs, and actively transfer technology and management resources.

In 1959, Xerox was the first company to launch a xerographic office copier, the 914 model. Based on the 'electro-photography' process invented by Chester Carlson, the machine used an electrostatic process to transfer images from one sheet of paper to another. The technology was very successful and allowed Xerox and Rank Xerox to expand rapidly and dominate the market with shares approaching 100% during the 1960s and early 1970s. The success resulted in explosive growth. During 1964, the corporation recruited more people than there were on the payroll on the 1st of January of the same year.

The technology was protected by a large number of patents. However, these patents began to expire during 1970s allowing many new players to enter the market. Before 1980 more than 70 companies had produced their own copiers. There were companies from the office equipment industry, the electronics industry, the reprographics industry and the photographic equipment industry. Some of them were not successful and were soon forced to exit the business. But others, like Ricoh, Kodak and Canon, challenged the dominant position of Xerox and Rank Xerox in the market. Canon in particular gained substantial ground by attacking the low-volume market, a market segment which was initially considered unimportant by Xerox. The situation was worsened by the ability of the Japanese firms to offer low-priced high-quality products in the market.

Although its revenues continued to increase, Xerox's market share fell from 93% to 40% in 1985. To recover its leadership position, David Kearns, then chairman and CEO of Xerox, adopted the quality approach and implemented a new strategy called 'Leadership Through Quality'. Initially, the strategic focus was on product quality and cost, but the emphasis soon moved to incorporate all aspects of operations.

As Rank Xerox was progressing in the implementation of the new strategy to change the culture of the company to one of quality, new challenges were emerging. Customer requirements were changing, becoming rapidly more demanding. Increased competition had changed the nature of the traditional markets. Rapid developments of product technologies were directing Rank Xerox towards greater integration with computer-based technology.

In response to these challenges the company launched new products across the complete market range, putting itself ahead of the competition in laser printers, centralized electronic printing, high-volume colour printing and in systems reprographics. The Docutech system broke new ground in merging reprographic and computer technologies so that document information could be managed from a whole range of networks and workstations. Perhaps the most visible change from outside was the positioning of Rank Xerox as 'The Document Company', a name that emphasized the shift from the copiers to a broad range of document management solutions.

The results of these efforts could be seen by 1986. The erosion of the market share slowed down and the company started to climb back to leadership (see Exhibit CS2.1.1).

Rank Xerox France was one of the biggest operating units of Rank Xerox. Like the other operating units, RXF had close relations with the company's headquarters where it reported regularly and from where it received directions for the company's future strategy.

THE PRODUCT RANGE

Didier Groz, the director for quality for Rank Xerox France, described the nature of customers served by the organization:

> Rank Xerox France has around 150 000 customers out of a total of approximately 500 000 customers in Europe served by all Rank Xerox units. All these customers are very different. They range in size from one-person businesses to organizations of several thousands of employees, vary in nature from large private companies to governmental departments and agencies, and hail from all sectors of industry. Accompanying this diversity are large differences in their specific needs for document management.

As 'The Document Company' of choice for organizations, Rank Xerox France offered a range of different products to meet the needs of its diverse customer base:

- First, there were commodity products such as facsimile machines, electronic typewriters, and the like. The majority of these products were sold through indirect channels (dealers). The stockholding policy for indirect channels had not been fully established, but it was expected that sufficient stock should be held at the dealer's or distributor's site to meet a few days' demand.

- The second type of products were copiers and laser printers. The copiers were divided into high-, mid- and low-volume, representing different market needs. Low-volume copiers were copiers with speeds from 1 to 30 prints/minute, mid-volume copiers had speeds from 30 to 90 prints/minute and high-volume copiers had speeds of over 90 prints/minute. The copiers also had different requirements in installation and maintenance. These products accounted for 85% of the total production.

- A third type of product was copiers of very-high volume, high-end reprographic electronic systems and generally products that required a special customization in order to meet customer requirements. They were very heavy, consisted of more modules than ordinary copiers and were assembled at the customer's site. They usually cost upward of $200k.

The installation procedures were not the same for all types of products. Commodity products did not require any special installation and the customer was able to use the product immediately. For low-volume copiers, the driver who made the delivery was required to install the product and demonstrate the key operating features to the customer. For mid- and high-volume copiers and customized products, special installation and training were required by the technical support staff of Rank Xerox France. Exhibit CS2.1.2 shows the nature of product deliveries with respect to the different types of products.

Table CS2.1.1 Order-to-install times

Type of product	Order-to-install time	
	Minimum time	Maximum time
Commodity	2 days	5 days
Copiers and laser printers	1 week	3 weeks
Customized	3 weeks	1 month

Also the order-to-install[1] time was not the same for all types of products. Table CS2.1.1 provides information about the order-to-install times for different types of products.

MANUFACTURING

Rank Xerox France received products from four manufacturing sites in Europe: Mitcheldean (UK), Venray (Holland), Lille (France) and Coslada (Spain). There were also manufacturing sites outside Europe: in the United States and Japan. Furthermore, modules could be received from Canada, Brazil and Mexico.

The European manufacturing sites were mainly assembly sites, since around 80% of the production costs corresponded to purchased material. Mitcheldean assembled low- and high-volume copiers. Venray assembled mid-volume copiers. Printers were manufactured in France, while the factory in Spain was used for the production of toner.

Production plans were prepared in the following manner. At the beginning of each year, all operating units in the various countries were asked to provide the headquarters of Rank Xerox with their forecasts for the demand in their markets. At the headquarters these forecasts were analyzed, compared with historical data and a consolidated production plan was released. This plan organized the production of all Rank Xerox products in manufacturing sites all over Europe. It was a 12-month-horizon plan with a quarterly and a monthly revision. This meant that small adjustments to the initial schedule were possible to achieve a better fit between supply and demand. Claude Joigneault noted:

> Though we tried our best, our forecasting system was never wonderful. We had to predict not only the demand for products, but also for product configurations. This was complex as we had a wide range of product offerings.

The products came out of the factories already customized to meet the needs of the local market. For example, products which were produced for France could not be sent to the United Kingdom or Spain without modifications. Claude Joigneault elaborated:

> Errors in forecasts often created havoc with our ability to meet customer needs. If we had underestimated the demand for certain products for a specific month, there was no easy way to meet the demand by moving goods from other countries. Moreover, situations of shortages and oversupplies in neighbouring European markets tended to occur simultaneously. This increased our order-to-install times to unacceptable levels.

---

[1] Order-to-install time defines the period between the acceptance of the order and the installation of the equipment.

The manufacturing units' target plans were to produce and deliver a fixed number of units per month. This resulted in manufacturing pushing products, especially at the end of each month.

SUPPLIERS

The extended use of purchased material made Rank Xerox heavily dependent on its suppliers for the quality of its products. Starting in 1982, Rank Xerox embarked on a JIT manufacturing capability development programme in order to ensure the quality of the raw materials and the components that were purchased and the dependability of the supply deliveries. As part of this programme, the supplier base was reduced from 5000 to 300 and quality improvement of parts from suppliers was emphasized. The results were impressive. By the mid-1980s the number of parts being rejected at the assembly lines reduced from 10 000 to 250 parts per million (or from 1% to 0.025%). While previously all parts were delivered to the receiving dock, more than 70% of parts started being delivered directly to the assembly lines when assembly operators needed them. In the manufacturing sites, the throughput increased three times, the inventory of raw materials reduced by 80% and the transportation cost reduced by 40%. The overall manufacturing lead time decreased by 46% as a result of these changes. Exhibit CS2.1.3 provides a graphical summary of some of the changes enabled by the JIT manufacturing programme.

Furthermore, in 1984 Rank Xerox built on the JIT manufacturing initiative by introducing the Continuous Supplier Involvement Process, whose main features were to:

- work in full partnership with the manufacturing suppliers in order to pursue continuous improvement
- provide the suppliers with training and support
- involve them in product design and manufacturing activity at every stage
- recognize their contribution through an award called the Certified Suppliers' Award.

WAREHOUSING AND DISTRIBUTION

Finished products, spare parts and supplies from the manufacturing sites were stored in four supply centres located near the European manufacturing plants. Within these intermediate stocking centres, all materials were received at the receiving dock, inspected, repacked and stored. The items were shipped to the national warehouses for each operating company according to the predetermined supply schedule. Prior to shipment, it was necessary to pick and sort the items, and pack and consign them appropriately to ensure correct delivery. Exhibit CS2.1.4 describes the flow of products within the distribution network of Rank Xerox.

For example, Rank Xerox France had two warehouses, one in Garonor and another in Gonesse.[2] The finished goods, which were produced for France according to the consolidated production plan, were shipped from all manufacturing sites to the national warehouse of Garonor, where

---

[2] The size of Garonor was 11 000 m² while Gonesse had a size of 6300 m².

they remained in stock. When an order arrived at Garonor it was entered into a logistics information system (Accord) which scheduled the dispatch of the products according to the stock availability (as determined from a stock information system called Mistral). The products were checked and packed by the Rank Xerox France personnel at Garonor and then they were dispatched to the field logistics centre. Rank Xerox France had 21 such centres which were called regional platforms. The regional platform crew, which was subcontracted, was responsible for contacting the customer and arranging the delivery and installation of the product. Garonor also kept stock of spare parts required by Rank Xerox France's service organizations which provided technical support—repair and maintenance—to end customers.

The other national warehouse, at Gonesse, was used for the carcass management process: Rank Xerox France gave its customers the option to withdraw an old piece of equipment and replace it with a more modern one. The ratio was two new copiers (delivered) to one old piece of equipment (withdrawn). Depending on its condition, the old equipment could be used for refurbishment, asset strip or scrap. Refurbishment meant that, through parts and spares replacement, the product was reoffered to the market. The refurbished equipment had a lower price, but the same guarantee of three years. Asset strip meant that some good parts of the old equipment were selected and used in maintenance. Finally, a piece of returned equipment was characterized as scrap when it was very old or had a different technology (a competitor's product). In fact, in terms of copiers, approximately 80% of products contained some refurbished parts and spares.

The carcasses followed a reverse route from the customer site to the regional platform and then to Gonesse. A similar network was operated for all national operating companies in Europe. A logistics manager recalled the pressures within the distribution network:

> The target for the distribution network was to deliver a specific quantity of product each month at Garonor according to the manufacturing decision to push the stock there. We usually received the order for this quantity at the end of the month. But many times we had difficulties with the peaks, for example sometimes we had to receive 10 trucks at Garonor on the last two days of the month. The personnel at Garonor had to handle the products, put them into stock and at the same time fulfil the contracts they had.

TRANSPORTATION

The transportation activity was outsourced to a number of different transportation companies. On a European level, about 25 different transportation companies were used to move products between the manufacturing sites and the supply centres. The further movement of products to the national warehouses was handled by around 80 different transportation companies.

Rank Xerox France used 16 transportation companies by as suppliers of transportation services from the national warehouse through interim transit points to the end customer. The transporters were also controlling the regional platforms and, furthermore, were responsible for the delivery and installation of the products. The number of the transportation companies caused Rank Xerox France many problems. Claude Joigneault comments: 'We had many difficulties in managing them; for example, each company had its own tariffs. Also, the communication between them was not the best.'

Case 2.1

SALES

Sales orders originated in two different ways. The first was through a direct salesforce who negotiated contracts with customers. The salespeople validated the order, checked the customer's credit-worthiness and input the order into an ordering information system (Sophia). The other type of orders came from dealers who sold exclusively Rank Xerox products. They could input their order into the Sophia information system directly through Minitel, a French teletext system. The dealers made 40% of the sales revenues but the trend was to increase their number and cut down the direct salesforce.

The sales staff could not guarantee delivery dates to the customers. A manager recalled:

> If stock was not available at Garonor, the customer had to wait for the next shipment. This could take from a few days to a few weeks. Product availability varied according to the type of product, the time period within the month and the demand. Of course this was not very satisfactory for our customers. Complaints were quite frequent and most of the time we could not do anything about it because manufacturing had no flexibility.

PLANNING FOR THE 1990s

The competitive environment of Rank Xerox was evolving rapidly. A major challenge for the company was to have a logistics infrastructure in place to satisfy customer requirements in the most cost-effective way in the coming years (Exhibit CS2.1.5 provides information about the total logistics cost and the level of assets for 1988). In 1988, a major study was initiated within Rank Xerox to define the Rank Xerox Logistics Business Strategy for the 1990s. Bernard Costiou, the project manager for this study, commented:

> We knew that the speed and accuracy with which we could meet customer requirements would be of vital importance for customer satisfaction and loyalty. We had made significant progress during the 80s, but much needed to be done. The organization had to be changed to pull together the different units into one coherent focused operation. The question was what to do and how best to manage the implementation?

Exhibits

Exhibit CS2.1.1 Regaining market share

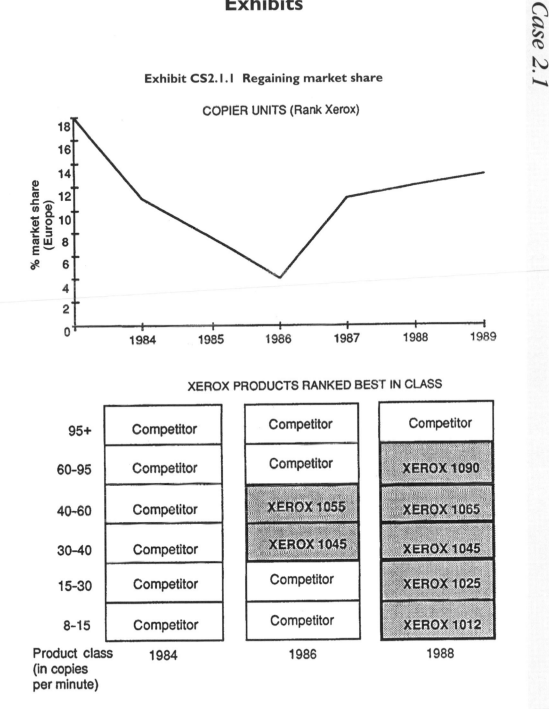

Exhibit CS2.1.2 Transportation/delivery service

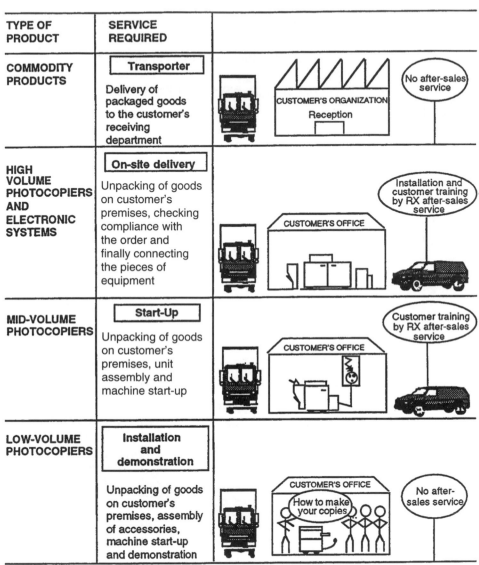

TYPE OF PRODUCT	SERVICE REQUIRED	
COMMODITY PRODUCTS	**Transporter** Delivery of packaged goods to the customer's receiving department	
HIGH VOLUME PHOTOCOPIERS AND ELECTRONIC SYSTEMS	**On-site delivery** Unpacking of goods on customer's premises, checking compliance with the order and finally connecting the pieces of equipment	
MID-VOLUME PHOTOCOPIERS	**Start-Up** Unpacking of goods on customer's premises, unit assembly and machine start-up	
LOW-VOLUME PHOTOCOPIERS	**Installation and demonstration** Unpacking of goods on customer's premises, assembly of accessories, machine start-up and demonstration	

RX = Rank Xerox

Exhibit CS2.1.3 JIT manufacturing

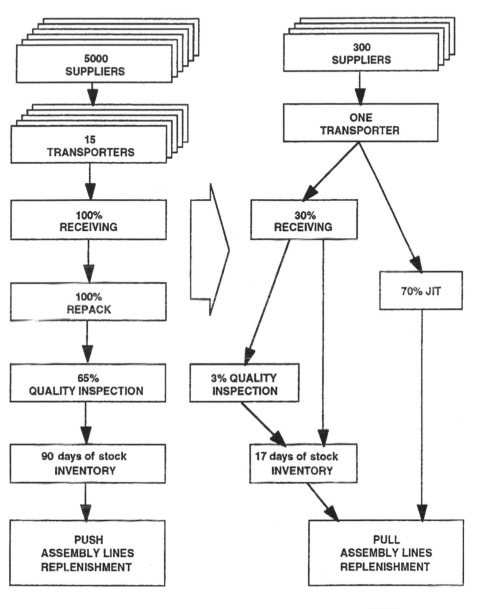

Case 2.1

Case 2.1

Exhibit CS2.1.4 Distribution flow (1988)

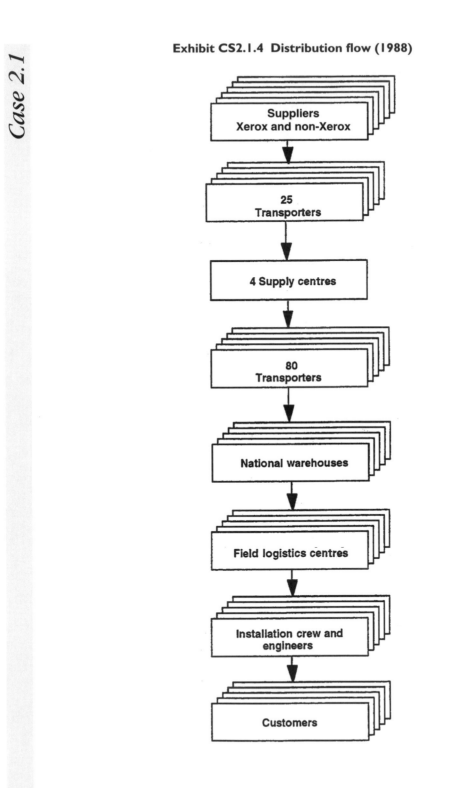

Exhibit CS2.1.5 Logistics costs and assets

COSTS ($ millions)

Transport	9.3
Personnel (salaries, expenses, temporary staff)	4.2
Warehouse (surface area, maintenance)	1.6
Information systems (RX France's contribution to logistics)	0.9
Others (equipment, external services)	0.7
Total	16.7

ASSETS

Type	$ millions	DOS (days of supply)
New	18.3	30
Renovated	4.3	6
Returns	20.5	31
On loan/on trial	4.3	6
Total	47.5	73

Case 2.2

This case was written by Nikos Christodoulou, Research Associate, and Soumitra Dutta and Enver Yücesan, Associate Professors at INSEAD. The case is intended to be used as a basis for class discussion, rather than to illustrate either effective or ineffective handling of an administrative situation. Financial support from the INSEAD Alumni Fund European Case Programme is gratefully acknowledged.*

Rank Xerox France (B):
Designing the Integrated Supply Chain

INTRODUCTION

In 1988, Rank Xerox France realized that its existing logistics strategy could no longer meet the corporate objectives of improving customer satisfaction and increasing market share and return on assets. The stock accounted for 73 days of supply (i.e., assets of more than $50 million), the warehousing area of 11 000 m² at Garonor seemed inadequate and the average order-to-install time was above 15 days. Delivery dates required by the salesforce were being frequently missed, and Rank Xerox seemed unable to respond to the ongoing changes in the competitive market environment. It was obvious that a radical change in Rank Xerox's logistics strategy was required.

In response to these challenges, a Europewide project aimed at redefining Rank Xerox's logistics strategy was launched in 1988. An internal Rank Xerox document described the objectives of the project as:

> . . . to shorten the Rank Xerox supply chain, enabling a far better flow of information to and from the field than is possible today, allowing echelons of stock (and thus warehouses) to be eliminated, and enabling very quick delivery to the customers 'point of need' from stocks that have been assessed to be in the ideal location for the business.

Bernard Costiou, director strategic projects and project leader in France, commented on the project:

> We did not really have a clear understanding of the essentials of the logistics process in Rank Xerox. There were many functions involved – manufacturing, information

management, warehousing, distribution and sales – and there was no systemic overall view of their interactions.

We had to improve performance, not only internally by improving quality and productivity, but also externally by increasing customer satisfaction and market share. So, we had first to define customer requirements and then to reconceptualize the whole supply chain through which the products flow from their source points to demand points. This involved determining which facilities would be used, where they would be located, which products would be assigned to them and which transport services would be used. Emphasis also needed to be given to inventory planning, order management and the effectiveness of the overall supply chain.

When the project finished, four years later, the results obtained were significant (Exhibit CS2.2.1). The level of assets had been reduced from 73 days of supply to 26 days. This meant a saving of about $35 million in assets. The level of customer satisfaction had been enhanced on many criteria, particularly on punctuality (+7%), reaching an overall 93.9%. The order-to-install cycle time had been reduced by 3.3 days, of which 2.7 days was due to savings in logistics and transportation times. Logistics costs had also been significantly reduced. Although the number of movements in 1992 was 13% higher than that in 1989, transportation costs were maintained at the same value which represented savings of $1 million. Non-transportation costs had been reduced by 29%, mainly due to a 50% cut in headcount leading to another $1 million in savings.

But more needed to be done. Bernard Costiou continued:

Despite this progress, we are not completely satisfied. In some areas, the performance has not improved sufficiently. For example, consider the order-to-install time: the average time has been reduced from 15 to 12 days, but this figure is clearly away from our target of 9 days. This result cannot be considered satisfactory.

THE ORIGINS OF THE PROJECT

The project—termed as the Core End Point Plan (CEPL)—started as an initiative from the information management function for improving configuration management (i.e., product description in terms of data) and echelon management (i.e., data requirements for supporting activities like ordering, delivery and invoicing). Due to the increasing complexity of products and markets, there was a need to communicate faster and to integrate the information flow between manufacturing, the supply centres and the distributors. But this was not easy as Rank Xerox had several different IT systems in the operating units in the various countries. Feasibility studies were conducted to investigate the opportunity of investing in a new central system which would improve communication, optimize the logistics cost and provide synergies. Bernard Costiou recalled:

Our first reaction was to build large centralized systems for each of the major functionalities we recognized within the supply chain, such as configuration, warehousing, finance and order management. But when we looked at the problem in more detail we realized that there was no such requirement from the customers' points of view. What they were seeing were not these individual functions, but rather flows of products, such as equipment, spare parts and consumable goods (e.g. paper, toners). What was needed was not another set of large software packages but a radical change in the way we perceived and organized the logistics process.

TEAM FORMATION

Once the scope of the project expanded beyond information management to include the entire logistics process, the next step was the formation of an international cross-functional team. Claude Joigneault, logistics director of Rank Xerox France in 1993 and a member of the team, described the selection of the team members:

A crucial factor for the success of the project was the creation of a cross-functional team to support the operation and look at all aspects: human, physical, organizational and systems. Members of the team were selected from all concerned functions: logistics, marketing, accounting and finance, information management and human resources management. The quality manager participated in the team to validate that everything would be in line with the corporate initiative for quality. Subcontractors such as transporters or suppliers were also included in the team. Everybody around the table had the same power and was equally respected whether he was a subcontractor or a functional manager. The team remained the same throughout the project.

Though the mandate of the team was to look at Rank Xerox's operations across Europe, France opted to be the pilot site for the implementation of the new strategy. Bernard Costiou, director strategic projects, was named the project leader in France.

PROCESS DESIGN

Since most of the team members were from different functions, they lacked an integrated view of the entire logistics process; they only knew their individual roles. A clear picture of the logistics process had to be formed and performance requirements had to be defined. Bernard Costiou elaborated:

We wanted to be sure that we had a clear understanding of the logistics process. We asked ourselves: Do we know exactly what the logistics process includes? Do we know who is doing what? Do we know where the activities begin and where they end? And then do we know how to measure the performance of each activity?

To gain a deeper understanding of the content of each task, meetings and interviews were conducted with people from different parts of the logistics process such as the supply centres, the warehouses and the information management department.

The team mapped all the physical and information flows. Process boundaries and interfaces with other processes were determined. Soon, a picture of the complete logistics process started to emerge: an integrated supply chain providing demand driven flow of goods linking logistics, manufacturing, suppliers and operating companies in a continuous operations cycle (see Exhibit CS2.2.2). The process started with suppliers delivering raw materials and parts to Rank Xerox factories and ended with the customers receiving Rank Xerox products.

Then, the team analysed the process by focusing on customer needs and the added value of each activity. Bernard Costiou commented on the guiding principle:

We were constantly thinking in the following manner. Which activities produce output? What is the added value of each activity to the end customer? For example, what is the added value of maintaining stocks and warehouses in each European country?

PROCESS REDESIGN

To design the new process, different alternative scenarios were assessed in terms of feasibility, cost and benefit. The prevalent scenario suggested the marriage of just-in-time (JIT) manufacturing (started in 1982) with JIT distribution (see Exhibit CS2.2.3). Stocks would be eliminated and flows would be simplified. The system would provide upstream and downstream integration. Within this framework all national warehouses had to close, since there was no need for most products to be held as stock at a national level. A central warehouse (the European Logistics Centre) would be needed to hold the stock at a European level. The flow of products out of the central warehouse would be triggered only by the customer order. An intermediate stage (transit point) would be necessary at the national level to substitute the national warehouses. Its role would be not to hold stock but to receive shipments from the central warehouse, match them with customer orders and ship the equipment to the regional platform. This intermediate stage would be called the dispatch centre level. Also, for each type of product exact logistics procedures had to be described (see Exhibit CS2.2.4):

■ For commodity products (e.g., typewriters, faxes) the order-to-install time should be less than two days. As the majority of these products would be sold through indirect channels, the dealer or distributor would need to hold stock and use appropriate distribution networks. In case the distributor ran out of stock, the operating unit would have to provide him with products within one day. Since this time constraint could not be fulfilled by the central warehouse, it was decided that for commodity products stock should be held at a national level.

■ For copiers and printers which were the bulk of RX operations (85% of the total unit sales) the standard lead time allowed sufficient time for orders to be met from the central warehouse. This eliminated the need for holding stock in the operating units. However, this required the elimination or minimization of several activities within the supply chain; for example, nationalization, preinstallation, quality testing. Equipment had now to be produced by the manufacturing plants in standard modular packages. Any specific local market requirements that could not be centrally met had to be added in the form of a kit at the final stage in the supply chain at or before installation.

■ For customized products (i.e. products manufactured upon customer's request) no change was required since they were build-to-order products and no stock was held.

Key performance indicators were also identified and agreed upon (such as order-to-install time, percentage of deliveries meeting the customer's requested date, total transportation cost). Then new performance standards were defined. Bernard Costiou elaborated:

Wherever possible we benchmarked ourselves against our direct competitors such as Canon and Kodak. Where this was not feasible, we benchmarked against the best companies in the field. Then we identified performance gaps and set our targets to be 10% better than the measured gaps.

CHANGE IMPLEMENTATION

A critical issue for the implementation of the changes was the delegation of power to the team. Bernard Costiou commented:

Before we proceeded with the implementation, we decided to validate the vision of the team with senior management. This is very important because when you change things you want to be sure that you have top management support. And the changes we proposed were not minor, for example in the new process the sales force had to negotiate contracts, orders and deliveries without available stock.

The members of the team were called implementation managers. This meant that while they still remained in their functions, their primary role was to implement the new process. Once a month there was a meeting between the team and the Rank Xerox France senior management committee to review the progress of the implementation. An effective method for implementing the changes was through internal contracts. This concept suggested that each job had an upstream supplier and a downstream customer. Bernard Costiou elaborated:

These contracts were considered as firm contracts and were as valuable as any other contract. They defined in detail what had to be done, within what time frame, who was the producer and who was the customer. Such contracts were initiated between different departments, for example, the information management department and the logistics department. These contracts were used to put people together, to agree on things and to follow their implementation.

To implement the changes a stepwise procedure, which comprised five major steps, had to be followed. The duration of the implementation phase of the project was two years (Exhibit CS2.2.5).

- The first step was the optimization of the transportation network. Since, for Rank Xerox, transportation did not constitute one of its core competencies, it was outsourced. A new call for tenders was released providing details about the new organization and tariffs and describing explicitly the obligations of the new contractors. As a result, the number of transportation companies used by Rank Xerox France was reduced from 16 to 2 (a number chosen deliberately to retain some competition and reduce risk). The emphasis, however, was on building a long-term partnership with the transportation companies.

- The second step was the implementation of JIT distribution for the copiers and the printers. This had to be done product by product and was completed after one year. This step involved the introduction of the European Logistics Centre (ELC), the new central warehouse to handle the stock from all European manufacturing plants at a European level. A new place was chosen, mainly for its privileged geographic location, close to the manufacturing site of Venray in Holland.

The ELC was implemented following a concept which Rank Xerox called 'rationalization', i.e., the elimination of data entry duplications, the handling of products using the most efficient methods, the reduction of inspections to a small sample of products and finally, the classification and storage of products according to their different characteristics (fast movers, slow movers, large parts, etc.).

■ The third step was the introduction of the dispatch centres which were subcontracted. A pilot system was tested for two months. A 24-hour operations cycle was finally implemented: during the night, the orders of the day were sent to the ELC, during the following day the equipment was prepared at the ELC and by the following night the equipment was shipped to the dispatch centers. At the dispatch centres the equipment was matched with the customer's name and address and shipped accordingly to the regional platform. So, within two working days the equipment had already been sent to the order entry point.

■ The fourth step was the management of the commodity products at the dispatch center level. Since the lead time was a crucial factor for the distribution of the commodity products, Rank Xerox France had to keep some stock at national level. After a two-month pilot implementation, the subcontractors were able to manage the commodity goods at the dispatch centers.

■ The final step was the management of the remaining products—mainly office systems products—via the dispatch centre level. This phase was very important because it was the end of the old organization, the closing of the national warehouses and the transition to the new system.

Later, the carcass management process (consisting of the removal and recycling of old equipment—see Exhibit CS2.2.2) was also implemented at the dispatch centre level. Accompanying these changes, there was a significant reduction in headcount of the non-manufacturing personnel associated with the supply chain. While in 1989 there were about 100 staff associated with the logistics of equipment (in Rank Xerox France), only about a quarter of them remained by the end of the implementation of the above changes. About a third were reemployed by the subcontractors, about another third left voluntarily on early retirement and the rest found other jobs within the organization.

THE NEW ORGANIZATION

From the start, it was clear that as the new organization emerged, people's roles and responsibilities would have to be redefined. Claude Joigneault explained:

> In the previous organization, manufacturing was not responsible for the level of service to the end customer, but only for delivering the planned production. But now, you said to the manufacturing manager 'You are in charge for the level of service to the end customer, not only to deliver finished goods to the European Logistics Centre, but to

Case 2.2

the end customer'. This required a change in the attitude and responsibility of the manufacturing plants. Now there was a complete supply chain and everybody added a portion to it. Everyone had to feel much more involved and committed to do a good job.

Bernard Costiou provided another example:

The new organization brought changes in the job content of many employees. They had different roles and responsibilities. For example, an important part of their new job was to cooperate with the transporters to achieve full loading of the trucks, in order to control the transportation cost. Fortunately, the required skills were not different from the previous ones, and the employees, after a short training period, were able to adapt themselves to the new environment.

However, many others felt uncomfortable with the change as familiar working patterns were altered. Claude Joigneault recalled:

At the beginning the salesforce was frightened since they had to change the way they were dealing with the customer. Previously they could negotiate contracts with customers with the knowledge that the product was in stock at Garonor. Now, they had to negotiate contracts on the faith that the organization would deliver on the committed date.

These challenges were anticipated and addressed. The policy deployment process (Exhibit CS2.2.6) communicated the company's policy and strategy to all concerned employees. Didier Groz, the director of quality for Rank Xerox France, commented:

The policy deployment process was a vehicle to make every team and each individual work to support the objectives of the company during the change process. It served as a great tool for focusing the minds and energies of the entire unit on the strategic priorities. The process ensured that each individual had clear personal targets and goals, could relate his activities to other people and understood how to contribute to the success of the company.

Communication was also achieved through employee involvement in the change process. In all operating units meetings involving all managers were followed by meetings at which these managers cascaded the information directly to their own staffs. Communication was regularly reinforced at quarterly and monthly review meetings.

At the end the salespeople were very happy with the new system. The redesign of the supply chain gave them additional flexibility and a better negotiating position with their customers. As the pilot within France was a success, all other operating units in Europe also shifted to such an organization.

PROCESS MANAGEMENT

Continuous process improvement was considered essential for sustaining the new logistics strategy. Central to these efforts was a dedicated focus on process management. A key role was that of a process owner—a person who, in cooperation with the customer and in line with the corporate objectives, defined performance goals and measured goal attainment. He or she also had the task of ensuring that all process personnel had a clear view of the goals and had been given the necessary tools, authority and responsibility to affect changes in performance. To manage the logistics process, Rank Xerox France developed two types of performance measures: internal and external. The former measured the performance of the process in terms of metrics such as order-to-install time, cost per order, cost per cubic metre of transportation, total costs and total revenues.

For the external measures, Rank Xerox France started carrying out several rigorous surveys. For example, twice a year an independent agency records the attitudes of customers and potential customers to the company and compares it to its competitors. Also Rank Xerox surveys every customer who buys any of its products 90 days after delivery (90-day-post-installation survey). Customers are interviewed by telephone and asked about the performance of the company's equipment, its responsiveness as an organization, the level of service provided by its sales and support staff, and so on. Didier Groz commented:

> Each year we set targets and we try to meet them. Then through audit procedures, we mark our performance. A performance can be satisfactory, which means that the targets have been met, or unsatisfactory. We never mark a performance as excellent because we never say that something cannot be improved.

ORDER-TO-INSTALL TIMES

By the end of 1992, progress was clear and the obtained results were very significant. While the logistics process functioned in a more effective and seamless manner as compared to the late 1980s, there were still some problems, for example with the order-to-install times. Didier Groz elaborated:

> We had improved our order-to-installation times, but we only seemed able to meet the specified delivery dates in about half the orders. Our surveys showed that customers were not very happy with our performance regarding on-time delivery.

> We had enlarged our vision of logistics to an integrated supply chain, but had we gone far enough? Perhaps we had to seriously question the entire organization to see how the other processes functioned. Perhaps the problem lay outside the supply chain.

Exhibits

Exhibit CS2.2.1 Business results

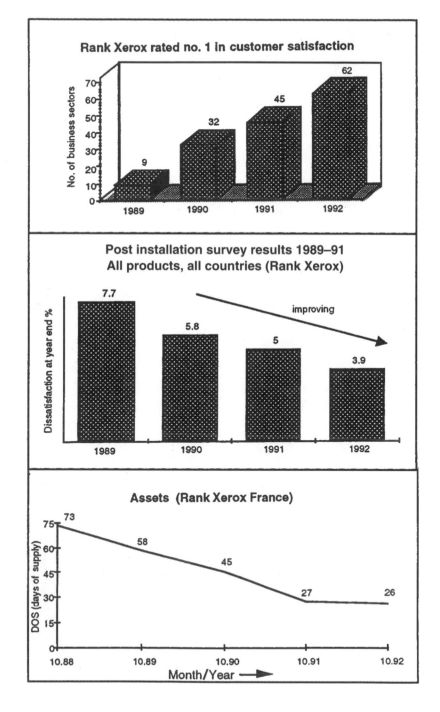

Rank Xerox rated no. 1 in customer satisfaction

Post installation survey results 1989–91
All products, all countries (Rank Xerox)

improving

Assets (Rank Xerox France)

Case 2.2

Exhibit CS2.2.2 The integrated supply chain vision

STRATEGY RELATIONSHIPS

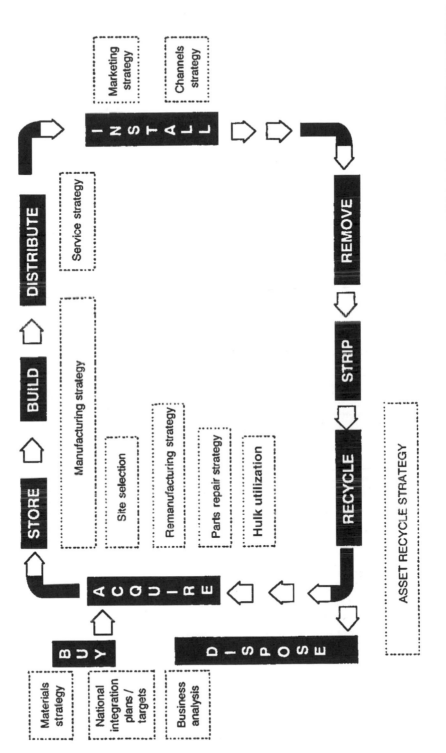

Case 2.2

Exhibit CS2.2.3 Supply chain integration

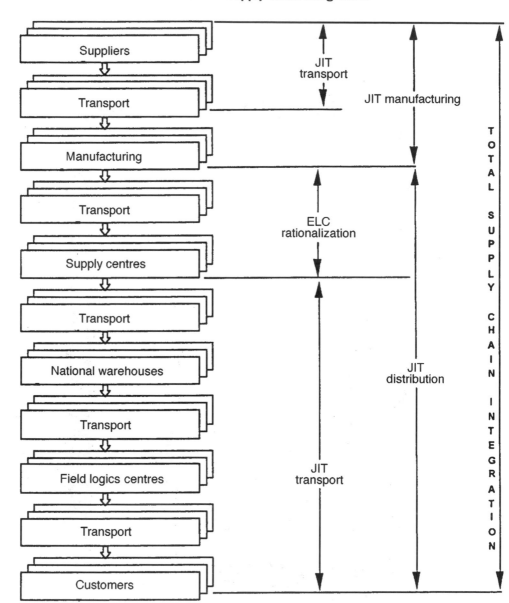

Exhibit CS2.2.4 Supply chain 'pull'

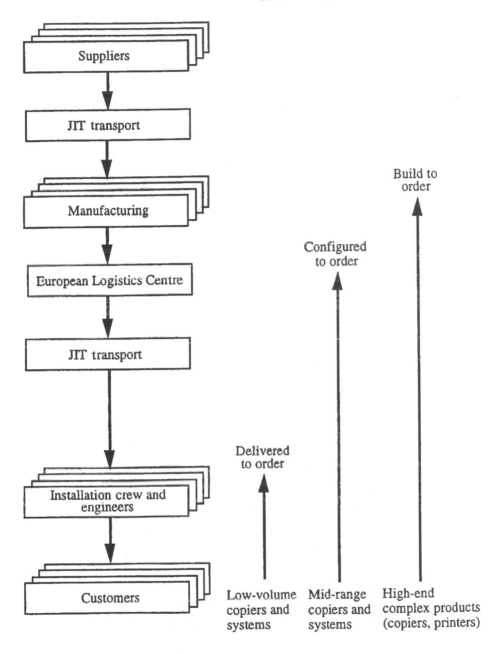

Case 2.2

Exhibit CS2.2.5 Project time schedule

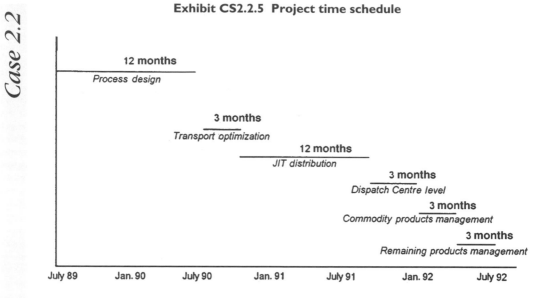

Exhibit CS2.2.6 The policy deployment process

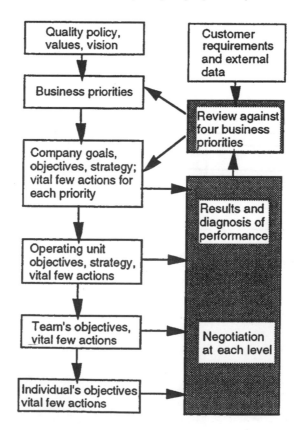

Case 2.3

This case was written by Francesca Gee, Research Associate, under the supervision of Soumitra Dutta and Jean-François Manzoni, Professors at INSEAD. It is intended to be used as a basis for class discussion, rather than to illustrate either effective or ineffective handling of an administrative situation*

Transforming TSB Group (D): The Second Wave (1992–95)—Focusing on Quality and Processes

In 1991, TSB Bank, TSB Group's retail banking division, found itself in a quandary. While its profitability was improving (the cost/income ratio had fallen from 76% to 59% in two years), customer service was not displaying the same auspicious trend. Both the bank's ability to attract new customers and its reputation for good service were declining. In 1990, only 36% of customers thought that TSB was 'the kind of bank people would want to move to', down from 43% in 1987. Also, only 28% felt that TSB provided high-quality service against 37% three years earlier.[1] One out of four TSB customers reported experiencing a problem with their current account every year. According to an internal document:

> *Independent market research into TSB customer perceptions shows that the bank is seen less and less as one that offers what the customer wants. One survey indicated that over half of those customers who left TSB did so because they were dissatisfied with the service they received.*

Yet improving customer service was of key importance with research showing that companies with a high level of customer service could at least double the length and profitability of their relationship with each customer.[2] It also showed that attracting new customers cost about five times more than retaining existing ones.[3] TSB's new chief executive, Peter Ellwood, had introduced a Quality drive in his previous job as head of Barclays' credit card division and wanted to make Quality a cornerstone of his strategy at TSB:

[1] *Quality News* (internal newsletter), Issue 1, 1991.
[2] Heskett, J.C., Sasser, Jr., W.E., and Hart, C.W.L., *Service Breakthroughs*, Free Press, 1990, p. 33.
[3] TARP, Washington DC, quoted in Binney, G., *Making Quality Work*, Ashridge and The Economist Intelligence Unit, 1992, p. 39.

It was clear to us that service quality had to be at the top of our agenda for the 1990s—not only because customers were clamouring for it or because it was a proven factor in improving profitability, but because we could see it would be the means of acquiring a sustainable advantage in a fiercely competitive market. With a large number of competitors in an overcrowded sector, products are easily copied. Good quality service distinguishes a bank in a way that customers can appreciate and that cannot be copied overnight.

TSB

TSB was the outcome of the protracted amalgamation of dozens of non-profit-making Trustee Savings Banks. In 1986, TSB Group (which included four separate banks: TSB Scotland, TSB Northern Ireland, TSB Channel Islands and TSB England & Wales) was floated on the London Stock Exchange and acquired several million private shareholders. In 1989, the new Group chairman, Sir Nicholas Goodison, merged the four banks into a single division, TSB Bank, and hired Peter Ellwood with a mandate for radical change.

Peter Ellwood's vision was to turn TSB into a leading financial services retailer, and he launched a series of far-reaching projects aimed at correcting the historic underinvestment in the branch network, an unwieldy multilayered management structure and the poor quality of management information systems. Both TSB's head office and branch network were to be redesigned.

One project, aimed at simplifying branch procedures, involved making an inventory of best practices throughout the network and encouraging the adoption of these practices by other branches. Selling space in branches was increased and the range of products widened, branch managers focused more on sales and less on administration, and staff began systematically to search customer records for potential leads.

These changes, particularly the reorganization of the branch network, resulted in significant reductions in the headcount (from 30 000 in 1989 to 25 000 in 1991). While most other major UK financial institutions reduced their staffing levels later, TSB was the first UK bank to announce large-scale layoffs in 1989, and they were the largest ever seen in British banking.

LAUNCHING A PROGRAMME OF QUALITY

The objective of TSB's programme of Total Quality Management (TQM) was to simultaneously improve customer service, cut costs by reducing waste *and* improve staff morale. While previous initiatives had tackled these objectives separately, the TQM drive was the bank's first attempt to deal with all three at once. In the autumn of 1990, TSB decided to use the method developed by American consultant Philip Crosby, which was based on the premise that all work is a process. The Crosby method revolved around four so-called 'absolutes':

- The definition of quality is conformance to requirements: understand what the customer wants and then align the organization to provide it,

- The system of quality is prevention: we prevent problems by improving processes rather than by trying merely to cure the symptoms,

- The quality performance standard is zero defects: we must get things right first time, every time,

■ The measurement of quality is the cost of non-quality: we measure how well we're doing in establishing TQM by measuring and reducing the cost of what goes wrong.

A 'Quality team' of two dozen members was set up in the personnel department and its brief was to make training in Quality matters as relevant as possible to TSB's staff, who would ultimately deliver the expected improvements. The goal was to train all 25 000 staff to use tools and techniques such as those shown in Exhibit CS2.3.1. TSB decided that they would be trained by their own line managers using customized Crosby materials, rather than by external consultants.

Training began in 1991, when three TSB managers were sent to the Crosby Institute in Florida for three months. The training was cascaded down through TSB so that two years later all staff had been trained. TSB felt that the risks involved in the cascade approach—losing control over the process and the message being distorted—would be more than compensated by the emergence throughout the business of 'Quality champions' whose enthusiasm would generate the critical mass of support needed to make the programme a success. The length of training varied from four days for members of the executive committee to six 90-minute sessions for branch staff. The entire effort cost £7 million plus £12 million worth of staff time, and it took the equivalent of 150 work-years.

The exercise also involved a major communication effort. All staff, including senior managers, were told of the benefits that other companies had derived from TQM and the Quality team produced newsletters and videos to inform support staff. TQM was presented as a practical, commonsense approach aimed at resolving hassles and problems. Staff spent 25% of their time correcting errors, following up complaints and doing things over again; TQM would enable them to identify problems that prevented them from doing their jobs, then to work together to develop and implement solutions.

Implementation began with 15 'fast-track projects' such as processing insurance applications faster, cutting the number of emergency orders for stationery or increasing the availability of cash from cash dispensers.[4] A simple change, such as stopping the bulk counting of cash in branches, saved over £1 million annually.[5] Director of Quality Roger Cliffe commented:

> The fast-track projects were successful in that 10 of them showed major business benefits while the other five yielded marginal returns. Overall, the projects demonstrated to the organization that it had the capability to make huge improvements in customer satisfaction and profit through the vehicles of small- and medium-sized improvement projects. It also demonstrated that a more radical process reengineering approach was needed where improvement involved large cross-functional projects and major technology spending.

---

[4] Unavailability of cash from dispensers was one of the top causes of customer dissatisfaction, especially among younger customers who were of particular importance to TSB.

[5] TSB, like other banks, received cash from the Bank of England in bundles. Counting these notes was a wasteful activity because if notes were missing, they could not be claimed back from the Bank of England. Stopping the bulk counting of cash saved about £1000 in each of the 1300 branches annually.

Peter Ellwood made his own commitment as visible as he could. In October 1993, he described his personal 'Quality plan' in a meeting with a small group of staff:

> I have started to log when I am late and also when I overrun a meeting. Then I measure it in [financial] terms. It is getting better, but I still start too many meetings late . . . I've also asked all the team that work for me what I do that causes them hassles. For instance, they tell me that I ask for a thing to be done instantly, and ask why don't I give them more notice. I now try to give them more time, and they say I'm getting better! I also have regular lunches with staff at all levels at which I encourage them to raise issues and concerns.

QUALITY IMPROVEMENT TEAMS

Quality improvement teams (QITs) were formed in 18 functional areas at the start of the Quality drive. Their main role was to provide leadership and to coordinate training and administration of the Quality programme. Each team was headed by a senior director and included about 15 members of varying seniority. Teams met monthly, with a member of the central Quality staff in attendance. Some QITs decided to identify Quality champions or 'sponsors' in their area. For example, each branch had its own Quality sponsor and sponsors from a cluster of 8 to 12 branches met monthly.

The teams also ran a staff suggestions scheme known as the 'Corrective Action Process'. When staff identified a problem that they could not solve themselves, they could complete a form (known as a 'snag form') which their manager sent to their QIT. The QIT then took appropriate action. Staff sent thousands of suggestions, and many resulted in cost savings, but many were perceived by management as unrealistic or not focused on business needs. A review of this process in December 1993 showed that it was slow and that issues occasionally became lost in the system. Staff felt it was unreliable, and they often failed to receive feedback on the issues they had raised, or when they did, it took too long.

After consulting staff and studying other companies, TSB designed a new system called Opportunities for Improvement (OFI) inspired by the programme of Milliken, past winner of the coveted Baldrige U.S. Quality Award and European Quality Award. Staff members could suggest improvements by filling in a form that required the proposal to meet one of three criteria: improving customer service, generating income, or reducing operating costs. The process guaranteed feedback: OFIs were processed by a small team and passed on to nominated individuals who had to acknowledge each suggestion within 24 hours, and give a decision within 72 hours. OFI forms were only submitted to the central team if the idea could not be implemented locally. The scheme was piloted in 1994 and rolled out from November. By January 1995, the OFI team had received over 1000 OFIs, of which about 40% were implemented. By May 1995, more than 650 OFIs were being processed each month. Exhibit CS2.3.2 shows some examples of OFIs.

SHARING IDEAS AND SUCCESSES

TSB immediately realized that it was important to share good ideas and successes. Information was passed on in various ways such as through regular meetings of QITs and Quality sponsors, dedicated newsletters or the weekly bank update sent to all branches, and through videos.

Recognition was another important aspect of the programme. After much soul-searching, TSB decided against financial rewards to stress the point that Quality was part of everyone's job, not something for which one got paid extra. Quality manager, Sue Meadows explained:

> We held many meetings to discuss how we would recognize Quality successes and whether we should pay people for suggestions. We wanted a fair process that would give all staff equal opportunities to be recognized. We also wanted to stress that quality improvement is not just about a few great big projects that yield enormous cost savings; it is about every individual taking responsibility and initiative in their work-place. That led us away from monetary rewards. Some people, because of the nature of their work, have more opportunities to think of changes that generate large savings—in technology, for example. We wanted the process to recognize effort and participation, and we wanted something open to all levels of staff.

The recognition programme involved newsletters and recognition events. Individuals or teams could be nominated by their colleagues or managers for outstanding achievement in the area of Quality. They were given Q-shaped silver or gold pins and invited to local and national recognition events. Over 400 staff attended the seven national events held between August 1992 and February 1995, all hosted by Peter Ellwood. Sue Meadows explained:

> Focus group discussions suggest that some senior managers may overestimate the importance of monetary rewards for junior staff. When we ask people how they would like to be recognized, cash comes up, but what they really want is appropriate recognition from their line manager, 'Thank you and well done'.

ASSESSING THE PROGRESS MADE

TSB measured the progress achieved through its Quality drive in various ways, for instance through the price of non-conformance (PONC), which expressed the cost of waste in monetary terms. By May 1994, PONC savings stood at £34.3 million, including £8.6 million in bottom-line savings. The Quality effort also involved developing more sophisticated measures of customer satisfaction, as Peter Ellwood explained:

> We needed to know what our customers really wanted and what they were prepared to pay for. We now conduct regular research that involves questioning up to 70,000 people, and the results of this research have had a significant impact on our strategic and operating plans.

TSB surveyed customers through a variety of mechanisms including focus groups, telephone interviews and questionnaires mailed to large samples of customers. The key indicator to emerge from this research was the customer satisfaction index (CSI), a weighted average of TSB's performance on the ten service features that customers cared most about. The CSI, which was measured annually until 1994, is now assessed twice a year. TSB also tracks its competitors' performance through internal and external industrywide studies. Roger Cliffe described the evolution of his department's attitude toward measurement:

> When Quality was first introduced, the Quality team used to keep massive central records of cost and service improvements made throughout the organization. As

Quality developed, the ownership for measurement was passed into the line. With this, the role of the central team changed from being the 'cop' to being the 'coach'. Our job is now to set policy, to teach and support and to coordinate any central activity such as OFI and recognition but not to police Quality in the business.

In work units such as branches or transaction processing centres, service-related performance measures were an important component of variable compensation programmes for both staff and managers. Customer Service Centres (CSCs) for example used four main measures:

- Service measures: average error rates (timeliness, accuracy) on various types of transactions, estimated monthly by staff on a sample basis

- Branch response: a quarterly feedback form completed by managers of each branch that is supported by the CSC

- Mystery shopper: periodic visits to branches by trained staff, acting as customers to review service performance

- Inspection reports: quarterly internal audits completed by audit staff

A regional manager in charge of CSCs pointed to the limitations of some measures:

We are currently revising the way we measure CSC performance. Branch response, for example, has not been a very good measure. Branch managers are not close enough to the support provided to their staff by the CSC, so we are relaunching this measure in 1995, giving a group of individuals in each branch the opportunity to give feedback to their CSC. We also have some problems with service measures which are estimated internally and are thus subject to different interpretations. For example, we have an error rate of about 1% on all standing orders, which is not large, but we also receive thousands of complaints each year on standing orders across the bank, so the two don't quite match. We don't think there is anything untoward—all staff know that any manipulation of the figures is a disciplinary offence—but it does mean that we have to understand better what gives customers cause to complain.

In the branch network, staff performance was measured along four service-related dimensions: availability of cash at dispensers, accuracy of mortgage paperwork completed by the branch, adherence to credit-scoring guidelines and the error rate on transaction processing. While these were measured in each branch, bonus payments were based on the performance of a cluster of branches.

Alan Gilmour, head of marketing research, commented on TSB's approach to customer service measurement:

I think it is fair to say that we probably have not, until recently, made customer service a 'meaningful obsession' throughout senior and middle management. But we should not be measurement-led, we shouldn't be thinking 'Let's keep measuring and measuring so that we develop an obsession', we should be action-led.

In 1993, TSB introduced a widely used method of self-assessment which enabled the 18 QITs to measure achievements in their respective functions and compare themselves with the best-in-the class as defined by the European Quality Award. The model, which is more fully described in Exhibit CS2.3.3, measured achievement along nine dimensions:

Enablers	Results
Leadership	People satisfaction
People management	Customer satisfaction
Policy and strategy	Impact on society
Resources	Business results
Processes	

Roger Cliffe explained:

The self-assessment for a QIT normally takes the form of a one-day workshop in which members of the QIT, acting under the guidance of a trained facilitator, review the strengths and areas for improvement in each of the nine EQA model criteria. Feedback from this workshop is used to develop a Quality plan for the QIT. The QIT also calculates a score for its achievement in each area using the guidelines provided by the model. The scores are used to prioritize improvement activities and as a basis for internal comparison between QITs and for benchmarking with external organizations. The EQA model has helped managers to focus on using Quality to improve business results.

REENGINEERING PROCESSES

The TQM programme focused on 'little victories' at the local level and on convincing staff that Quality would make their daily work easier and improve TSB's image through better service. The bank also turned its attention to large-scale processes that crossed functional boundaries. There is more to good service than the application of Quality, explained Peter Ellwood: 'A 'smiles and uniforms' campaign is hollow mockery of the customer if the processes delivering the underlying service aren't right.' (Exhibit CS2.3.4 has a graphical representation of the scope of quality and process management at TSB).

This exercise, which forced a reexamination of the fundamental ways in which TSB operated, was known as business process reengineering (BPR). Group strategic development director Archie Kane was convinced that 'If you can't get the processes right, you will never get customer service right'. The effort began in 1992 when TSB merged its banking and insurance retail sales under the newly created post of branch network director. Archie Kane, then operations director, was given the task of integrating all the bank's back-office processing. He realized that although there seemed to be many differences in the way the various back offices operated, they were also quite similar in many ways:

I approached the whole problem from an activity-based, costing standpoint. It then evolved to a process perspective. This was the start of a long story that took about 18 months. We spent a long time conceptualizing processes and workshops were long and painful. They involved fairly senior people from all over the organization, including the retail side and they met half-a-dozen times over a period of six months.

But BPR was a fairly new concept and the lack of an established methodology forced TSB to proceed through trial and error. Archie Kane set up a small department of business process analysts and retained consultants to teach them modelling techniques. The analysts started mapping different types of activities using IDEF methodology (described in Exhibit CS2.3.5) which shows an activity or a series of activities as simple blocks with inputs, outputs, controls and mechanisms. Archie Kane said:

> Six months later, we ended up with 74 processes for the whole business. We grouped them into three categories: Direction-setting, which includes strategic planning, budgeting, and top management control, support, and operations, which is the heart of the business including initial contact, transactions, operational centres, branch, network, marketing. I thought, That's brilliant! I have 74 processes in three boxes, this is going to be a lot of use to top management! That was probably our darkest hour.

The BPR process was difficult and time-consuming because of the magnitude of the change that was envisaged. Archie Kane explained:

> We spent a lot of time thinking and working it out. It was not [just] taking a little support process and reengineering it . . . We didn't do it by reaching on to the shelf and pulling down a textbook . . . It involved reading all there was on the subject, talking to all the consultants, finding out that they didn't know the answers, then trying to think our way through it. Neither was it a bolt of lightning; it took us a lot of time thinking through these things, and a lot of people were involved developing the vision with many long workshops and much going back and forth to senior management.

More workshops were held until a new concept emerged, that of focusing on customer needs and customer-facing processes. Archie Kane said:

> It was pretty radical for us. We had to get ourselves out of supply-side thinking, which is very difficult for banks. We had a whole range of people from traditionally separate businesses like the credit card business, the pensions business, the branch network, and we had to get them to think in terms of customer-facing processes. What does the customer need? If he wants to buy a home, there is a homebuying process, which will involve a bundle of separate products handled in different parts of the organization, but it is one process; if anything goes wrong in any of those places, the customer will be unhappy.

In this way, five customer-facing processes were defined: homebuying services, payment services, insurance services, savings and investment services, and lending services. Exhibit CS2.3.6 shows the path that led from mapping activities to identifying these five processes.

> That was the first big breakthrough. But while the customer-facing processes were very useful, we don't manage the business that way. So we took all the operations processes and mapped them across the customer-facing processes. Across all these customer-facing processes, we have the same types of activity: we generate leads, we approach customers, we sell products, we process those products.

Three generic processes were thus identified: the new business process, the transaction process and the change process. In the case of credit cards, for instance, creating the account when a new card was sold was a new business process; handling monthly payments was part of the transaction process; replacing a lost card, registering a change of address, or increasing the spending limit all came under the change process. Customer-facing processes and generic processes could be represented as a matrix on which distribution channels were mapped. The result, shown in Exhibit CS2.3.7, looked like a Rubik's cube.

As a pilot project, TSB reengineered its home buying new business process. Until 1993, a customer who wanted a mortgage was interviewed four or five times and had to fill in five forms with 167 questions; an average 30 days would elapse before a formal offer to lend was made. The redesigned process reduced costs by more than 25%, had a single application form, one interview and a formal offer was made after an average of seven days, with a long-term objective of reducing cycle time to five minutes.

PROCESS MANAGEMENT

The three generic processes which were the main focus of the reengineering effort were entrusted to dedicated teams working full-time on BPR. Each of the five customer facing processes was assigned to a process-owner team (POT) responsible for identifying and implementing process improvement and reengineering opportunities. Each POT included about 10 senior managers who represented all the key departments involved in the particular process. POT members also headed cross-functional teams composed of lower level staff. For example, the homebuying process relied on 10 such teams, one for each of the bank's geographical areas. There was a lively internal debate as to whether the POT and team members should keep their line positions. Said Archie Kane:

> We came to the conclusion that making POT membership full-time would do more damage than good, because if members left their line departments they would no longer have any influence on them.

Archie Kane estimated that POT members devoted about one day a week to their BPR activity. For the same reason, POTs were initially given very few resources, but this changed later on. Archie Kane:

> There was a lot of resistance initially. Success varied depending on the quality of the people involved. Some people saw it as a good thing, a way of getting things done; others saw it as something that wasn't in their objectives and refused to become involved. Reengineering means managing the business in a very different way— horizontally rather than vertically.

Roger Cliffe expressed a similar view:

> We did not have the confidence to move away from the functional, 'vertical', management structure to a horizontal process-based structure in a single step. Instead we opted for a matrix management approach with the functional structure supported by strong process management teams. We believe this style fits best into our current management style, but this will undoubtedly change as we become more experienced in managing business processes.

The POTs realized that process costs had to be identified accurately. Domain controllers or reengineering champions from different functions were identified and asked to look at the cost base of a particular function and to distribute that cost base across the processes which had been defined. Roger Cliffe described the process:

> Costs in this business are a very sensitive issue. People found it difficult to apportion costs in a different way. When we saw the numbers we wondered, 'Does this process really cost this amount of money?' We had not realized how much cross-functional processes were costing. The immediate reaction was that we'd got the numbers wrong. In some cases we had, in other cases we hadn't.

Commenting on Exhibit CS2.3.8, which presents different levels of reengineering, Archie Kane summarized TSB's approach:

> Most of the examples you will find are people reengineering a minute part of the process: Ford Motor Company redoing their payables cycle. Big deal, you can do that in the morning and play golf in the afternoon! Most examples consultants use are pretty underwhelming, they are not deep enough or wide enough. We're talking about something different: fundamental reengineering of the corporation, and real processes that cut across the organization's boundaries.

SOME MEASURABLE RESULTS

The impact of the BPR effort was only beginning to be felt. Archie Kane felt that one problem shared by many banking institutions, both in Europe and in the United States, was that customers were referred to different business units for each product:

> That's a structure constructed by managers because it's convenient to manage that way. It is not designed for the convenience of the customer. I call it the 'Filofax syndrome' — customers need a Filofax to keep the addresses and telephone numbers of all the business units that they have to deal with! When customers think of TSB, they want to think of one organization. They don't think of TSB Life and Pensions, TSB General Insurance, TSB Trustcard, TSB Homeloans . . . In fact, they get really irritated by that.

His vision was to give customers a single interface at four contact points: over the telephone, face-to-face (branch and home visits), by mail, and through interactive terminals using a sound and picture video technology that could handle virtually any transaction. Reengineered processes were due to be rolled out across the bank by the end of 1996, but redesigned processes had already saved the bank £1–2 million. Archie Kane expected long-term savings of up to £60–70 million, but that, he added, was not the only objective: 'Any fool can get costs down; the trick is to get customer service up at the same time, and if you can't get the processes right, you will never get customer service right.

The success of the Quality effort varied across the organization, with many pockets of excellence but much depended on the managers in charge of training staff. By 1995, measurable results had come mostly from processing centres, such as the self-contained Brighton credit card division, which had a strong culture. It was not so easy for branch staff to see serving the customer as a process, although they did make many small improvements that had an impact on customer service, but which could not always be measured in financial terms.

Quality-related performance measures painted a mixed picture. TSB's customer service index (CSI) performance improved significantly from 65% in 1991 to 70% in 1994, as did its ATM cash availability, from 89.6% in 1991 to 95% in 1994. By 1994, industrywide surveys on customer service typically positioned TSB among the best banks, although still significantly lower than the best building societies.

On the other hand, a staff attitude survey showed that, in 1993, staff were still unsure about TSB's commitment to quality, and external studies conducted in 1994 showed that while TSB had made good progress, it could not yet claim a lead over competitors on a number of service dimensions (Exhibits CS2.3.9 and CS2.3.10).

Trends in European Quality Award self-assessment scores also reflected mixed results. The 1994 scores showed that all units had improved with some reaching fairly high scores, but the whole bank's average still fell short of European 'Best-in-class' benchmarks. Roger Cliffe said: 'Although the scoring is not as important as identifying opportunities for improvement, it is an indication of how much more we have to improve before we can regard ourselves as world class.

Nevertheless, TSB's operations had improved significantly, as shown by the performance of its processing centres and it expected these improvements to keep pushing up the CSI. Alan Gilmour added:

> Banks have tremendous power to make a mess of someone's life: by not paying a bill on time, they can get the electricity cut off. A negative experience colours for a long time your experience of that brand, so shifting customer perception takes a long time.

Peter Ellwood:

> Letters of complaint that I receive from our customers show we have a long way to go before we reach zero defects. However, we see this challenge as an opportunity and one which will create tangible improvement for the customer. We have to accept it as a major change and success cannot be achieved overnight.

THE IMPACT ON THE ORGANIZATION

Staff enthusiasm for Quality was often high initially, but then tended to fade, said Sue Meadows. Telltale signs were a fall in the number of QIT meetings and reports of improvements. In early 1993, the drive was relaunched under the name 'Quality in Action' with a video. Managers also received a document applying TSB's structured 'Troubleshooter methodology' to the problem of insufficient staff involvement (see Exhibit CS2.3.11). The Quality drive received another boost in 1994 when the branch network was restructured and the number of its QITs increased from 3 to 10.

Sue Meadows saw various reasons for a loss of momentum. Redundancies among senior branch network staff set back the programme: they had been seen as Quality champions and had trained their branch managers. More generally, the Quality effort had been introduced while TSB was undergoing radical restructuring that involved thousands of redundancies. At first Peter Ellwood had worried that it would be difficult to introduce Quality while such changes were taking place. 'But I actually think it has helped offset some of the downsides of change', he said

Case 2.3

later. 'It helped improve staff morale because it touched everyone, got staff involved and more able to have their say.' While visible commitment from senior managers was considered essential, their perceived degree of involvement varied. Roger Cliffe remarked:

> Management commitment as perceived by staff varies across the business. Some staff perceive their managers as being very committed to Quality, but a fair number say the opposite. We made it a management imperative to apply Quality. Peter Ellwood is passionate about it and he's a great disciple. He went through the training and he hosts all the recognition events himself.

As Peter Ellwood remarked,

> One of the main barriers to successfully implementing Quality is people's incredulity—they simply don't believe that it will survive. Real Quality improvement within an organization takes time to achieve and is a difficult job even in the absence of other change taking place in the business.

Staff also remarked on the difficulty of applying Quality principles in an environment of scarce resources. Sue Meadows explained that over time the Quality department placed progressively less emphasis on financial measurement of savings. She also put the variations of momentum in perspective:

> This happens at all companies. Even in the absence of major changes in the organization, Quality starts off on an upward hill with enthusiasm and activity levels high at the initial stage. After a while, it levels off and then slips.

By 1995, TSB's best performing unit in terms of Quality was Mortgage Express, a subsidiary which the Group wound down because of its huge losses in the early 1990s.

Roger Cliffe explained:

> Mortgage Express is an excellent example of business success through the application of Total Quality. Their small size and unique position within TSB has allowed them to move further and faster along the Quality journey than most other parts of TSB. Their staff morale is higher than anywhere in the business,[6] customer satisfaction shows excellent positive trends and last year they made record profits. Possibly they have the best of both worlds because they have all the support of the central team, but the independence of a small business unit.

Like most firms which pursued TQM and BPR simultaneously, TSB also faced the problem of integrating the two initiatives. Roger Cliffe explained his view:

> The intention, and it seems to work pretty well, was to give the organization the view that change is not only something that is inflicted on them—it is something they can

---

[6] A survey carried out at the end of 1994 showed that 81% of Mortgage Express staff were satisfied and 74% reckoned that the company was well managed. (*Daily Telegraph*, 3 January 1995).

take part in. The Quality programme has helped in telling people, 'It's up to you to make it work', 'You can improve things at your own level'.

Another objective of the Quality effort was to make individuals more aware that they were part of a process. Archie Kane, however, felt there was no real need for the staff who dealt with customers directly to be aware of the process architecture:

People spend too much time worrying about that sort of thing. I don't think it matters. If you're serving McDonald's hamburgers at the counter, you don't need to understand how the burgers are made. All we need to do is to supply robust service and process platforms that remove the propensity for errors and train staff to use them. Once we have them right, we can position people on top of them.

Roger Cliffe concluded:

People always say you should never implement Quality when you're making radical changes. But in this business that means you'd never implement Quality at all! The challenge is to implement quality at the same time as you are changing the organization quite dramatically.

Case 2.3

Exhibits

Exhibit CS2.3.1 Some tools and techniques used to train TSB staff

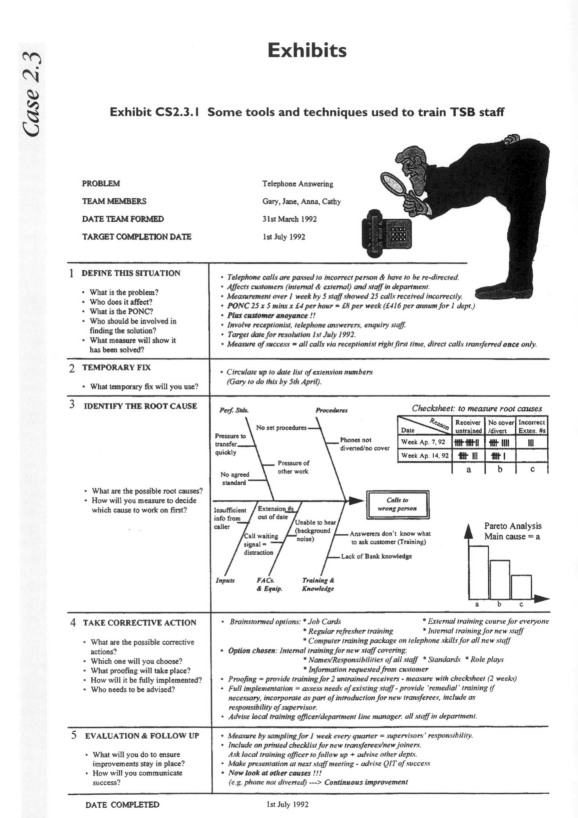

PROBLEM	Telephone Answering
TEAM MEMBERS	Gary, Jane, Anna, Cathy
DATE TEAM FORMED	31st March 1992
TARGET COMPLETION DATE	1st July 1992

1 DEFINE THIS SITUATION

- What is the problem?
- Who does it affect?
- What is the PONC?
- Who should be involved in finding the solution?
- What measure will show it has been solved?

- *Telephone calls are passed to incorrect person & have to be re-directed.*
- *Affects customers (internal & external) and staff in department.*
- *Measurement over 1 week by 5 staff showed 25 calls received incorrectly.*
- *PONC 25 x 5 mins x £4 per hour = £8 per week (£416 per annum for 1 dept.)*
- ***Plus customer anoyance !!***
- *Involve receptionist, telephone answerers, enquiry staff.*
- *Target date for resolution 1st July 1992.*
- *Measure of success = all calls via receptionist right first time, direct calls transferred **once** only.*

2 TEMPORARY FIX

- What temporary fix will you use?

- *Circulate up to date list of extension numbers (Gary to do this by 5th April).*

3 IDENTIFY THE ROOT CAUSE

- What are the possible root causes?
- How will you measure to decide which cause to work on first?

Perf. Stds.
No set procedures
Pressure to transfer quickly
Pressure of other work
No agreed standard

Procedures
Phones not diverted/no cover

Checksheet: to measure root causes

Date \ Reason	Receiver untrained	No cover /divert	Incorrect Exten. #s
Week Ap. 7, 92	‖‖‖‖ ‖‖‖‖ ‖‖	‖‖‖‖ ‖‖‖‖	‖‖‖
Week Ap. 14, 92	‖‖‖‖ ‖‖‖	‖‖‖‖ ‖	
	a	b	c

Calls to wrong person

Insufficient info from caller
Extension #s out of date
Unable to hear (background noise)
Call waiting signal = distraction
Answerers don't know what to ask customer (Training)
Lack of Bank knowledge

Inputs   *FACs. & Equip.*   *Training & Knowledge*

Pareto Analysis
Main cause = a

a b c

4 TAKE CORRECTIVE ACTION

- What are the possible corrective actions?
- Which one will you choose?
- What proofing will take place?
- How will it be fully implemented?
- Who needs to be advised?

- *Brainstormed options: * Job Cards   * External training course for everyone*
  *   * Regular refresher training   * Internal training for new staff*
  *   * Computer training package on telephone skills for all new staff*
- ***Option chosen:** Internal training for new staff covering:*
  *   * Names/Responsibilities of all staff * Standards * Role plays*
  *   * Information requested from customer*
- *Proofing = provide training for 2 untrained receivers - measure with checksheet (2 weeks)*
- *Full implementation = assess needs of existing staff - provide 'remedial' training if necessary, incorporate as part of introduction for new transferees, include as responsibility of supervisor.*
- *Advise local training officer/department line manager. all staff in department.*

5 EVALUATION & FOLLOW UP

- What will you do to ensure improvements stay in place?
- How will you communicate success?

- *Measure by sampling for 1 week every quarter = supervisors' responsibility.*
- *Include on printed checklist for new transferees/new joiners.*
  Ask local training officer to follow up + advise other depts.
- *Make presentation at next staff meeting - advise QIT of success*
- ***Now look at other causes !!!***
  *(e.g. phone not diverted) ---> **Continuous improvement***

DATE COMPLETED	1st July 1992

Exhibit CS2.3.2 Examples of Opportunities for Improvement (1994)

1. Up to 1000 customers a week were inconvenienced while waiting for their Speedbank cards to be returned to them after having it retained by a non-parent branch Speedbank machine. Procedures have now been changed with waiting times being cut from two weeks to three days.

2. Branches using old-style front-office workstations have been unable to process certain enquiry transactions (account details, recent transaction history). In order to obtain the information that the customer required, the cashier would have to close the till and do the transaction on a back-office machine, this would result in the customer waiting unnecessarily at the counter for five minutes. Measurement in one branch showed that this could happen over ten times a day. Network Systems have been able to identify the root cause of the problem and have notified branches of the action they need to take to resolve the problem for their branch.

3. Joint account customers whose accounts are monitored by consumer credit department (CCD) were being overcharged for letters issued to them. CCD would send two advices through the branch for charging, one for each named customer, even though only one charge should be passed. Measurement at one branch showed that 260 such advices are received each year.

4. During the course of the day, commercial business centres (CBC) are required to process computer transactions, many of which a require manager's override. The old procedure meant that a transaction slip had to be completed for every separate transaction. This amounted to 190 000 slips across all CBCs per year. The procedure has now been reviewed, and each CBC is now able to list all the transactions on a single sheet.

5. A licence has been obtained by CBCs to enable prepaid envelopes to be included with facility letter documentation that is issued to customers. This has led to customers returning the documentation far sooner than previously, enabling CBCs to set overdraft limits in time to meet customer requirements.

6. Many passbook customers do not have a linolite signature in their book. When they use a branch other than their own, that branch has to request a copy signature from the customer service centre. If the signature matches, the branch is able to pay the customer cash, but was not permitted to insert a signature in the book, which would resolve the problem. This had to be done at the account holding branch. Measurement showed that these delays for customers were happening three times a day in every branch. Procedures have been changed to allow any branch to insert a signature into a passbook once CSC confirmation has been received.

Case 2.3

Exhibit CS2.3.3 The European Quality Award model

The underlying philosophy of the European Quality Award model is as follows:

Leadership is the driving force that links *policy, people, resources* and *processes* to produce not only *business results* but also measurable results in *customer* and *people satisfaction* and *society*.

The European Quality Award assesses corporate achievement along two sets of dimensions: Enablers and Results. 'Enablers' are the tools to achieve results. 'Results' show how well a particular business, or business area is doing.

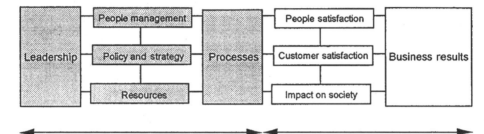

Category	Points	Sample relevant areas
Leadership	100	■ Visible involvement (communication, accessibility, role models) ■ Consistent culture (assess awareness, review progress) ■ Provision of support (funding and championing improvements)
Policy and strategy	80	■ Vision (how Quality is reflected in mission) ■ Business plans (how plans are tested and aligned with organization) ■ Communication (how policy and strategy are communicated)
People management	90	■ Continuous improvement (planning, management) ■ Performance objectives (negotiation, appraised and review) ■ Involvement (suggestion scheme, empowerment)
Resources	90	■ Financial and information (cost data, availability of data) ■ Materials and technologies (goods supplied, fixed assets)
Process	140	■ Key processes (how defined, interface issues solved) ■ Innovations (how new ideas are discovered and used) ■ Review/improve (use of feedback and measurement)
Customer satisfaction	200	■ Customer perceptions in respect of product and service Quality ■ Direct and indirect measures of customer satisfaction
People satisfaction	90	■ What employees think about their business area ■ Direct and indirect measures
Impact on society	60	■ Perception of company within society at large ■ Active involvement of company in the community

Exhibit CS2.3.4 The scope of quality and process management at TSB

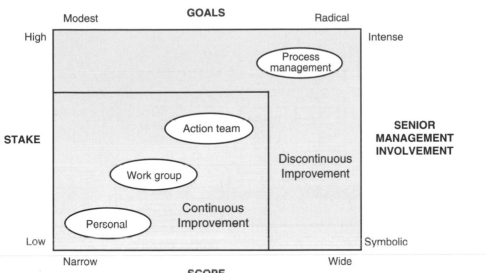

Exhibit CS2.3.5 IDEF methodology

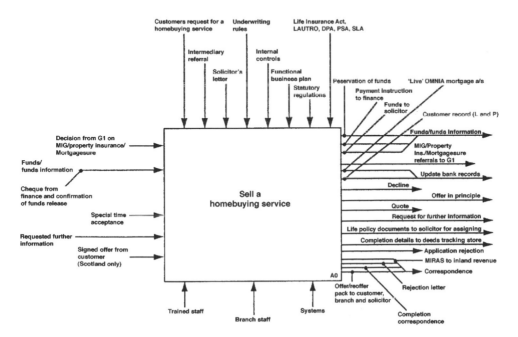

Case 2.3

Exhibit CS2.3.6 TSB's customer-facing processes

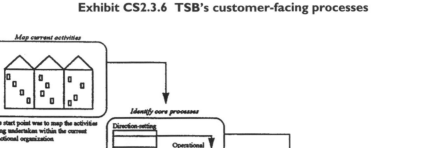

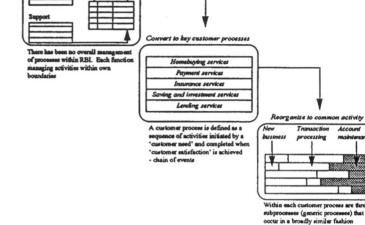

Map current activities

The start point was to map the activities being undertaken within the current functional organization

Identify core processes

Direction-setting

Operational

Support

There has been no overall management of processes within RBI. Each function managing activities within own boundaries

Convert to key customer processes

Homebuying services
Payment services
Insurance services
Saving and investment services
Lending services

A customer process is defined as a sequence of activities initiated by a 'customer need' and completed when 'customer satisfaction' is achieved - chain of events

Reorganize to common activity

New business | Transaction processing | Account maintenance

Within each customer process are three subprocesses (generic processes) that occur in a broadly similar fashion

Exhibit CS2.3.7 TSB's Rubik's cube

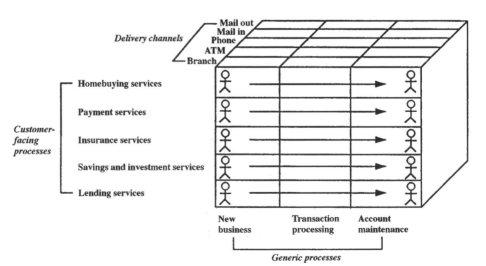

Delivery channels
Mail out
Mail in
Phone
ATM
Branch

Customer-facing processes
Homebuying services
Payment services
Insurance services
Savings and investment services
Lending services

New business | Transaction processing | Account maintenance

Generic processes

Exhibit CS2.3.8 The scope of process management

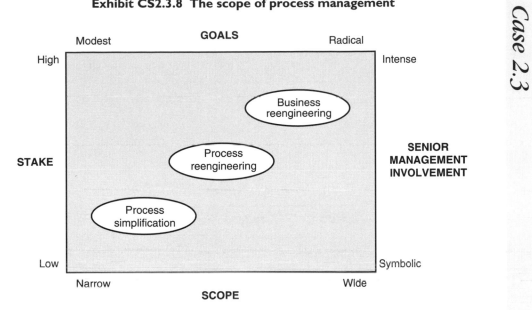

Case 2.3

Case 2.3

Exhibit CS2.3.9 Quality-related indicators at TSB

Customer service index (%):

	1991	1992	1993	1994
TSB	65	67	69	70
Competition	73	72	71	73

Cash availability at cash dispensers (%):

	1991	1994
TSB	89.6	95

Customers experiencing a problem in last 12 months (%)

	1992	1993	1994
TSB	77	71	76
Competition	47	59	55

Customers who are very likely to recommend TSB (%)

	1992	1993	1994
TSB	30	34	36
Competition	n/a	38	39

Exhibit CS2.3.10 TSB brand survey results

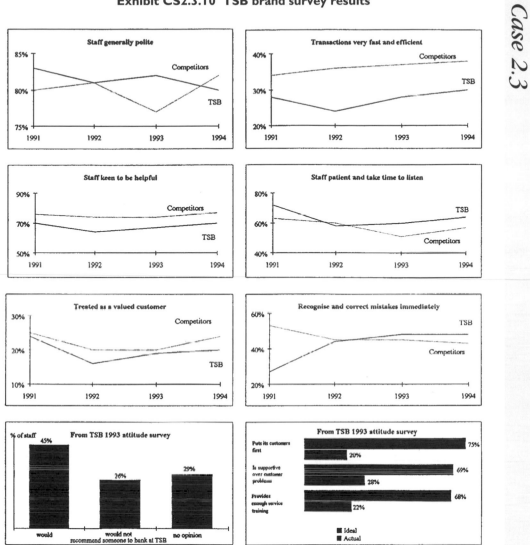

Case 2.3

Exhibit CS2.3.11 TSB's troubleshooter methodology

MANAGERS' ACTION POINTS

*The notes below have been prepared using the Troubleshooter chart format to help
you to present this programme to your staff and answer their questions. They give
an overview of why the programme has been developed and how it fits in with other
Quality initiatives. After watching the video with your staff, a team action plan
should be agreed and implemented. This could include:*
- *Holding a brainstorming session to identify service improvement opportunities*
- *Prioritizing these and using Troubleshooter charts to make local improvements*
- *Setting a target of a minimum of one local improvement project underway at any time*
- *Ensuring that all successes are documented and communicated to your QIT*

---

① **DEFINE THE SITUATION**
- *Quality training has been very successful in heightening awareness and there have been many successes. However, there are still wide differences across the business in the degree to which people are actually applying their training (some pockets of excellence, some areas of inactivity).*
- *This affects everyone in the TSB—the potential PONC is the investment which has been made in the training and the missed opportunities to improve customer service.*
- *To find the solution it was necessary to talk to staff from all levels around the business.*
- *The problem will be solved when there is evidence that Quality tools and techniques are being used on an everyday basis across all parts of the business.*

---

② **TEMPORARY FIX**   • *None appropriate*

---

③ **IDENTIFY THE ROOT CAUSE**

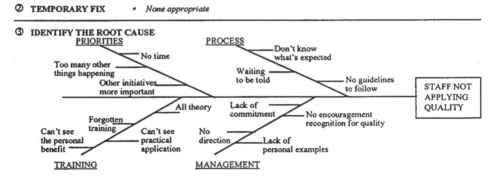

Research with staff showed main root causes to be:
- *perceived lack of management commitment and recognition for Qualit achievements*
- *more direction and practical help needed to apply Quality tools*
- *Lack of awareness of successes and benefits*

---

④ **TAKE CORRECTIVE ACTION**
- *Troubleshooter charts and tools and techniques booklets launched via QITs to give practical help.*
- *Local newsletter developed by some QITs to highlight successes and local workshops/meetings arranged.*
- *National recognition events held (hosted by Peter Ellwood and John Elborne to demonstrate senior management commitment and reinforce the importance of Quality).*
- *Quality in Action communications programme developed comprising of:*
    - *personal audio tape message from Peter Ellwood to all senior managers to highlight the importance of management participation*
    - *video to be shown by managers to all staff to stimulate activity and remind people about the help and support available.*

---

⑤ **EVALUATE AND FOLLOW UP**
- *Monitor activity and use of tools by:*
    - *number of local success stories fed to QITs*
    - *random surveys of staff on effectiveness of programme (telephone, face to face, questionnaires).*
- *Communicate results and successes (local newsletters, on-line, other appropriate channels).*

DATE COMPLETED   *All staff to have seen video by end of April.
Action plans to be agreed and implemented by end of May.*

Information Technology and Performance Innovation

Chapter

3

The ubiquitous potential of technology

The strategic importance of information technology (IT) has long been understood by managers. Not only does IT form the backbone of major industries, such as banking, airlines and publishing, it is an increasingly important value-adding component of most consumer products, such as television sets, cameras and cars.

This is very different from the early decades of computing, when was IT largely relegated to 'invisible', back-office functions. To see a computer, managers had to actively seek out a 'glasshouse' where large mainframes hummed away in antiseptic environments. Today, they do not even have to leave their chairs to locate the computers. They probably have more computing power in their briefcases than they could find in their entire organizations a couple of decades ago. Their cars have more computing horsepower than was contained in the lunar landing craft that was used to achieve the 'great step for mankind'. Even their portable phones, which are fast becoming a mandatory part of modern managerial attire, each contains about half a million lines of computer code within its small case.

Increasingly the value added component of products and services is enabled by IT. Consider a very mundane product: the elevator. There are about six major elevator manufacturing companies throughout the world. How does an elevator manufacturing company, such as Otis, differentiate itself from its competitors and deliver value to customers? Outsourcing of many of the mechanical components to a few global suppliers leaves little room for value differentiation. On the other hand, each building into which Otis installs its product has a specific usage pattern. What if an elevator could be smart and learn (over time) the specific usage pattern of that building? It could, as a result, change its scheduling algorithms to minimize the waiting times at each floor. For example, if an elevator were able to 'learn' that on the 17th floor a particular office closes every Thursday at 3 p.m., it could position all elevators at 2:55 p.m. on the 17th floor. Guaranteed maximum waiting times at each floor are critical for adding value to customers, be it in attracting high rents or saving time in critical environments, such as hospitals. This value innovation is being delivered today with the help of IT.

In addition to the ubiquitous nature of software, IT is often a core part of the business fabric and innovation strategy for organizations. Consider, for instance, developments occurring in the marketplace created by the Internet. Figure 3.1 depicts the home page for Security First Network Bank (SFNB), the first FDIC insured virtual bank. SFNB has a dramatically lower cost structure than traditional banks. Traditional retail banking has been built on the fundamental premise of bricks and mortar— branches which exist in communities and to which customers go for different services. Now SFNB is able to offer a radically different concept of a bank: a virtual bank, open 24 hours a day, that is accesible from the comfort of a customer's home (or office) and is able to offer distinctly competitive products and services. All this is enabled by IT.

The current era is often called 'the information age', and is estimated that at the present, global volume of information doubles every fifth year. (This global volume of

Figure 3.1 Security First Network Bank home page

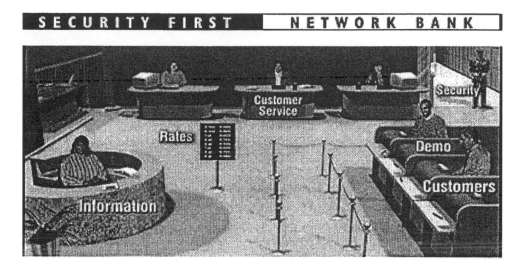

information includes traditional text-based information, but also consists of sound, images, animation, film and futuristic virtual reality simulations stored in both electronic media and networks.) This is a dramatic change from when the amount of global, written information first doubled in the 300 years from 1447, when the printing press was invented, through to 1750. The global volume of information doubled again by about 1900 and then again by 1950. Based on current trends, some experts believe that by the year 2020, the global volume of information will double every 72nd day.[1]

This exponential growth poses tremendous challenges for an organization. On one hand, it can open the doors to inestimable information resources and opportunities for growth, while on the other, it can spell chaos and a possible breakdown in an organization's ability to process information and conduct business.

Companies that are overwhelmed by this will cease to exist and be consigned to history. Those that passively participate in the information explosion will struggle for survival and eke out a meager existence in the shadows of market leaders. True winners of the information age will be those that actively build on relevant external and internal information to create and sustain leadership positions in their chosen markets.

Trends in the impact of technology

During each of the last few decades, major technology shifts have caused fundamental changes in the impact of technology on business. For example, the 1970s saw the advent of the personal computer (PC) that revolutionized the distribution of computing

---

[1] Arthur Andersen (Denmark) *Knowledge as a Competitive Factor,* 1995, p.8.

within organizations. IT was no longer restricted to centralized glasshouses, but started appearing within departments and eventually on employees' desks.

Progress in telecommunications in the 1980s led to the creation of organizational networks and client-server architectures, and people could start using their computers for communication and coordination. This was a major shift away from traditional 'computation-oriented' applications of IT systems, which now started to have a more direct and lasting impact on the way people worked together in organizations.

The 1990s is seeing another major shift—the creation of a global information space, most clearly exemplified by the Internet—and the following sections outline some recent trends that are having a major impact on the way in which technology is being used by organizations for performance innovation.

Increased connectivity

Today, we are witnessing the rapid emergence of virtual information spaces (VIS), which is created by a major increase in the degree of *connectivity* across organizations and within society at large.

Historically, organizations have created shared information spaces by colocating employees, forming teams and increasing communication between different parts of the firm. The power of teamwork for creating new products and services and delivering higher value is well accepted. However, organizations have had to face real physical constraints in creating the right level of teamwork and communication. A simple limitation was that a person could not physically be in two different locations at the same time. This made it difficult to get employees to participate in teams which were not colocated temporally and geographically. Such constraints were particularly acute for global, geographically spread organizations that relied on leveraging the skills of key personnel for their competitive success in the marketplace. Consequently, organizations have traditionally built local information spaces on the backs of effective teams and relied on a central communication function and mechanisms, such as job rotation, to bridge gaps across local information spaces.

The situation has changed dramatically over the last few years with the maturing of networking technology. For more than a decade, organizations have been investing in electronic links (such as EDI) with strategic partners. The widespread adoption of the Internet is now creating a unique global VIS which is shared by both individuals and organizations. Mirroring the growth of this global VIS, are several intraorganizational VISs which are spurred by the current growth in intranet technology. Indeed, the rapid diffusion of the Internet and intranet makes it possible today to imagine a world in which every individual and organization has access to a shared VIS.

Some countries have long enjoyed shared VISs, a good example being France with its Minitel videotext system. Nearly every French home has a Minitel system and uses some of the 25 000 different services available through the Minitel. The ability to perform routine transactions, such as to order groceries and purchase train tickets, or to access information from different sources and communicate using e-mail, has long been available via the Minitel system. Several entrepreneurs have set up innovative services on the Minitel, and the French public has reaped the benefits of their national virtual space for more than a decade.

While the Minitel experience in France provides a good preview of life in a VIS, there are some fundamental differences that make the emerging VISs of today more interesting. The first of these is that the emerging global VIS is technologically open and transcends both organizational and national boundaries. The momentum of the global VIS is so pervasive that even providers of 'closed' virtual spaces, such as France Telecom (for Minitel) and America Online, have been forced to provide connectivity to, and plan for the eventual absorption of, their closed virtual spaces within the global VIS. Today, organizations cannot think of implementing their local VISs with intranets without providing connectivity to the global VIS constituted by the Internet.

Another difference is the *increased bandwidth* of connectivity. While the telephone has always provided an international virtual space, its bandwidth has been very limited. The bandwidth of connectivity within VISs has increased significantly today and is poised to increase dramatically in the future. This will enhance the richness of VISs by making it possible to transmit easily real-time video and virtual worlds.

A final difference is the increase of local memory and intelligence within the emerging VISs. A telephone network does not allow the storage of significant information at any one telephone set. The Minitel is essentially a simple terminal which only provides access to the network.[2] The emerging VISs are built with PCs that can not only store and process information locally, but can also respond to the needs of others in the VIS.

Increased accessibility

A direct and immediate impact of VISs is a radical increase in the degree of accessibility to information. It is now a relatively mundane task to surf the Internet, to move across organizations with complete disregard for national boundaries and to retrieve information from remote locations. Such fluid and rapid access to global information would have seemed a pipe-dream only a few years ago.

Coupled with increased accessibility has been a dramatic reduction in the cost of access. Most organizations distribute information on the Internet free. There is often competition to attract customers by giving away as much relevant information as possible. Time and distance no longer have any relevance to the cost and accessibility of information.

The hurdles to accessing information have been reduced significantly, this has increased the degree of freedom in search strategies. These strategies are no longer

---

[2] Some more intelligent forms of Minitel that integrate the Minitel with a PC are now available.

constrained either to a predetermined route or by those designed by others. Individuals are free to pursue multiplicative search strategies in which they are able to design their own options, explore different search paths and switch search modes when appropriate.

Professional organizations, such as accounting and consulting firms, are at the forefront of exploiting the increased accessibility of information. For example, Arthur Andersen has built an impressive array of databases of global best practices and a shared groupware platform that enables employees worldwide to access and exploit firmwide information freely. A partner from Arthur Andersen (Denmark) provided an example of how his team leveraged the increased accessibility to firmwide data:

> *We have on-line the Proposal Toolbox which contains details of all proposals submitted by Arthur Andersen worldwide. In one instance, we were required to submit an urgent bid in an industry sector which was completely new for the Danish office. Using the Proposal Toolbox, we found all other proposals made by Arthur Andersen worldwide in this industry sector, reutilized parts of proposals prepared by other offices, located and called upon other Arthur Andersen employees who had actually worked on the proposals, and submitted a professional, complete bid within three days!*

Increased interactivity

Interactivity is the essence of services.[3] Certain fundamental trends are spreading this sense of interactivity to all products, even those which traditionally have had low levels of interactivity associated with them.

Typically, the interactivity of products arises from embedded software. All consumer products have varying degrees of embedded software. This can range from around a million lines of code for a top-of-the-line mobile phone to a few hundred lines of code for a room heater. The amount of embedded software in most products is doubling every two years, which leads to a systematic increase in the intelligence and information content of products. While it is easy to see how a mobile phone interacts with its user, even mundane appliances such as room heaters 'interact' with us by, for example, learning the usage patterns of specific rooms and adjusting their heating patterns accordingly.

This increase in the information content of products and services is forcing many organizations to reevaluate their strategies and core competencies. For example, consider Alcatel, a major telecommunications equipment manufacturer. which traditionally thought of itself as a hardware company. Today, as up to 75% of the value

---

[3] The concept of interactivity has been defined in the literature in different ways. J. Deighton ('The Future of Interactive Marketing', *Harvard Business Review*, November–December 1996, pp. 151–162) identifies three components of interactivity: the ability to address an individual, to remember the response of that individual and to address that individual repeatedly in a way that takes into account his or her unique response.

added by its key products is based on software, it views itself more as a software company. As the bulk of the company consists of hardware engineers, it is now in the throes of a major cultural change to disseminate the required skills and attitudes for success in the fast-moving software world.

The increased information content of products is also changing the way we view products. A manager who recently bought a new GM car received a note stating that he should bring the car in to a dealer to fix a defect in the engine. He went in expecting a mechanical part replacement, but found instead that the software controlling the valves of the engine needed to be replaced—there was a bug in the software!

Besides the gradual impregnation of intelligent software into products, a more radical, broader phenomenon is under way—traditional 'hard' products are becoming 'soft'. For example, consider Autodesk, a major Silicon Valley producer of CAD (Computer-aided design) software. Autodesk has an entire division that functions much like a publishing house: it produces software descriptions of building parts such as doors and windows. These 'soft' representations of physical objects are accessed by architects for incorporation into the electronic design of buildings.

With increased levels of interactivity, the concept of products and product delivery changes. Customer interactions with an organization become more participatory as opposed to reactive in nature. Software producers, such as Mircosoft, are well known for involving large groups of their customers in testing and debugging different versions of software prior to market launch. Similar principles of customer participation and involvement are now being explored by other organizations, for example, Levi allows customers the option to design jeans on-line

Changing paradigms

The VIS is changing the paradigm for the creation and utilization of information. For decades, the basic paradigm of information processing has been hierarchical. Certain entities have held privileged positions and have taken on the responsibility of creating, processing and disseminating information. This 'one-to-many' model of information processing is breaking down and is giving way to a radically different, non-hierarchical, 'many-to-many' model of information management. Participants in the emerging VISs are more equal and better able to express their unique contributions.

The publishing industry provides a good example of this change. Traditionally, newspapers have held privileged positions in society with immense power due to their ability to access, process and disseminate information. Today, several on-line newspapers have been started, often by new ventures with a negligible fraction of the resource base of traditional publishing houses. The changing information paradigm has forced major publishing houses to go on-line and give information away for free or for a fraction of the cost of a traditional publication.

Recently, there has also been a radical shift in the notions about how technology is used within organizations. Until the mid-1980s, most applications of technology could be classified as 'automation' of *existing* procedures. These existing processes were typically built on the assumption of limited information access. For example, departments within organizations traditionally did not have immediate access to information within other departments. These information asymmetries led to the creation of elaborate interdepartmental transfer procedures (with their accompanying delays and quality problems). Automating such inefficient procedures often led to little organizational benefit.

The situation is very different today, with intranets and similar technologies spreading information widely across organizations. Critical information is suddenly more transparent and more readily available throughout the organization. This leads to a fertile ground for the *rethinking of business procedures* with the overall goal of performance innovation. Organizations such as Hewlett Packard (HP) and Silicon Graphics have merged the concepts of 'free information' and 'process redesign' with impressive results. For example, HP has developed an intranet application to help employees 'self-authorize' the purchase of petty items, thus drastically reducing the associated administrative overhead.

To take advantage of the pervasive nature of intranet technology within different organizations, HP is now working on redesigning the way it interacts with external business organizations. An ongoing application is in the area of using the Internet/intranet technologies to manage customer support, which has been outsourced to different organizations: via the Internet, partner organizations are able to access internal product-related information and human expertise from different HP sites and provide an overall higher level of support to HP's customers.

Excellence in IT

Technology excellence within an organization can conceptualized at three levels (see Figure 3.2):

Figure 3.2 Levels of IT excellence

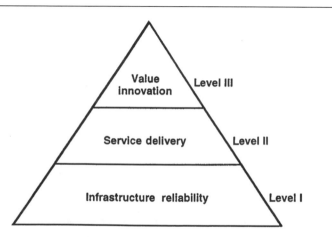

■ *Infrastructure reliability (level I)*: This is the fundamental level of technology excellence within an organization. Every organization needs a good and reliable technology infrastructure. While not the most exciting aspect of technology management, a lack of excellence at this level will hamper an organization's ability to leverage technology in a sustained manner.

Donald R. Walker, senior vice president and chief information officer (CIO) at United Services Automobile Association (USAA) notes that the lowest level of infrastructure reliability presents some of the thorniest issues for technology managers:

The hardest decision are infrastructure decisions. In and of themselves, there's no return on investment. There isn't an ROI until you get a business application that uses the infrastructure.[4]

Although USAA is often quoted as one of the leaders in the innovative use of information technology, it has not shied away from actively investing in its technology infrastructure. In fact, USAA is currently engaged in a $105 million effort to upgrade the company's entire technology infrastructure.

■ *Service delivery (level II)*: This level defines the minimum expected value from technology investments within an organization. Excellence in service is the motto for IT units striving for excellence at this level. Customer satisfaction with the IT is often the best measure of success here. As long as the IT organization is efficient, responsive and credible in the eyes of the customer, there will be a strong partnership between the line functions and IT units, and business needs will be satisfied.

Robert Walker, former Vice President and CIO at Hewlett Packard, describes it in the following manner:

To be really effective, [an organization] needs to have credibility . . . [Credibility is] a mathematical product of two terms. One is empathy, that is to say, understanding what users need and being able to convey that understanding. The other is performance, or delivering on what users think they ought to be doing. Mathematically speaking, if either term is zero, you have zero credibility.[5]

■ *Value innovation (level III)*: Technology units operating at this highest level of excellence focus relentlessly on creating value for the organization. Their key imperative is to leverage their credibility to forge strong partnerships with top management. These organizations have a passion for leveraging the enormous potential of IT for creating business value. For example, several leading financial services organizations have been noted in the literature for their uncanny ability to align their IT and business strategies. Focusing in particular on the

[4] Mayor, T. 'Ensured stability' *CIO*, August 1997, p. 62–68.
[5] Hildebrand, C. 'The nature of excellence', *CIO*, August 1997, pp. 46–58.

Chase Manhattan Corp. and Charles Schwab, Michael May, managing partner of strategy in Andersen's Financial Services Industry practice in New York, noted that

Both companies start with technology as a strategic weapon, not just a business strategy enabler.[6]

The same has been said about British Airways, which is the subject of cases 5.4 and 7.2.

At this level of IT excellence, the focus shifts from the efficiency-oriented measures to more intangible value-based benchmarks. Larry Prusak, a manager in IBM's consulting group notes that

[Value is] usually measured by narrative or story rather than hard numbers because that's how innovative ideas are born.[7]

Organizational challenges

Organizations face several challenges in exploiting the enormous potential of IT. The problem begins in the 'IT Factory' with the fundamental production of software, and extends to the strategic linkage of IT systems to organizational goals.

The notion of a 'software crisis' has been a recurring theme in the literature. Software projects are often reported to exceed budget and time constraints by orders of magnitude. Examples of software disasters abound in the literature. The *Scientific American* described the experiences of California's Department of Motor Vehicles in a 1994 article:

In 1987, California's Department of Motor Vehicles decided to make its customers' lives easier by merging the state's driver and vehicle registration systems—a seemingly straightforward task. It had hoped to unveil convenient one-stop renewal kiosks last year. Instead the DMV saw the projected cost explode to 6.5 times the expected price and the delivery date recede to 1998. In December, the agency pulled the plug and walked away from the seven-year, $44.3 million investment.[8]

While California's DMV could walk away from a non-critical application, others are not so lucky. The American Internal Revenue Service (IRS), for example, had to concede in 1997 that despite having spent $4 billion developing modern computer systems, the systems 'do not work in the real world'.[9] Arthur Gross, an assistant commissioner of internal revenue, added that dysfunctional as some of those systems might be today, the IRS remains wholly dependent on them. More generally, some industry observers claim that as much as three-fourths of all large systems are 'operating failures',—i.e. they consume resources, but do not generate value.

[6] Field, T. 'Banking on bonds', *CIO*, August 1997, pp. 94–100.
[7] Hildebrand, C. *op. cit.*
[8] Gibbs, W.W. 'Software's chronic crisis, *Scientific American*, September 1994.
[9] Johnston, D.C. 'IRS admits its computers are a nightmare', *San Francisco Chronicle*, 31 January, 1997.

Even organizations with a distinct history of IT achievement are not immune from such crises. American Airlines built its reputation for IT excellence during the 1980s on the back of its famous SABRE airline reservation system. The same American Airlines were humbled in a few major disasters, for example while attempting to build the CONFIRM reservation system in 1992 for hotel and car rental companies such as Hilton and Budget. In 1992, which was a huge loss-making year for the US airline industry, American Airlines had to take a $165 million write-off due to the failure of the CONFIRM project. American Airlines was also the butt of several jokes in the European media when its efforts to build a reservation system for the French railway system led to a major fiasco during the traditional peak summer tourist season of 1994. Software errors caused havoc with the traditionally efficient French railways and either grossly overbooked or underbooked trains leading to a national outcry.

Problems in the 'IT factory' are so widespread that the *Scientific American* notes that

for every six new large-scale software systems that are put into operation, two others are cancelled. The average software project overshoots its schedule by half; larger projects generally do worse' (pp. 72–73).[10]

Compounding these problems is the trend for the amount of software code in most consumer products and systems to double every two to three years. Consequently, software developers are scrambling to cope with the pressures of developing systems that are not only a couple of orders of magnitude bigger and more complex than those developed a few years ago, but also need to meet ever-increasing demands for higher quality and superior performance.

The story takes a turn for the worse when one considers the degree to which IT is integrated with the organization's strategy. Not only is the 'software factory' producing systems that are overbudget and late, these systems often do little to advance the organization's strategic goals. Thomas Theobald, the chairman of Continental Bank, describes how the choice of 'important' applications by his bank's IT department baffled him:

It always fascinated me . . . that left to their own devices, technologists will solve internal problems. . . . I can remember a couple of years ago when I was trying to find out the [priority list] for software development, [I noticed] one of the things we did real quick was a new system to order tickets for sporting events. Now, I'm sure it saved a nickel, but why the hell that should show up on the screen of things that are important to the future of Continental Bank, I don't know—except it's typical.[11]

Similarly, in a 1996 survey of 350 European Directors conducted by EDS in collaboration with MORI and INSEAD, over 80% of the senior managers surveyed across different industry sectors wanted their IT strategies to be linked more closely to their business strategies.

---

[10] Gibbs, W.W. 'Software's chronic crisis, *Scientific American*, September 1994.
[11] Fitzgerald, M., 'Still searching for the IS Holy Grail, *Computerworld*, **25**, 34, p47.

A decade ago, when IT was relegated largely to back-office systems, general managers were content to let technologists rule the IT world. But today, with IT positioned as a critical part of the value innovation strategy of organizations, general managers are realizing that they have little choice but to reign in their IT departments and exert more direct control over IT strategies. As they try to do so, they face numerous challenges.

The most obvious challenge is the 'fear of the unknown'. Technology is progressing as a breath-taking pace—a pace with which even technical specialists find it difficult to keep up. Industry observers claim that if other popular technologies, such as automobiles and planes, had progressed at the same rate as computers, a Rolls-Royce would be available today for less than a nickel and a Boeing 747 would cost around $5 and circle the globe five times on a gallon of fuel. Faced with this inevitable technological progress, business managers have often watched in horror as each generation of technology has whizzed by, and how their children and grandchildren often seem more comfortable with computers than they themselves. A manager once aptly described his and his peers' helplessness by the statement, 'we are constantly trying to catch up with the past'.

Technological progress would have been a shade less intimidating had technical specialists not taken it upon themselves to generate acronyms and technical terms at an equally hectic pace. Faced with impenetrable jargon, even technology-sensitive managers have been discouraged from active involvement in strategic IT planning. Only a brave few have persisted and resisted the pressures of tech-talk. An especially persistent senior manager once confided that when he was frustrated, he limited his technical specialists to answer his questions in two words: either 'yes' or 'no'.

This pervasive abdication by general managers of their role in IT management has had unfortunate consequences. Many organizations find themselves helpless captives of run-away technology. Pulled in different directions by rapid technology shifts, organizations have often invested significantly in IT without corresponding returns. It is not uncommon for managers to claim that their top three concerns with IT are 'cost management, cost management and cost management'. In fact, a focus on costs now dominates general management thinking about IT. On one hand, this is understandable because 'cost ' is one element of IT projects that managers 'understand' and can exert some control over. Unfortunately, it has also distracted attention away from the more important 'benefit' side of IT projects. Managers in organizations have paid scant attention to the organizational and human aspects of the implementation of IT systems; aspects which are critical both for ensuring adequate return from IT projects and represent elements of IT management which are directly under the control of general managers.

IT innovation strategies

The model depicted in Figure 3.3 can be used to illustrate the potential of IT for performance innovation. The horizontal axis titled 'Technology Excellence' represents the adequacy of technical skills within the organization. The vertical axis titled

Figure 3.3 A model for IT innovation

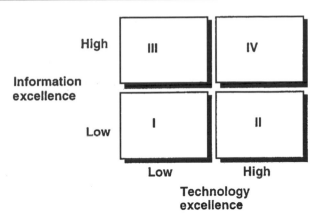

'Information Excellence' represents the ability of the organization to plan and manage its business information needs.

An organization needs to have appropriately high levels of both technology and information excellence in order to use IT successfully to innovate and improve performance. Good information management skills will be of no use if the organization's IT skills are so poor that it is unable to provide a reliable technology infrastructure or develop quality IT products. Similarly, a state-of-the-art technology infrastructure and the best technical skills will be of little use if the business does not have a good handle on its strategy and evolving business needs.

For example, banks had traditionally invested large sums of money in sophisticated IT systems. Regretably, however, their information planning was product-line oriented and was driven by the stove-pipe nature of their organizations. About a decade ago, banks realized that they did not have an overview of all relationships with a customer. Not only was this unacceptable in terms of not satisfying customers, who increasingly wanted a comprehensive view of their relationships with their banks, but it also severely limited the bank's ability to cross-sell products. Banks, while investing in state-of-the-art IT systems, had forgotten to pay adequate attention to planning how their business information needs, such as the need to have an overview of all customer relationships, would evolve. They had positioned themselves in quadrant II in Figure 3.3 and, as a result, were not in a position to innovate using IT.

Depending upon the organization's position in the IT innovation matrix of Figure 3.3, different strategies for performance improvement are possible.

The technology imperative

This is the desired strategy for an organization which is positioned in quadrant I. If the level of technology excellence is low within the organization, the primary focus of the

Figure 3.4 The technology imperative strategy

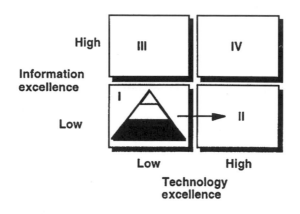

organization's technology strategy should be to increase technical skills and upgrade to a technology infrastructure with the desired levels of quality and reliability. John Gerdelman, the president of network MCI Services summarized this imperative as follows:

> *Reliability is the foundation. If the data center isn't operating and the network isn't up, you don't get to get in there and talk about innovation.*[12]

The desired movement within the IT innovation matrix in the technology imperative strategy is horizontal as shown in Figure 3.4. The emphasis within the three level model of IT excellence is on level 1, infrastructure Reliability (shown by the shaded figure in quadrant I in Figure 3.4). The focus of the organization has to be on technology-related aspects including the following:

- building an infrastructure that is technologically up-to-date and adaptable for the future;

- maintaining a high percentage of uptime of both the data centre and networks;

- supporting a work environment where a majority of workers have access to PCs and networks;

- controlling the total cost of IT systems across the organization.

Technology infrastructure projects usually have a long lead time before any business value can be generated from them. As a result, an organization's top management needs to coordinate technology infrastructure investments in relation to future plans. For example, within USAA, the CEO mandated a decision-making structure called 'the Information Technology Acquisition Process', which requires executive council members to sign off on major projects at key milestones, such as concept definition, concept validation, full-scale development and implementation stages.

---

[12] Hildebrand, C. *op. cit.*

Note that the technology Imperative strategy is also appropriate for organizations positioned in quadrant III of the IT innovation matrix. Realistically, however, few organizations will find themselves in this quadrant. If the organization has an adequate level of excellence in business information management, it is quite likely that it would have paid adequate attention to improving its poor technology infrastructure, which would have prevented it from meeting its business information needs.

The service imperative

This is the desired strategy for organizations positioned in quadrant II of the IT innovation matrix, which involves an appropriate technology infrastructure and adequate technology skills, but a low level of excellence in business information management. Such organizations have to move up vertically to quadrant IV of the IT innovation matrix as depicted in Figure 3.5.

Once the technology infrastructure is stable and reliable, the technology unit has to focus on judging internal and external customer needs, and strive to satisfy and, if possible, exceed them. Technology units need to develop good partnerships with end-users, agree on mutually acceptable levels of cost versus service trade-offs and deliver on their promises with the highest level of operational and project management skills. The emphasis within the IT Excellence Model has to be on level II, service delivery, as shown in Figure 3.5. This strategy demands that the organization focus on best practices for excellence in operations support and service delivery, which include the following:

- knowing the business and understanding the business impact of IT actions;

- partnering with business users to develop and implement best-in-class applications;

Figure 3.5: The Service imperative strategy

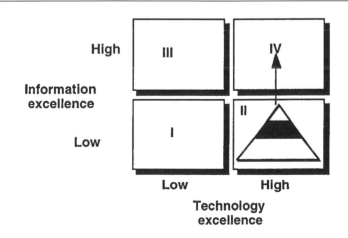

- providing a rapid-response, cost-efficient help desk;
- measuring and baselining service levels for continuous improvement;
- setting good standards and establishing sound management practices;
- devising project and risk management strategies.

The value imperative

For organizations positioned in quadrant IV of the IT innovation matrix, the emphasis has to be on level 3, value innovation, as depicted in Figure 3.6. To use IT as a strategic enabler for business innovation, CIOs and top management have to form an active partnership. CIOs have to ensure that top management is aware of the potential impact of IT for value innovation, while top management has to articulate its vision for the organization in cooperation with the CIOs. Constant contact with top business managers helps technology managers to understand and appreciate the latest trends in business thinking. With a passion for constant learning and innovation, technology units serve the ultimate need of business functions: to create market value by leveraging IT.

Best practices for technology units in the Value Imperative Strategy include the following:

- The CIO forms a close team with the CEO and other top managers.
- IT management is involved in strategic planning for the business as a whole.
- IT strategy supports business goals and can adapt to changing business conditions.
- The IT unit applies resources towards functions that are critical to competitive advantage.
- A managerial group explores and links new technologies with strategic business applications.

Figure 3.6 The value imperative strategy

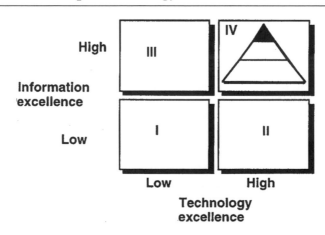

Management involvement in IT innovation

Across countless surveys of critical success factors for IT within organizations, the issue identified most frequently is probably the need to align IT with business strategy. In 1972, Adams claimed that 'the successful implementation of an MIS depends on the active and informed participation of executive management'.[13] More recently, different authors have repeatedly identified the alignment of IT with business objectives as one of the top few determinants of IT success in organizations. For example, Nath noted that for both senior general managers and MIS managers 'upper management commitment is deemed critical' for aligning MIS with their organization's goals.[14] In their survey of 55 CEOs, Jarvenpaa and Ives found that the CEOs who participated in the management of IT were more involved in it, which in turn led to their firm being more progressive in the use of IT[15]. These and other studies make it clear that participation and involvement of senior managers in IT planning and management is beneficial for aligning IT with business strategy.

While there is consensus among researchers and practitioners on the importance of managers participating in and getting involved in IT management, many companies find themselves in a situation where business managers do not consider IT to be an area in which they need to get personally involved. Jarvenpaa and Ives mention that 'few nostrums have been prescribed so religiously and ignored as regularly as executive support in the development and implementation of management information systems'.[16] Lederer and Mendelow note that their research has shown that 'top management still needs to be convinced of the potential strategic impact of information systems'.[17]

An interesting question in this regard is 'Why does management not participate willingly and proactively in IT management?' Lederer and Mendelow attribute this to the fact that top management often lacks awareness of the potential of IT, and/or takes a traditional operational view of computing, and/or lacks adequate faith in their MIS staff, and/or does not view information as a resource to be managed.[18] Age and familiarity with computing are also often mentioned as factors relevant for determining the degree of executive participation and involvement.[19]

The story of Continental Bank (CB) is particularly relevant in this context. The management of IT was perceived to be a problem area by CB's management team. First, the bank's legacy systems were seen as largely unable to cope with the new need

---

[13] Adams, W., 'New role for top management in computer applications', *Financial Executive*, April 1972, p. 54.
[14] Nath, R., 'Aligning MIS with the business goals', *Information and Management*, **16**, 1989, p. 71.
[15] Jarvenpaa, S.L. and Ives, B., 'Executive Involvement and Participation in the Management of Information Technology', *MIS Quarterly*, **15**(2), June 1991, 205–227.
[16] Jarvenpaa, S.L. and Ives, B., *op. cit.*, p. 205.
[17] Lederer, A.L. and Mendelow, A.L., 'Convincing Top Management of the Strategic Potential of Information Systems', *MIS Quarterly*, **12**(4), December 1988, p. 525.
[18] Lederer, A.L. and Mendelow, A.L., *op. cit.*
[19] Jarvenpaa, S.L. and Ives, B., *op. cit.*

for nimble and quick response to customer needs in changing business environments. Secondly, CB had chronic staff problems; the IT unit employed about 500 people, which was too much to carry out the daily operations and maintenance of the systems, but not enough to conduct major system projects. There were also questions about the level of technical skills of the IT staff. Finally, there was the distance between the IT and business staff. A lack of adequate communication between the two groups created frustrations for CB's management. R.L. Huber, who was responsible for CB's back-office and data processing operations, described his concerns:

> *The more I learned about the situation, the clearer it became that the bank was already effectively outsourcing technology services to a group of people who happened to work at the same company I did but whose work was different.*[20]

Faced with these concerns, CB's management struggled with the issue of whether they could manage IT themselves or whether it was in the bank's interest even to try to do so. Eventually, they decided that it was probably best to outsource it.

Strangely, after outsourcing CB's managers started getting involved actively in managing IT. They established a management structure to ensure that CB remained in control over its technology strategy. The management structure performed the functions of balancing the technological requirements of individual business units with the entire bank, ranking IT proposals in a bankwide priority list and ensuring the alignment of the bank's IT and business strategies. One year into the outsourcing relationship, Huber made the following observation on the changed attitudes of CB's business units:

> *Another obvious success is the changed behavior of the bank's business units. They are now active and disciplined participants in the IT process. With virtually all IT work on a hard-dollar, contract basis, they are devoting time on the front end of projects to define clearly and carefully their technology needs and how they want to spend their budgets . . . The process is more demanding now, but it's paying off in reduced technology costs and improved quality.*[21]

How to explain the changed attitudes of CB's management? The same group of managers moved from a passive to a more participative and involved role after outsourcing. There is little evidence that newly acquired skills or increased familiarity prompted this transition. The answer probably lies simply in top management's will and desire to get involved in the management of IT. As long as CB's management treated IT as someone else's (i.e. the in-house IT department's) management problem, IT was a problem orphan for the bank. When forced by the hard-dollar outsourcing contract and the lack of an in-house IT department to delegate IT management to, CB's management started taking a more participatory role in IT management with positive results.

---

[20] Huber, R.L. 'How Continental Bank outsourced its "crown jewels"', *Harvard Business Review*, January–February 1993, p. 123.
[21] Huber, R.L. *op. cit.*

In the light of CB's story, one can question whether issues like age and lack of familiarity with technology are anything but excuses for executives to be not involved in managing IT. No one expects senior business managers to be masters of technology, and they do not need to be so to effectively align IT with their businesses. The emphasis is on management understanding and involvement which makes IT planning and management neither special nor abnormal, but simply a part of the normal business planning of the organization.

Case 3.1

Case 3.1

*This case was written by Soumitra Dutta and Arnoud De Meyer, Professors at INSEAD. It is intended to be used as a basis for class discussion, rather than to illustrate either effective or ineffective handling of an administrative situation.**

Knowledge Management at Arthur Andersen (Denmark): Building Assets in Real Time and in Virtual Space

INTRODUCTION

$K = (P + I)^s$. This simple equation is the basis of our efforts to manage knowledge effectively and be competitive in the information age,

explained Jesper Jarlbaek, the managing partner of the Business Consulting Practice at Arthur Andersen (Denmark). He continued:

We see knowledge (K) as being captured by people's (P) ability to exchange information (I) by utilizing technology (+), exponentially enhanced by the power of sharing (S). The power of this simple equation is tremendous. It is clear and it has universal appeal. When you explain it to someone, everything makes sense, everything falls into place!

Andersen Worldwide SC is the world's largest professional services organization with 1995 revenues of $8.1 billion. It consists of member firms, operating from over 360 locations in 76 countries and has over 82 000 employees. Arthur Andersen (AA) is the business unit of Andersen Worldwide SC that provides audit, business, tax advisory and specialty consulting services.

An ability to organize and communicate knowledge seamlessly on a worldwide basis is viewed as the critical competitive factor in the auditing and consultancy tasks performed by AA (see Exhibit CS3.1.1). Compounding the challenge is the realization that the global volume of knowledge is doubling every fifth year and is predicted to double every 72nd day by the year 2020 (see Exhibit CS3.1.2). AA, more than many other organization, has long realized the importance of

---

*Copyright © 1997 INSEAD, Fontainebleau, France.

managing organizational knowledge effectively to be competitive in an ever-changing, intensely competitive information age. An internal document described the issues thus:

> Our objective is to develop and maintain our knowledge capital so that, as a company, we will command the greatest, best structured and most valuable knowledge capital in the knowledge society . . . . Our unique strength lies in the fact that we have always worked with intensive knowledge sharing. Since Arthur Andersen's origins in 1913, we have had centralized knowledge databases, to which employees all over the world have contributed Best Practice examples . . . Arthur Andersen was applying the science of the knowledge society long before the notions of information and knowledge societies were invented.

The Danish offices of Arthur Andersen have developed a distinct competence in knowledge management which has been widely quoted and has been copied as an example of best practice in knowledge management within the larger AA organization. However, despite extended experience in knowledge sharing, the intricacies of managing knowledge in the information age were only starting to be understood. Jesper Jarlbaek continued:

> We thought we understood the equation, $K = (P+I)^S$, but as we go forward we continually discover new issues, interpretations and problems inside the equation. We are finding that knowledge management is not simply a task of building technological systems or collecting information. Questions are raised about our entire organization and about how people relate to each other. Our journey has only begun!

A TECHNOLOGY PLATFORM FOR KNOWLEDGE SHARING

AA Denmark (AADk) has always been a pioneer in the use of information technology (IT). Close links with AA technology groups in the USA and trusted partnerships with local suppliers in the Danish market have been key factors in AADk's technology leadership. It has often been ahead of its suppliers, as Carsten Sorensen, the previous manager of IT for AADk, described it:

> Twelve years ago, our permanent global PC network was created, years ahead of other firms in Denmark. We set up our first stationary IBM PC in Denmark one year before they were being retailed by IBM Denmark. We put the first portable Compaq PCs into service two years before Compaq Denmark was started. About eight years ago we began using CD ROM systems. In Denmark this meant having to import CD ROM drives in order to be able to start using the new technology!

The Business Systems Consulting (BSC) practice stream (comprising three consultants) was started within AADk in 1991. Initially, the focus of this stream was divided equally between consultancy services in computer risk management and PC-based applications. However, the focus soon shifted to groupware with the acquisition in early 1992 of a small company comprising five experts in networking and a fortuitous demonstration of Lotus Notes to Jesper Jarlbaek by another partner who was passing through Denmark. Jesper Jarlbaek elaborated:

> I was very impressed by Lotus Notes and recognized the tremendous impact it could have on both our business and those of our clients. Though Lotus Notes had been announced as being part of the Arthur Andersen Worldwide Office Automation Platform[1] in 1990, Denmark was not scheduled to get it till late 1993. As we had the

Case 3.1

necessary technical skills in-house, we decided to implement a test site of Lotus Notes within the Business Systems Consulting practice area. We started experimenting with Lotus Notes and doing things which were beyond the scope of the Office Automation Platform, but we did know this then!

Soon after the implementation of the test site of Lotus Notes in AADk, a central European Lotus Notes server for the BSC practices in Europe was set up in Denmark in the winter of 1992. Jesper Jarlbaek continued:

The BSC consultants did not want to wait for the roll-out of the Office Automation Platform to be completed. So with support from the worldwide BSC practice stream management, it was decided to establish a Europewide network on Lotus notes for information sharing within the BSC practice streams in Europe. I lobbied hard to have the server based in Denmark, and, in retrospect, 1992 was the year of learning for us. Our 'playground' in BSC gave us ideas which we normally would not have had without the initial experimentation.

In parallel to the experiments with Lotus Notes within BSC, AADk was preparing for the scheduled roll-out of the Worldwide Office Automation Platform[2] in late 1993. The Office Automation Platform was aimed at giving everyone within AA a PC, a uniform set of tools and links to a common technology platform. This was unique in the history of the company. In anticipation of the new computing platform, an IT committee in the Copenhagen office was established in Spring 1993 within AADk to prioritize applications, decide upon investments and manage the accompanying organizational change. To help prepare employees for the coming change on 1 September 1993, or 'D-Day' as it was called, the IT committee sent out monthly newsletters to everyone and organized a number of training sessions.[3] The need to establish a 'natural dependency' between users and the new platform was identified as a critical success factor for a successful roll-out. Jesper Jarlbaek, the partner leading the committee, elaborated:

As no user-applications were being rolled out from Chicago on the new platform, we chose four applications for development and implementation within Denmark on Lotus Notes: Sign-In, Office Memos, Bulletin Boards and Cash Receipts. Each replicated a daily manual task with the additional advantage that the information presented was simpler, more friendly and real time. All four applications[4] were easy

---

[1] The Arthur Andersen Worldwide Office Automation Platform was announced in 1990 and comprised five applications: Microsoft Word (a word processing package), Excel (a spreadsheet package), Powerpoint (a presentation package), Microsoft Access (a database package) and Lotus Notes (a groupware package). The roll-out of this platform was synchronized and managed on a worldwide basis over a five-year period by an Arthur Andersen technology group in Chicago.

[2] The Office Automation Platform was a major enhancement of the existing global e-mail and data network, AA Net, which had been operational since 1984 on Wang minicomputers.

[3] In total, there were 56 training sessions of half to one day each for all 200 employees of AADk in small groups of six to eight staff per training session.

[4] Office Memos was a one-to-many e-mail application, which eliminated the need for secretaries to photocopy memos, number them (for archiving and security) and then distribute them in different mailbags. Bulletin Boards was a many-to-many discussion forum which allowed AADk employees to conduct several business and non-business discussions simultaneously. Both of these applications were not possible in the older e-mail (one-to-one) system which existed within AA worldwide previously. Cash Receipts allowed partners to check the status of all payments received in the mailroom by the office on a daily basis. Previously, all incoming payments were logged in a manual register and photocopied for distribution to all partners.

to understand, made people comfortable with technology and gave them a glimpse of the power of groupwork on the common platform. For example, Sign-In was a location-logging system to enable each employee to specify the location at which s/he could be reached during the day. Previously, each employee phoned the switchboard at the start of a day and gave their location for the day. This list of locations was photocopied at 9:30 a.m. each morning and distributed to all partners and practice streams.

During 1993 and 1994, Lotus Notes and the use of the new Office Automation Platform spread from the BSC practice stream to the entire Danish AA organization. A number of innovative applications (described later) were developed within AADk for groupwork and knowledge management. A survey in February 1994 showed that about 87% of all employees were logging into the new platform daily. This reflected a very high degree of acceptance as a large number of employees are based at client sites on a daily basis.

By 1995, AADk had also evolved into the largest Lotus Authorized Education Centre for groupware applications on Lotus Notes in Denmark. Though it was not selling the Notes system, it was regularly organizing training and information sessions for increasing awareness of the potential of groupware and leveraging these sessions for building its image, generating new client contacts and selling additional services.

SPACE MANAGEMENT TO ENCOURAGE GROUPWORK

In October 1994, AADk moved to a new building which allowed it to implement an open office concept to encourage teamwork and sharing. Jesper Jarlbaek elaborated:

Having started only in 1962, Arthur Andersen in Denmark was relatively young. Hence the previous building we chose was a grand old building which gave us a sense of being 'established'. However, staff were dispersed in small offices across floors throughout the building and it was not supportive of what was needed in the 90s: networking among people with very different skill sets. Also, we had grown by over 50% since 1989. Thus, when a decision was taken to change buildings, we decided to opt for a new building which would both be 'modern'—in accordance with our desired image for the coming decade—and at the same time be supportive of knowledge sharing and teamwork.

Strict office-space regulations from Chicago[5] meant that the new building would have about half the office space as in the old building. As no one wanted to share desks and the space available was dramatically less than before, the options were, as one manager put it, 'either to have very small desks or to rethink the concept of offices!'

A committee of several staff members was set up to design the new office space. The committee conducted several brainstorming sessions and visited the locations of five other

---

[5] Given the burgeoning cost of real estate worldwide, Arthur Andersen was trying its best to limit rising office costs accompanying its rapid growth. Whenever an office changed buildings, it had to follow strict regulations about the maximum allowable space per employee.

acknowledged 'advanced' offices of leading Danish firms. After numerous discussions, a vision of the new office concept started to emerge. Key aspects of the vision were:

- Open offices: All boundaries would disappear allowing teams to sit and work together in open spaces.

- Total mobility: No one would have a preassigned desk or a fixed phone. Rather desks would be shared by individuals on an as-needed basis when they were not based at client sites. Incoming telephone calls would be redirected through a digital exchange to their desk location for the day.

- Clean desks: Desks would not have any drawers or storage areas. This meant that all work had to filed away in the appropriate place at the end of the day.

- Easy access: Professional staff would have all required support facilities within 10 m of their desks.

- Electronic networking: All desks would have a PC/docking station which would be electronically networked into the AA Office Automation Platform.

- Single card access: One card would be used by all employees to access all functions such as security, restaurants, parking garage and library.

Jesper Jarlbaek commented on the new office concept:

Everyone loved their desks! We had to promise that even though they would not get their own desk, their desks would be bigger and more functional! To help ease the transition, we sent out regular newsletters and created a mock-up of the new office in our old building months prior to the actual move. However, there was still quite a bit of anxiety and frustration prior to the move.

In preparation for the move, support services (e.g. for printers, office supplies and reference works) were clustered to enable the internal administrative functions to support a high level of service in the new building. All local manuals and reference documentation were also moved on-line in order to save on shelf space and to enhance access and maintainability. This meant that staff would no longer need to change floors to consult documents or get a printout. Interestingly, partners were reluctant to give up their private offices, even though they agreed to have their individual office space halved!

The move to the new premises was completed successfully in November 1994. Jesper Jarlbaek continued:

Three months after the move, we conducted and published the results of a written survey to gauge the level of satisfaction with the new building. The strategy behind a written survey was to enable the 'silent majority' to express their views. The results were very satisfying—80% of the staff reported to be 'satisfied' or 'very satisfied' in 36 of the 37 measured areas. By publishing the survey, we were able to ensure that the 'dissatisfied minority' did not distort the positive atmosphere and we had a documented basis to conclude that the move had been successful.

ARCHIVING AND KNOWLEDGE STORES

A number of different archives (such as a general correspondence archive, a working papers[6] archive and an office copy[7] archive) were maintained centrally in different locations within AADk. Adding to this dispersion was the fact that most employees had their own 'personal archive' in order to avoid walking across floors to retrieve documents! The overall process of storing and retrieving information into/from the archives was wasteful and often inaccurate.

The introduction of the new office concept provided the opportunity to rethink the organization of archives and knowledge stores:

■ First, the entire archiving process was redefined around the client. It was decided to store all information relating to a client in one logical location. As the new desks did not have drawers, all work had to be filed in the appropriate client's archive at the end of the day. This was a change from the previous practice of each individual/team storing-client related documents in a personal archive, a habit which impeded sharing. The new discipline encouraged the philosophy that a client belonged to the firm and not to a team or an individual.

■ Second, rather than assigning physical shelves of equal sizes to all clients, small partitions were created and assigned flexibly to each client on an as-needed basis. A Lotus Notes application was designed to keep track of the physical locations of different client partitions. This change dramatically reduced the amount of required physical archiving space.

■ Third, an electronic client folder was created on Lotus Notes for each client and it contained electronic filings of all client-related documents generated in-house and all incoming and outgoing mail/faxes. The aim was to start electronic filing for low volume, high value-adding documents[8] and then progressively move on to other high volume documents such as working papers. A simple work-flow application within the electronic client folder both facilitated sharing and enhanced security.

■ Fourth, many manuals, both local (such as the employee policy manual and address directory) and AA firmwide (such as audit procedures), were moved on-line and updated centrally. Though some AA firmwide manuals were available on-line previously, their usage was poor because not all employees had PCs and not all desks had docking stations connected to the worldwide network, AAnet. But this changed in the new offices where access to AAnet was easy and common. Exhibits CS3.1.3 and CS3.1.4 summarize some of the AADk local and AA firmwide knowledge-bases available on-line. The

---

[6] These are the intermediate working papers produced by auditors during the audit process. They are not sent to the client.

[7] The office copy contains the original financial statements and the corresponding audit opinion (which is the end result of the auditing process) and various quality assurance documents. While the original audit opinion is sent to the client, the other documents are for internal use only.

[8] All correspondance with clients, of both final and intermediate results, is very important as AADk is liable to external stakeholders of the client. For example, if a client were to suddenly go bankrupt, then AADk could be required to prove in court that it did an honest and fair job in auditing the client's accounts.

Case 3.1

widespread availability of useful knowledge further stimulated the use of the Office Automation Platform.

Carsten Sorensen, the designer of the new archiving system, commented on the impact of the changes in the archiving process:

> A typical large client may be served by 10 different service lines. Previously the Client Relationship Partner had a patchy overview of our relationships with the client. Now we have one knowledge store per client for the partner to consult. This allows better activity tracking of clients, timely retrieval to provide quick responses to client queries, and supports our mission to make each client a full-service client.[9]

> Our outputs are deliverables produced by the teamwork of several members of our organization. The process is inherently iterative with many documents existing in a dozen or more versions of which perhaps only one or two are sent to the client for comments at intermediate stages. The new archiving system has made it possible for us to support the sharing necessary for effective teamwork and also to build in appropriate quality control procedures during intermediate stages. This is important because AA can be held legally responsible for its deliverables such as audits. In such cases the contents of working papers and intermediate documents shared with clients becomes very sensitive.

The technological and physical archiving innovations implemented at AADk have been formalized as COCOA (Copenhagen Correspondence Archiving) and is being rolled out in the larger AA organization as a best practice.

ADVANCED DECISION SUPPORT

Global knowledge sharing via the on-line knowledge-bases often provided a unique competitive advantage. Jesper Jarlbaek provided an example:

> We have an on-line knowledge-base called the Proposal Toolbox which contains details of all proposals submitted by Arthur Andersen worldwide. In one instance we were required to submit an urgent bid in an industry sector which was completely new for the Danish office. Using the Proposal Toolbox, we found all other proposals made by AA worldwide in this industry sector reutilized parts of proposals prepared by other offices, located and called upon other AA employees who had actually worked on the proposals, and submitted a professional, complete bid within three days!

The constant need to render client-oriented processes at AA more effective and efficient has led to the development of a number of advanced IT systems for decision support. Exhibit CS3.1.5 provides details on decision support and expert systems which have been developed within the global AA organization to aid the auditing process. These systems have had a major impact on improving the quality of the auditing process and improving client satisfaction.

---

[9] That is sell a variety of audit, tax and consultancy services to the same client. Traditionally, the different streams of AA have been fairly protective of their own client bases.

A CULTURE OF TEAMWORK

On paper, AADk was a very hierarchical organization, particularly on the audit side. There were several levels between a first-year assistant to partner with fairly fixed roles and responsibilities at each level, especially during the first five years before an entering employee became a manager. However, there was significant teamwork and informal networking within the organization.

Carsten Dalsgaard, the partner responsible for external knowledge management services at AADk, described the prevailing culture of teamwork and networking:

> One of the best things about Arthur Andersen is the 'one-firm' concept. Arthur Andersen is able to leverage its international expertise in an unrivaled manner. We are only able to do it because we work through teams—teams of all kinds: across practice, across offices and across countries if necessary. As we move from team to team across projects, we form our own informal networks. These networks are reinforced periodically at annual firmwide or regional meetings and at training courses. You just cannot survive within Arthur Andersen without a good people network!

The general consensus was that teamwork was far more established in the consultancy practice streams of AADk where projects and clients changed rapidly and required a fluid passage of employees from one project to another. In contrast, more than 70% of audit clients were the same from one year to the next. Consequently, audit teams for a particular client were less volatile and some team members worked with the same client for a number of years.

A proliferation in the number of practice streams since the late 1980s has accentuated a need for flexible cross-practice teams bringing together different skill sets. The number of cross-practice teams is expected to increase sharply through the late 1990s as AA gradually reorganizes its activities worldwide along industry groups. Cross-office teams are less common, and are often formed either in response to the specific needs of an international client or to leverage the special expertise of a particular AA office. AA invests more than $15 thousand in education for each consultant per year. This both provides a common framework for cooperation across practices and offices, and eases project skill mixing by ensuring that an employee at a certain level has certain proven skills.

Though each team consisted of employees from different hierarchical levels, the group dynamics worked well. A junior employee commented as follows:

> Even when I was fairly new in the organization, I found that my opinion was respected. I could influence the project and I learnt a lot from the chance to collaborate with several experienced partners and managers on different projects. I feel that I have a particularly close relationship with colleagues at my level worldwide as we all went through the same training programme at the start and have kept meeting frequently at different reunions in the subsequent years. I find that the people network within AA is a deliberate antidote to the hierarchy.

Jesper Jarlbaek added his view on the relation between the hierarchy and the sharing of knowledge:

Case 3.1

Hierarchies are not a prevalent part of Danish management philosophy! However, in my opinion, effective knowledge sharing is aided by certain structures. Our organizational hierarchy can be an example of such a structure which makes the sharing of knowledge more manageable and effective.

Even though the technological support for knowledge sharing was becoming more sophisticated over the years, the role of people networks was seen as central to AA's knowledge management efforts. In the words of a manager:

Tools are useful for finding facts; but they don't tell you how to do something. So even if you find the names of contacts through the on-line databases, you have to call them to learn about the details of doing the job.

Indeed, collaboration is seen as a necessary way of life within AA. A manager described it as follows:

Especially from my level up, we are required to be entrepreneurial and to constantly seek out work. So there is this implicit pressure to talk to others—you never know where leads may come from! There is this 'collaboration-thing' on the grapevine. You would not want to have a reputation of being unhelpful!

Jesper Jarlbaek added:

We are trying actively to promote sharing across offices. During recent meetings of the heads of management consulting practices in Europe, we have dedicated the first couple of hours to providing examples of and discussing means of enhancing cooperation across offices.

A Focus on Knowledge Management

Each service line within the worldwide AA organization has a knowledge manager who is responsible for the collation and dissemination of knowledge related to that service line on a worldwide basis. Each AA office also has a manager responsible for knowledge management. Carsten Dalsgaard, the partner responsible for knowledge management within AADk elaborated on his role:

I have to ensure that our Danish offices are plugged into the AA global knowledge network. This means two things. First, we should be making effective use of the global AA knowledge-base. Second, we should be making our contributions to the development of the knowledge capital within AA. This latter aspect is particularly important—if you see value in and use knowledge created by someone else, but you do not add to it, it is unfair.

After the completion of each project, project team members were required to fill in a report summarizing their key benefits from the assignment experience and also mentioning any best practices which they may have observed within the client's organization. These observations were summarized and shared locally, and also transmitted to the global AA knowledge managers.

The global knowledge managers filtered through the 'incoming knowledge', distilled the key messages and incorporated them within appropriate firmwide knowledge bases.

The degree of knowledge shared also influenced the evaluation of personnel at all levels of the organization. Carsten Dalsgaard elaborated:

> *Project team members are required to search for all possible sources of relevant knowledge within the global AA organization. They are rated by project leaders on the degree to which they have used firmwide resources. The effective leverage of global AA resources is one of the components in the annual ratings of staff. Even partners are subject to a peer evaluation of the degree to which they have been helpful and cooperative.*

Advanced decision support systems also contributed to the ongoing knowledge management processes. For example, Win SMART, a risk assessment expert system (see Exhibit 3.1.5), was used more than 100 000 times a year worldwide. The results of all these analyses are centrally consolidated and used to update the weightings of the risk drivers. This knowledge is incorporated and distributed in new versions of Win SMART and forms the basis for the computation of the updated risks for assignments.

ISSUES AND CONCERNS

A particular concern was related to the realization that different practice streams exhibited different attitudes to knowledge sharing. This was particularly evident in the differential rates of adoption of groupware for knowledge sharing within the practice streams. Jesper Jarlbaek elaborated:

> *The business systems consulting practice stream has been very successful over the years in sharing knowledge with groupware on a global scale. They have a natural inclination to share. In contrast, the audit practice streams while being good users, have been poor adopters of groupware for knowledge sharing. I think that the difference is mainly due to differences in the nature of the skills used. Knowledge within the business systems area is specialized, technical in nature and fluid. Auditing, on the other hand, relies on well documented rules and models, much of which were developed several decades ago.*

While AA has developed a special competence in the use of technology for top-down sharing of knowledge through an impressive array of on-line knowledge-bases, the overall 'write/read ratio' is low, i.e., the amount of new knowledge contributed by employees is low as compared to the use of existing knowledge by employees. A partner speculated that the reasons for this were probably rooted in the culture of the company:

> *Most employees of Arthur Andersen, especially at the senior levels, are focused on chargeability—as this is a key figure of their performance evaluation. This may lead to a natural inclination to not share information with each other, to read more and to write less! However, everyone has to be writer and not only be a reader for the network to succeed.*

A manager thought that the current profit center structure was not most conducive to sharing:

> Today, we have profit centres per office and per practice stream within each office. While this sharpens our focus and increases accountability, it does not always create the right incentives for sharing. I do not believe that we really understand how to measure knowledge management and create the right incentives for effective knowledge sharing.

Knowledge management within AADk largely mirrored the different practice streams. This was at odds with the industry trend of moving towards an organization structured along industry groups, such as retailing. An organization structured along industry groups would not only require a higher degree of integration across employees with different skill sets, but also need a different structuring of knowledge across the organization globally.

In addition, there were challenges along the international dimension. Jesper Jarlbaek continued:

> You need an environment of trust and openness for effective knowledge sharing. But the reality is that there are large differences in management styles and trust levels across European countries. For example, there was significant resistance to giving junior staff access to Lotus Notes within the international tax practice stream in a neighbouring country. It was feared that the staff may send the wrong information to other AA offices and increase legal liability. The answer is not to shut off the use of the enabling technology; rather there should be organizational mechanisms to ensure that people have adequate skills, talk to each other and trust each other.

Practical issues like differences in languages also posed major hurdles as explained by Carsten Dalsgaard:

> Summaries of past projects are entered in our on-line database, EDGE, in English. But the whole project proposal is only in Danish. At present, it is not possible to translate the proposals into English or other languages. This makes it difficult to share all details with colleagues in the global AA organization.

While employees universally praised the global knowledge management infrastructure of AADk as unique, many were critical of information overload in the existing system. A manager commented:

> There is too much information today. I spend a lot of time reading and listening, not enough doing. At our billing rates, this can be deadly! While technology makes it easier for me to accomplish certain things, it does not help in differentiating us from our competitors.

Another junior staff member added:

> There is information overload. There is a need to structure knowledge better to make it more useful. There needs to be a higher degree of quality control into what goes into our knowledge bases. For example, the reporting of key learnings required after the

completion of each project occurs in a relatively ad hoc manner. Thus the quality of knowledge input varies widely from project to project.

Technological progress and trends in knowledge sharing was also threatening to change the very nature of AA's business. Jesper Jarlbaek commented:

The trend is to share our knowledge directly with the client. For example, for the first time we have packaged our knowledge about accounting principles and started selling it to clients as a CD-ROM/Internet based knowledge product. If you start sharing everything, what do you have left to generate revenue with? Market trends are forcing us—and everyone else too—to start moving further up the value ladder!

This could have implications on the nature of clients sought by AADk. Jesper continued:

We have traditionally sought clients who could obtain a competitive advantage from our core competencies. But now we may have to target clients more carefully with an additional explicit focus on being challenged and developing new competencies in the process.

THE KNOWLEDGE EQUATION FOR THE FUTURE

Clearly AADk has invested significantly over the years in setting up an advanced infrastructure for the management of knowledge. While the equation, $K = (P + I)^S$ captured the essence of what AADk had tried to achieve, the organization was becoming aware of further changes and fine-tuning that were needed to continue building its competence in knowledge management.

Jesper Jarlbaek outlined some of the questions which were foremost in his mind:

The knowledge equation has been widely publicized in our organization. I wonder if it is deeply embedded within the ingrained culture of our firm and behaviors of our staff?

The formulation of the knowledge equation is simple and that has been one of its strengths. But is it robust enough to capture the intricacies of knowledge management? Are we missing some critical aspects?

Technological shifts such as the Internet and the global market space will require radical new ways of organizing knowledge and competing. Are we prepared to meet this challenge? Or do we run the risk of becoming a corporate dinosaur?

We believe that our people are our most valuable knowledge assets. Technology is now enabling us to create different kinds of knowledge assets— in real time and in virtual space! What are the long-term implications of these changes on the recruiting and career development of our staff? What kind of skills do they need and how do we develop them? How do we measure the effectiveness of their performance?

Exhibits

Exhibit CS3.1.1 Knowledge—The competitive factor in the auditing and consultancy business

The ability to collate and communicate knowledge is crucial to an auditor's efficiency and to the quality of both the auditing and consultancy tasks performed by AA.

AUDITING

Statutory auditing is performed with a view to allowing society to obtain an independent, state authorized, public accountant's statement concerning the financial position of the company, the result for the period under review and continued operations. The auditor's report is the conclusion of many hours of analysis and consolidation work. The shorter the report, the better.

The purpose of the auditing process is to build up a stock of knowledge enabling the auditor to evaluate the correctness of the client's recorded financial transactions and the company's ability to continue its operations.

The general turbulence in the markets, combined with the companies' multifarious and changing applications of information technology, partly in connection with administrative systems and partly in its integration with product/market-oriented systems, is making increasingly greater demands on the auditor's know-how and professional breadth of view. Moreover, frequent changes in client demands result in the need for technological innovation on the part of the auditor.

Personal experience base and access to relevant information, and the ability to assimilate and make conclusions on the basis of the same, are of the essence in the speed, quality and hence the effectiveness of the auditor's services.

CONSULTANCY

AA provides a range of consultancy services ranging from tax and financial planning to business systems consulting and computer risk management. Consultancy services arise from specific identified needs of clients and require the quick and cost-effective delivery of customized solutions based on leading-edge business practices.

The range and diversity of consultancy tasks done worldwide by different AA offices provides an enormous fund of global knowledge capital that is potentially available for its employees for any specific project. However, this also means that the organization has to have a unique capability to collate the knowledge garnered from experiences in different projects, structure this knowledge for dissemination worldwide and allow for its employees to leverage this knowledge capital for new projects in a quick and cost-effective manner.

The pressure of managing AA's cumulative knowledge capital effectively is also increased by the global nature of most consultancy projects and the desire of all clients to be at the leading edge of worldwide best practice.

Source: Internal Arthur Andersen documents.

Exhibit CS3.1.2 Growth in global volume of knowledge

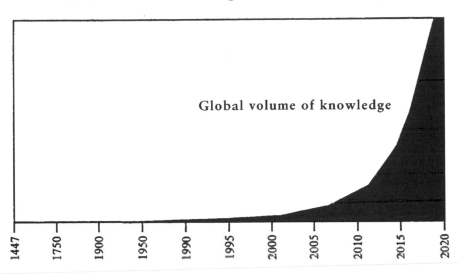

Historically, the development of the global, written volume of knowledge gained impetus in the year 1447, when Johann Gutenberg invented the printing press. The printing of books, among other things, gained ground, and in the space of 300 years or so (by about 1750) the global volume of knowledge had doubled. The global volume of knowledge had doubled again by about 1900 and had redoubled further by the year 1950.

Today, it is estimated that there is a doubling every fifth year. On the basis of the exponential constituent S in the knowledge formula $K = (P+I)^S$, one can expect a doubling of the global volume of knowledge *every 72nd day* by the year 2020.

In addition to traditional text-based information, the global volume of knowledge will consist of sound, images, animation, film and futuristic virtual reality simulations, etc. This observation includes not merely information on the local and global networks, but also stored and printed information on electronic (e.g. computer disks, tapes and CDs) and paper-based media.

This information overflow poses a challenge that will place demands on the ability of every company to make constant adjustments. The considerable volume of information will not only imply an inestimable knowledge resource, but also spell possible information chaos. Controlling information processing resources (including the filtering, structuring and validation of data) will be of great importance in selecting the right information when it comes to performing a large number of actions, procedures, decision support and so on.

Source: Internal Arthur Andersen documents.

Exhibit CS3.1.3 Sample Arthur Andersen Denmark local knowledge-bases available on-line

Employee Policy manual

Employee rights and contracts rules are summarized in this Lotus Notes application. Updates are seen from the number of unread documents on each employee's local PC in Lotus Notes.

Computer manual

The computer manual is used as an introduction to new employees or as reference information. The computer manual provides a general overview of most of the tools in daily use in the auditing, tax and consultancy departments.

ISO manuals

Arthur Andersen is currently the only ISO 9001 certified firm among State Authorized Public Accountant in Denmark. For the purpose of being certified, AADk developed and employed a series of electronic manuals, which are updated with procedures and methods of relevance to its ISO 9001 certification. The electronic manuals are updated according to standard ISO 9001 rules, with a closed panel of authors, a centralized approval procedure, and so on.

Private addresses

The staff database contains private address information, a photograph and information concerning department or division, post, and so on.

Telephone directories

Client telephone lists, short codes in the telephone system, telephone numbers to other Arthur Andersen offices, etc., are maintained in a centralized database.

Client list

The client list contains an index of Arthur Andersen Denmark's clients including address information and telephone/fax numbers. It also indicates the person within Arthur Andersen in charge of the client's account.

Note: There are more than 50 different local knowledge-based applications which have been developed locally within AADk and are in widespread, regular use by the Danish staff.

Source: Internal Arthur Andersen documents.

Exhibit CS3.1.4 Arthur Andersen firmwide knowledge bases available on-line

AA On-line

AA On-line is a Lotus Notes application through which the whole of Arthur Andersen can communicate globally on the basis of categories divided according to function or topic. The information in AA On-line is divided into various forms of communication:

Publication of announcements (one to many): Announcements are used partly for the purpose of information from the management to the employees and *within* the individual function areas, e.g. all staff in computer risk management, and partly in connection with the daily updating of international news from American newsbrokers and newspaper publishers. Publication is used for date-sensitive information, e.g. announcements of international meetings, new products and so on.

Conferences (closed, limited forum of users): Topics that call for international discussion, but cannot be directly referred or confined to the classification of functions or topics in AA On-line can be opened as a conference. Instead of, for example, having to gather partners, managers and other staff somewhere around the world every month with input for the project complete with all the associated travelling costs and interruptions to the individual's local work, information is exchanged via a conference on AA On-line.

Resource information (few to many): In AA On-line, resource information is kept on the various function or topic areas. In conjunction with the development of areas of application for data analysis in auditing, for instance, a resources review is kept in AA On-line, detailing the offices' present use of tools, the type of client, the local strategy selected for developing the business area and so on. At the individual offices, the employees in the function and topic groups share methods, facts and experiences with colleagues all around the world. 'Knowledge managers' within the individual function areas constantly monitor developments in this aggregation of resource information and in this way ensure that knowledge is shared in a structured and qualitative fashion.

Discussions and Ask Arthur (many to many): AA On-line not only grants the individual employee access to information already recorded, but enables him or her to seek advice and guidance from colleagues, regardless of their geographical location. A question is posted on AA On-line, and the replies received are tagged onto the question. The original problem is thus supplemented with proposed solutions, including references to the employees involved. In this way a consistent knowledge database is built up in AA On-line—again, dynamically.

Global best practices

This is the result of a centralized collection of basic data for auditing, methodological and procedural standards, and the systematic consolidation and analyses thereof for reporting Global Best Practices and useful benchmark data. Global Best Practices information is often generic and can be used on companies cutting across different trades and industries. The object of the exercise is to lend inspiration for creative development, of *inert alia*, procedures, methods and business routines. The databases are edited centrally and used together with Win Art Plus, an Arthur Andersen information search tool. Win Art allows for free-text searching and switching between reference material and cross-references with a click of the mouse—straight on words, terms and concepts.

Source: Internal Arthur Andersen documents.

Case 3.1

Exhibit CS3.1.5 Decision support and expert systems in the audit process

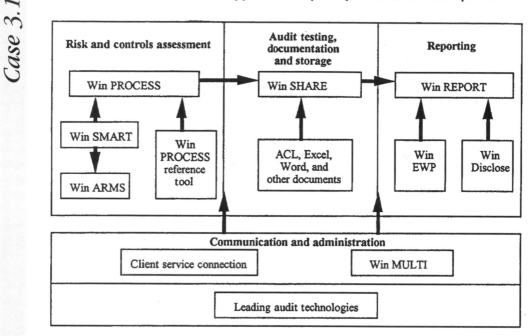

Win SMART

Prior to embarking on any auditing or consultancy services, a risk analysis of potential client arrangements is performed to sketch out and delimit the auditing process. Win SMART is a user-friendly expert system that optimizes and rationalizes a qualitative decision-making process. Win SMART is engineered as a work-flow, guiding the user first through questions ranging from the general to the specific. When all questions have been answered, the risk profile for the commitment is computed, resulting in completed forms ready for signing by the management of Arthur Andersen and a graded risk profile in the range of low to maximum risk. Depending on the risk computed, consultative comments are generated on the forms. Win ARMS is a related tool which facilitates the accept/retain approval process.

Win PROCESS

Win PROCESS is Arthur Andersen's most complex and sophisticated expert system. As its database, Win PROCESS takes the completed Win SMART and a wide range of benchmark analyses classified by industry in a work-flow-controlled application with the following objectives: (a) to comprehend, communicate and evaluate the client's expectations of the auditing process; (b) to provide the client with relevant suggestions for possible added value initiatives; (c) to match detailed, client-specific accounting data with relevant historical and benchmark industrial data to identify risk and variance and to form a basis for creative proposals for improvements; and (d) to generate cost-effective planning on major auditing assignments. The Win PROCESS Reference Tool is a resource and training tool which provides high-level and detailed overviews of Win PROCESS.

Win SHARE

Win SHARE organizes all electronic audit files and serves as a platform for electronic documentation. It receives all reports from Win PROCESS and works in conjunction with the ACL tool and other supporting Excel and Word documents. The ACL tool allows for detailed analysis of data extracted from a client's computer systems. Based on a knowledge of the client's administrative routines, transactional details are output from the client's administrative systems into an electronic feature-base for the company's financial position for the period. In addition to totalling the transactions per account and reconciling these with balance sheet, debtor, creditor and for example commodity group balances, logical and statistical analyses can be performed with the ACL data analysis tool.

Win REPORT

The Win REPORT is the definitive consolidation tool for use in the documentation and conclusion phase of the audit. It will make it possible to arrange the final accounts, and will also contain reporting and analytical facilities on the basis of Win SMART, Win PROCESS and Win SHARE data. Win REPORT works in conjunction with two supporting tools: Win EWP and Win DISCLOSE. While the former allows for complex consolidations, the latter supports the determination of financial statement disclosure requirements.

Win MULTI

This tool automates the multi-location audit administration process. It facilitates communication among offices.

Client service connection

This is a tool to promote electronic communication and knowledge sharing between engagement teams and clients, if desired.

Leading audit technologies

This is a marketing, training and recruiting tool. It describes many of the above tools.

Source: Internal Arthur Andersen documents.

Case 3.2

This case was written by Soumitra Dutta, Associate Professor at INSEAD. It is intended to be used as a basis for class discussion, rather than to illustrate either effective or ineffective handling of an administrative situation. Information on Continental Bank is taken entirely from published sources.**

Aligning IT with the Business: Banco Comercial Português and Continental Bank

INTRODUCTION

Starting on 6 May, 1986, Banco Comercial Português (BCP) grew in little over half a decade to become one of Portugal's largest and most profitable banks (see Exhibit CS3.2.1). In mid-1993, it had around 300 locations, 4300 employees and the equivalent in Portuguese escudos of over $11 billion in assets. BCP's extraordinary growth was marked by a continuous series of innovations, remarkable for the Portuguese banking environment in the late 1980s. None of BCP's competitors was able to match either the rapidity or diversity of its innovations. Technology played a critical role in enabling BCP's innovations. However, BCP's leadership position arose less from the use of cutting-edge technology as from a successful alignment of IT with its business. Today, BCP is an acknowledged leader in the strategic use of IT within European banks.

In the early 1990s, Continental Bank (CB) was recovering from the disastrous effects of failed large-scale energy-related investments made during the early 1980s. After a FDIC bailout in the mid-1980s, CB made several tough decisions including a dramatic refocusing of its core business away from retail banking to business customers. Thousands of employees had to go as the bank overhauled its internal structures and systems. Despite the changes, the early 1990s found CB with a weak reputation, strong pressures on profit and a continuous analysis of its every action by regulators, analysts and investors. Operating under such difficult circumstances, CB's management team aggressively started cutting costs in different areas. As part of its cost-cutting and refocusing moves, CB became the first money-center bank to completely outsource its IT when it signed an outsourcing agreement with IBM in 1991.

BCP and CB are two different banks on opposite sides of the Atlantic. Both of them have been successful in their own way. From a small start-up bank BCP has become one of the largest banks in Portugal within less than a decade.[1] On the other hand, CB has moved from a federal bailout to a focused, profitable operation which attracted BankAmerica's attention and led to a decision in January 1994 by BankAmerica to pay $1.9 billion to acquire CB. While BCP manages IT in-house, CB has chosen to out-source IT. Within these two apparently very different approaches to managing IT, both banks need to align IT with their businesses. This case describes how each bank approaches the issue with the aim of drawing some general conclusions from a comparative analysis of their experiences.

BANCO COMERCIAL PORTUGUÊS

The founding of BCP

The Portuguese banking environment was relatively immature until the mid-1980s. Real change only began in 1984 when the government passed legislation enabling individual investors to create private banks in Portugal. Soon after, in June 1985, BCP was founded as a private bank with a total of 204 shareholders and 3.5 billion escudos ($25 million).

From the outset, BCP acted in a professional, rigorous and thorough manner. One of its first actions was to develop a strong internal market research capability. It established a 'market segmentation' task force to analyse in depth target market segments, customer needs, existing products and business potential. High net worth individuals and medium-sized companies were identified as the most lucrative market segments for BCP to focus on initially.

In July 1985, soon after creating its market segmentation task force, BCP initiated a study to articulate a vision for using IT in support of the bank's overall strategy. The study came up with the following generic definition of IT at BCP: applications were to be real time, flexible, based on the client and independent of the branch;[2] and BCP would be a user rather than a developer of technology. The study also addressed the two important issues of the hardware platform and the software infrastructure.

The hardware decision was tied to BCP's aim to be a major Portuguese bank. If BCP were to opt for a mid-size solution, such as the (then) commonly used IBM system 38, the bank's ability to grow quickly might be stifled. The other choice was a large mainframe in the IBM 43xx series. A large mainframe was expensive, but it would allow BCP to meet the processing needs of rapid growth. The board deliberated in depth upon the choices and finally decided upon the mainframe alternative. A senior manager commented on the outcome:

> There is no technical solution to a management problem, but there is a management solution to a technical problem. The beauty of the decision was that it was a management decision, not a technical one. The industry, and some shareholders, thought we were crazy putting one-third of our start-up capital into a mainframe computer while opening only two branches.

---

[1] It should be mentioned that several other private banks started in Portugal at about the same time as BCP. None of these other banks has been able to replicate the success of BCP and most are struggling today (see Exhibit CS3.2.1).

[2] This would allow clients to switch branches as BCP grew and offered additional branch locations. At other Portuguese banks, in contrast, clients needed to set up new accounts with new numbers in order to change branches.

Case 3.2

As BCP's business strategy was built around customers, BCP adopted (after a careful study of the available choices) the Customer Information System (CIS) package from Hogan, a US-based banking software firm, as its core platform. The CIS organized information around customers and gave BCP managers a comprehensive view of relationships with their clients.

BCP's rapid growth

BCP recruited over 100 talented employees and opened its doors on time in May 1986 with an aggressive marketing campaign. From the beginning, excellence in customer service was explicitly established as BCP's key operating objective. A critical decision in assuring high levels of service was to assign primary customer service responsibility to the account managers.[3] The account manager was supported by the bankwide CIS systems, which provided data on all of BCP's products, as well as a comprehensive view of a client's financial position and dealings with BCP.

While other Portuguese banks paid zero interest on cheque accounts, BCP launched checking accounts paying 6% and 5% annually, for high net worth individuals and medium-sized corporate segments, respectively. Six weeks after BCP's opening, the Portuguese government, in response to lobbying from established retail banks, introduced a regulation limiting interest payments on checking accounts to 4%. However, due to BCP's flexible customer-oriented software structure, BCP was able to develop in less than a week a new product that could 'sweep' balances between client accounts. Each evening, checking account balances were swept into overnight treasury bills paying 8% interest and then the balance was reposted in the checking account the following morning. The market response, as one BCP executive explained, was enthusiastic: 'With the fantastic publicity the BCP dual accounts gave us, the fact that no one could copy us easily allowed us to grow quite nicely.'

By the end of 1988, BCP was well established with total assets of 296 billion escudos ($2.02 billion) and a network of 19 individual and corporate branches. Over the next few years, BCP launched a series of different business groups which catapulted it from a small start-up bank to a major player in the Portuguese banking market.

In early 1989, BCP launched two other business groups focused on private and corporate banking respectively. A major turning point in BCP's development came in November 1989 with the establishment of NovaRede, a banking group targeted at individuals in the middle income range. In May 1992, BCP established its sixth business group to serve small companies and independent businessmen. Exhibit CS3.2.2 provides an overview of the growth of BCP's banking operations over the years.

The launch of the new business groups was accompanied by a wave of new products and services, all very innovative for the Portuguese banking environment of that time. Exhibit CS3.2.3 provides a list of major BCP innovations over the years. BCP's flexible technology infrastructure gave it the unique ability to create and handle complex financial products, which consistently appealed to the Portuguese market. For instance, BCP produced an account paying interest at differing rates depending on the average daily outstanding balance. Technology also helped to differentiate BCP in the eyes of its customers by enabling it to launch innovative products and services with rapidity and flexibility.

---

[3] BCP was the first Portuguese bank to establish the role of account managers for clients.

BCP's IT infrastructure enabled NovaRede to maintain a minimum number of staff. NovaRede's maximum of five employees per branch contrasted with an average of 40 for the branches of its competitors. Cost per branch was also low. NovaRede branches took an average of just 18 months to break even. At NovaRede, BCP eliminated the personalized services of an account manager, but the former's customers had direct access to the IT infrastructure for their transactions. BCP established a direct line for telephone banking which customers could access round the clock. To overcome the competitive limitations[4] of Multi-Banco (a centralized ATM service owned by a consortium), BCP developed its own parallel network to provide additional services (not available from Multi-Banco) such as cheque dispensing.

BCP also developed a system of digitizing signatures that could be called up on-line by a teller in any branch to directly verify signatures on cheques (thus reducing dramatically the cost of back-office cheque processing). In addition, if there were insufficient funds to cover the cheque amount, the teller of the bank had the ability to send an electronic mail message to the account holder's branch to request special clearance. The whole procedure took a few minutes, compared with up to an hour at other Portuguese banks; it also allowed a saving of about a million fax messages a year.

A major innovation for NovaRede was the 'salary' account, which resulted from the initiatives of NovaRede's dedicated salesforce. The salary account allowed clients to receive access to their salaries on the 15th day of the month (instead of the traditional end of the month), an automatic personal line of credit equal to three months of salary, and free personal accident insurance. A BCP executive emphasized the importance of the salary account:

> *The salary account was something completely new for Portugal and relied on some highly specific software products. NovaRede had an entire year's head start before anyone else could match it.*

GROUP TECHNOLOGY STRUCTURE AND PLANNING

The central data-processing group consisted of about 200 staff members divided into four divisions: applications development; systems, communications and operations; business group performance and change management; and organizational quality assurance and help desk (see Exhibit CS3.2.4). The general manager of the central data-processing group reported directly to the board. In addition to the centralized IT department, a certain number of IT staff were distributed in each business group.

The central applications development team was further subdivided into several permanent project teams with each project team responsible for a set of major (related) software products. Each project team had a corresponding IT Users Committee which was headed by the general manager of a business group and comprised several business managers, front line users and some

---

[4] The centralized ATM service Multi-Banco gives Portugal a tremendous advantage over other countries in being able to offer a consistent network to all banking customers within the country. However, there are drawbacks in this for BCP. All ATM development plans must be channelled through an external company for developing new services. But BCP provides additional services which help to provide enhanced service to existing customers and to attract new customers. For example, BCP has contracted with a firm to refill its ATM machines on a 24-hour basis every day of the week. Thus, on a Saturday night, when the ATM machines of many other banks are empty, non-BCP customers can use BCP's teller machines as they are always full.

technical specialists from the associated project team. The IT Users Committee played a key role in integrating the technology staff with the users and overseeing the implementation of new projects.

Jardim Gonçalves, chairman of the board of BCP described the role of users in IT planning:

> At other banks, it is the data processing department that defines the information system. In BCP it's the users that decide it.

For requests needing new software to be developed, users started by filling in a one-page proposal form which contained brief qualitative descriptions of the proposed change, its objectives, impact on service levels, productivity and profitability, the name of the user 'owner' and a subjectively assigned priority. The business groups collected such proposals and channelled them to the IT users committees. Each IT users committee reviewed the assigned requests and determined relative priorities and redundancies among them. All conflicts were sorted out by informal meetings of the IT users committee members with the concerned users. After some negotiations, a list of proposed projects for the year was produced for each project team. This list was presented to the board which gave final approval after reviewing the projects in accordance with the strategic directions of the bank.

Each IT users committee also reviewed the progress of the different projects on a regular basis. If there were any changes which called for new projects or for modifications in planned projects, the committee decided upon ways to accommodate the change requests. The entire planning process was business driven, fairly informal and flexible, and relied on a close partnership between business and IT staff in the various users committees.

Information management

The effective management of both external and internal information was of high strategic importance within BCP. There were three important stores of information: the CIS, the Strategic Database and the Earning Analysis System (EAS).

The CIS stored information about current and potential customers and was accessed by the account managers. The Strategic Database stored detailed customer information profiles and relevant market information. The Strategic Database was used primarily by the marketing divisions of the different business groups. Information about key internal profitability parameters was available through the EAS. Based on EAS-reported information, BCP was able to quickly obtain a very clear picture of profitability for each business area.

A dedicated central staff unit was responsible for maintaining the information content of the CIS and the Strategic Database. The manager of this central unit explained his role:

> We maintain a clear distinction between information maintenance and technology maintenance. While the data processing department is responsible for the technical maintenance of the Strategic Database and the CIS, my unit is responsible for managing the information within them.

We act as an interface between the users and the data processing department. When our users have a need, they come to us. We translate their needs into action requirements for the data processing department. This is easy for us as we understand the needs of our users.

A separate central staff unit was responsible for maintaining the EAS. The manager of the staff unit responsible for the EAS commented on the role of his division:

During the initial years, the emphasis within line management was on volume. A few years ago, the focus shifted to volume plus profitability. The EAS has been critical in enabling us to refocus our business with the modified emphasis.

We view ourselves as information providers to our users. We have to ensure that the information we provide is reliable, clear and timely. We have to decide what we want to measure, how we want to measure and have to explain these measures to the users. This is our job. It cannot be done by the data processing department.

Both the central staff units were headed by senior business managers and each contained a small select group of technically literate business employees. BCP management clearly viewed these information maintenance units as business units and valued their role within the organization.

Group culture

The bulk of the information flows in BCP were informal and rapid, and it attached considerable importance to informal communications and motivating people. As one manager explained:

We don't have a real organizational chart. We might draw them from time to time. But we never approve one. When you approve something and then you want to change it, you have to make a decision to change it.

Cross-functional team work was valued highly within BCP since the start. Team members were encouraged to 'own' a project and often continued working on it after implementation.

With 1.2 terminals per employee, working with computers was an operational necessity for all employees in BCP. Training on computer systems formed a major part of the initial three-week training given to all new employees.

Frequent and close interaction between users and technical specialists was key to maintaining a strong technological awareness within BCP. There were frequent rotations of IT staff between the central department and the business groups. The general manager of the central data processing group commented on the job rotations:

The IT department does lose a lot of technical people, who move to other areas of the bank. However, this reinforces mutual understanding and ensures that there are good technicians in user departments like marketing and planning.

Further, a manager had the following comment on the attitude of BCP's top management towards technology:

The board is very aware of the strategic importance of IT to BCP. In fact, one of the board members was previously the general manager for IT. The top managers of the different business groups are also continually involved in IT decisions through the IT users committees.

CONTINENTAL BANK

Background

In its previous existence as a full-service bank, Continental Bank was at death's door. Today, as a slimmer, sleeker incarnate, the institution is a specialized banker catering to corporations and the wealthy.[5]

In the late 1970s, CB had aggressively participated in energy-related investments. When energy prices fell in the early 1980s, CB found itself with more than $1 billion in bad loans. By 1984, the bank was on the verge of collapse; it was haemorrhaging losses and facing an exodus of depositors. The gravity of the situation forced the Federal Deposit Insurance Corp. (FDIC) to bail CB out with $4.5 billion in the mid-1980s. As a result of its troubles, CB shrank from being the sixth biggest US bank to the thirty-third.

As the new majority owner of CB, the FDIC hired a new chairman in 1987: former Citibank vice chairman Thomas Theobald. In order to resuscitate CB, Theobald made some dramatic and drastic decisions, the most important of which was to simplify CB's diversified range of business lines and refocus on the core business of corporate banking.

A major asset for CB was its large base of loyal corporate customers. CB had over 2000 corporate customers including several publicly held companies such as Chrysler Corp., Deere and Co. and Allstate Corp., major private concerns and many prosperous smaller firms. According to a recent survey,[6] CB had a significant banking relationship with 16% of the 1000 largest corporations in the United States.

In accordance with the new strategic direction, among the first things to go was CB's retail banking operations, which made CB one of the first major banks with no retail operations. However, a bigger challenge for Theobald was to catch up with changes within corporate banking that had passed CB by while it was handicapped by the federal bailout and the ensuing tight supervisory regulations. By the late 1980s, corporate customers were not content with favourable loans, but were demanding several sophisticated products, such as cash management, mergers and acquisition (M&A) advisory services and interest rate swaps.

While the essence of CB's strategy lay in striking a balance between lending and fee-based activities with corporate customers, CB's continued survival depended upon modernizing its product inventory, tailoring products to fit its customer base, cross-selling and building strong customer relationships, being quick and nimble in effectively mobilizing bankwide resources to satisfy customers' needs, cutting costs aggressively and removing management distractions away

---

[5] Kiely, T., 'Continental gets down to business', *Bankers Monthly*, **109** (5), May 1992, p. 33.
[6] King, R.T. Jr., and Lipin, S. 'New Profit center: corporate banking; given up for dead, is reinventing itself', *Wall Street Journal*, Sec: A, 31 January, 1994, p. 1.

from the core business. Within such a context, CB's management undertook a thorough evaluation in the second quarter of 1990 to obtain a clearer picture of how to improve the core business and outsource or dispose of all other non-core business.

By the end of the year, CB had outsourced food, security, messenger services, property services and legal services. By 1992, the company had been slimmed down to about 5000 employees from a high of 12 000 a decade earlier. Richard Huber, CB's vice chairman commented on these spin-offs:[7]

> While we considered ourselves nimble bankers, we had no special expertise in other areas that make up a large organization—operating cafeterias, for example, or running a law firm.

Theobald also believed that this restructuring not only lowered CB's overhead but also improved its ability to respond nimbly and effectively:[8]

> We really are selling hand-tailored suits. The customers all get something different, and the larger you are, the harder that is to do. It's as simple as that.

The overall direction was to make CB a more 'scalable' organization, i.e., 'one that can deal with a doubling of the volume or a halving of the volume in a given product in a relatively short period of time'.[9]

MANAGEMENT OF INFORMATION TECHNOLOGY

> *Everything the information technology unit did took too long and cost too much.*[10]

The management of IT was perceived to be a problem area by CB's new management team. First, the bank's legacy systems were seen as largely unable to cope with the new need for nimble and quick response to customer needs in changing business environments. The business need was to take a new idea for a product, test it, tweak it, package it, and get it out to customers. In response to this need, Theobald had initiated a move towards deal teams, i.e., a loose, cross-functional coalition of product specialists from different areas and relationship managers. These deal teams needed consistent access to relevant information across the bank. The IT department tried to solve this problem by developing an integrated IT architecture, but their efforts were frustrated by the myriad assortments of partially compatible desktop systems and databases which had been allowed to proliferate in different parts of CB.

Second, CB had chronic staff problems. There were about 500 people in the IT unit, but this number was seen as too little to do major system projects and too much for carrying out the

---

[7] Huber, R.L. 'How Continental Bank outsourced its "crown jewels"', *Harvard Business Review*, January–February 1993, p. 122.
[8] Milligan, J.W. 'Can Continental prove Wall Street wrong?', *United States Banker*, **103** (6), June 1993, p. 34.
[9] Kiely, T. 'Continental gets down to business', *Bankers Monthly* **109** (5), May 1992, p. 33.
[10] Huber, R.L., *op. cit*, p. 123.

daily operations and maintenance of the systems. There were also questions about the level of technical skills of the IT staff.

Third, there were financial constraints in continually making the huge investments required to stay on top of the technology. These constraints were particularly acute given the weak performance of the company (CB suffered a loss of $71 million in 1991).

Finally, there was the distance between the IT and business staff. A lack of adequate communication between the two groups created frustrations for CB's management. Huber, who was responsible for CB's back-office and data processing operations, described his concerns in the following manner:[11]

> The more I learned about the situation, the clearer it became that the bank was already effectively outsourcing technology services to a group of people who happened to work at the same company I did, but whose work was different.

Theobald commented on IT awareness of business issues:[12]

> It always fascinated me . . . that left to their own devices, technologists will solve internal problems. . . . I can remember a couple of years ago when I was trying to find out the [priority list] for software development, [I noticed] one of the things we did real quick was a new system to order tickets for sporting events. Now, I'm sure it saved a nickel, but why the hell that should show up on the screen of things that are important to the future of Continental Bank, I don't know—except it's typical.

He was also critical of the degree of involvement from the business side:

> I think its [technology] has been overbought. . . . We who purchase have overbought through basically our own ignorance and lack of practice, lack of involvement.

Faced with these concerns, CB's management struggled with the issue of whether they could manage IT themselves or whether it was in the bank's interest even to try to do so. Eventually, the decision was made that it was probably best to outsource it. Though no detailed plan was prepared, it was believed by managers that outsourcing would give CB more ready access to the latest technology, cut the development time for new products and transform the bank's IT costs from fixed to variable. Competitive advantages from IT were not seen as accruing from having an internal IT division, but rather from having access to the best technology at the right quality/price ratio.

Outsourcing IT

> The one fundamental about change that we all forget is that once you begin the process, things will be different. But it may not be what you imagined it would be.[13]

---

[11] Huber, R.L., op. cit, p. 123.

[12] Fitzgerald, M., 'Still searching for the IS Holy Grail', Computerworld **25** (34), 26 August 1991, p. 47.

[13] Former CIO John Gigerich as quoted in Kiely, T., op. cit. John stepped down in the final weeks of the outsourcing negotiations.

Case 3.2

In March 1991, a committee of business managers was set up to analyse the risks in outsourcing IT and suggest ways to overcome them. A consulting firm was also hired to assist the committee in their 'reality testing'. Tom Gigerich the former CIO of CB recalled the first reaction of many of his staff to the outsourcing news:[14]

> *They wanted to work harder, do more. But there was nothing wrong with their performance. It was just—times changed.*

The next step was the formation of two councils to guide the bank through the outsourcing process. The business council was chaired by Huber and consisted of managers from CB's important businesses. It was responsible for strategic and business recommendations to the board of directors. The technical council consisted of selected technically knowledgeable staff from all the bank's businesses. Chaired by a credible, technically literate business manager the technical council had the mandate to conduct detailed analyses to evaluate and select vendors and make recommendations about which IT functions to outsource.

The technical council disaggregated IT into its distinct elements and examined the advantages and disadvantages of outsourcing each component. Despite some differences among members about the number of IT functions to outsource, the technical council turned in its recommendation in two months to essentially outsource all IT functions.

A call for proposals was issued and different vendors were evaluated. The major requirements on the vendor were to purchase all existing equipment, assume all responsibility for CB's hardware and software and employ CB's IT staff with comparable benefit packages. The last requirement was important not merely for altruistic reasons, but also because CB's IT staff were the only ones who really knew how CB's poorly documented systems actually functioned. The emphasis was on trying to develop a strategic partnership with a reliable outside vendor.

Responses from three vendors, Andersen Consulting, Computer Sciences Corporation and ISSC (an IBM subsidiary) were evaluated against several criteria including the ability to satisfy CB's specific technological needs, career plans and security for CB's IT staff and price for a 10 year contract. ISSC along with its partner Ernst & Young won after a two-round bidding process. Three months of intense negotiations followed with the winners to decide on the finer details of the outsourcing process. Huber, the chair of the business council, described the contract talks in the following manner:[15]

> *The bank was determined to gain much tighter management control over information technology by outsourcing. That required developing new, stricter methods to measure and document progress, then incorporating them into contract terms and pricing.*

The outsourcing contract was finalized in September 1991 and on 1 January, 1992 CB 'switched from building information technology to buying it'.

---

[14] Kiely, T., *op. cit.*, p. 34.
[15] Huber, R.L., *op. cit.*, p. 128.

Managing IT after outsourcing

One of the first steps was to establish a management structure to ensure that CB remained in control over its technology strategy. This consisted of a technical oversight group and a 20 person team of technically literate business employees.

The technical oversight group was composed of representatives from different business units and performed the functions of balancing the technological requirements of individual business units with the entire bank, ranking IT proposals in a bankwide priority list and deciding which projects had to be done first and which could wait. The 20-person team consisted of selected business people who understood both the business and the technology. They advised bank units on technology projects and assisted in communication with ISSC.

A year after the commencement of the outsourcing relationship, Huber made the following observations on the changed attitudes of CB's business units:[16]

> Another obvious success is the changed behavior of the bank's business units. They are now active and disciplined participants in the IT process. With virtually all IT work on a hard-dollar, contract basis, they are devoting time on the front end of projects to define clearly and carefully their technology needs and how they want to spend their budgets . . . . The process is more demanding now, but it's paying off in reduced technology costs and improved quality.

While Huber estimates that the outsourcing contract will save CB several millions of dollars each year, he also realizes that no one can truly predict the long-term cost savings (or losses) accurately. However, there is an overall feeling of increased satisfaction with IT at CB after outsourcing.

LESSONS FROM BCP AND CB

What kind of lessons are there for managers in these relatively mundane stories of BCP and CB? At first glance, they might seem to be two disconnected tales occurring in different continents within different contexts, but on a closer analysis they hold some simple but powerful insights for managers.

BCP is by most accounts the most successful story in European banking within the last decade. It is also an organization which has been very successful in aligning IT with its business. A clear proof of this is the rapid stream of technologically enabled innovations (Exhibit CS3.2.3) that it has successfully unleashed in the Portuguese banking market. Not only have these innovations formed the cornerstone for BCP's strategic success, but they have also given it an unassailable competitive advantage as none of BCP's competitors have been able to match them. What contributes to the alignment of IT with BCP's business?

In contrast to BCP, CB's predicament with IT prior to outsourcing represents an all too common situation in organizations. The alignment of IT with the business was poor as evidenced by the

---

[16] Huber, R.L., *op. cit.*, p. 129.

lack of the IT department's ability to respond to the business demands for nimbleness and speed of response. Why was CB's management having serious problems in managing IT and directing IT priorities? Why are such problems so common in many organizations?

Discontent with the management and alignment of IT led CB to outsource IT. After outsourcing IT, CB's management seems to be more satisfied with the management of IT. Why has this happened? What is the difference between CB's approach to managing IT prior to and after outsourcing? How does CB's approach to managing IT after outsourcing compare with BCP's approach?

Case 3.2

Case 3.2

Exhibits

Exhibit CS3.2.1 Growth of BCP

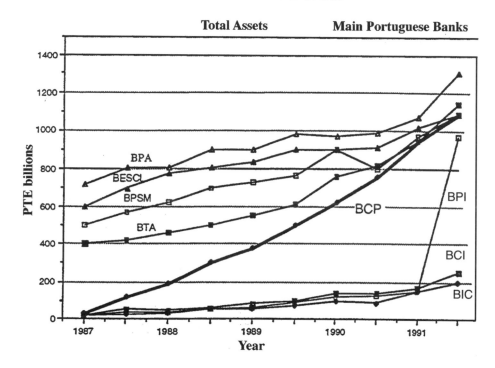

Total Assets　　　　　**Main Portuguese Banks**

In less than five years, BCP clearly outpaced the two other Portuguese commercial banks, which appeared in 1985 (BCI and BIC), and joined the group of Portugal's largest commercial banks (note that BPI merged with another local bank in 1991).

Legend:

BPA	Banco Português do Atlantico	BESCL	Banco Espírito Santo & Comercial de Lisboa
BTA	Banco Totta e Açores	BPSM	Banco Pinto e Sotto Mayor
BPI	Banco Português de Investimento	BCI	Banco de Comércio e Indústria
BIC	Banco Internacional de Crédito		

Source: Banco Comercial Português, by Pommes, C. de, Taubman, C., Doz Y. and Horwitch, M. INSEAD case study, 1993.

Exhibit CS3.3.2 Development of BCP's banking operation

Operation	Main phase	Target clients	Products and services	Operating systems
BCP Parent Group	1986–89	High net worth individuals	■ Cash management products (e.g. Conta Mais) ■ Mortgage credit ■ Personal loans ■ Prestige credit cards ■ Package of insurance	■ Autonomous branch network ■ Decentralized back office ■ Global range of flexible financial services ■ Account Managers ■ Personalized treatment
	1989–91	Medium-sized corporations	■ Cash management products ■ Short term capital financing ■ Treasury accounts ■ Investment accounts	■ Cross-selling to insurance and leasing ■ Corporate terminal ■ Risk assessment system
Corporate banking	1989–91	Portugal's top 500 corporations	■ Off-shore financing and cash management ■ Swaps ■ Loan syndication ■ Capital market services	■ Access to dealing rooms ■ Corporate terminals ■ Electronic banking ■ 'Tailor-made' products
Private banking	1989–93	Extremely wealthy customers	■ Investment products ■ Mutual funds ■ Insurance products	■ High level of discretion ■ Exclusive branches
NovaRede	Acquired 1990	Urban retail customers	■ NovaRede account ■ NovaConta salary account ■ Funds transfer ■ NovaRede Visa credit card ■ Insurance products	■ Fast service, low costs ■ Low-cost branch operations ■ Monthly combined statements ■ Automatic cheque dispensing ■ 'Direct Hotline' phone banking ■ Credit-scoring techniques
Merchant banking	June 1991–93	CISF	■ Medium & long-term financing ■ Valuations and M&A activities ■ Privatizations ■ Capital market business	■ Complementary business ■ Separate computing systems
International	March 1992–94	Portuguese émigrés in France	■ Retail banking facilities	■ Joint venture with Spain's Banco Popular
Small business		Small business	■ Hire purchase-type financing ■ Cheque management service ■ Loans	■ Specialised credit-scoring ■ Revolving line of credit

Source: Banco Comercial Português, by Pommes, C. de, Tauoman, C., Doz, Y. and Horwitch, M. INSEAD case study, 1993.

Exhibit CS3.2.3 BCP main innovations

Product	Category	Description	Date	Competitive Response
Account managers	New service	Personal service	May 1986	1990 Barclays, 1991 BCI(Santander)
Nova Conta Mas	New product	High-yield bank account	Late 1986	1990 Barclays, 1991 BCI(Santander)
Prestice Débico	New product	Visa premier gold debit card (first gold debit card in Portugal)	Late 1986	BCI launched gold card 1987
Conta Fitulos	New service	A more flexible securities account. Supported in BCP central system, it allows any customer access to the stock exchange at branch level	May 1986	No one followed
Direct marketing	New concept	BCP introduced direct marketing methods such as telemarketing and mailing. The first in the Portuguese banking industry	May 1986	Others followed late 1988
Image through premises	New concept	BCP premises designed in order to appeal to the segments they serve	May 1986	BPA and Barclays followed in 1990, BCI in 1991 and BFB in 1992
Market segmentation	New concept	BCP specified the product and service level for each of the tightly defined segments	May 1986	1990 Barclays and 1991 BCI. Others developed products aimed at special market segments. BCP still remains the only bank to follow a full market segmentation strategy
Financial group	New concept	In order to serve each customer's global financial needs, BCP established a group to co-ordinate such aspects as insurance, leasing, factoring, etc.	1987	Others have slowly developed a financial group
Tradelink	New product	An information service to businesses around the world	1986	A year later BESCL launched a similar product
Commercial bonds	New product	Innovative short- and medium-term bonds for medium-sized companies.	Late 1988	No one followed
Madeira 'off-shore' branch	New concept	Taking advantage of tax and operating conditions in Madeira's financial 'off-shore' status	Jan 1989	Others followed in late 1989
Monthly income account	New product	A new financial product combining a high interest rate with flexibility of withdrawal and the payment of monthly income	Mid-1989	Others followed a year later

Prestige Crédito	New product	Visa Premier ('gold') credit card. First ever gold credit card issued in Portugal	Mid 1989	Barclays launched a gold credit card in 1991
Short term bonds	New product	Securities issued by medium size companies at very competitive interest rates with 1 year maturity.	Late 1989	BPSM followed 2–3 months later
NovaRede distribution network	New concept	Small branches with reduced staff (5 maximum), automated and close to customers. ATM and CAT in all 200 branches (end 91)	Nov 1989	Other banks opened a few branches of that same size afterwards
Salary account	New product	An account based on automatic transfer of monthly salary, enabling early payment of salary from the 15th, an automatic credit line, and free accident insurance	Nov 1989	Others followed in 90–91 with similar products
Hot-line	New service	A telephone service through which clients may place orders, or request information	Nov 1989	BPA launched a hot-line in late 90. Barclays and BFB in 1992
Monthly combined statement	New service	A document which summarizes all financial transactions and states all the client's assets	Nov 1989	No other bank offers such a complete service
Mortgage credit	New product	BCP moved into an area that was an 'exclusive' of special credit institutions	1990	Others followed in the same year
ADR program	New concept	BCP was the first Portuguese company to have their share capital listed in a foreign market	Aug 1990	No other Portuguese bank is listed outside Portugal
Sponsoring TV show	New concept	Sponsoring a popular TV show	Oct 1990	Others followed in 1992 (BESCL, BFB, etc.)
Personal credit	New product	BCP launched credit facilities for individuals just after the removal of legal restrictions	1991	Other followed with 4–5 months lag
Managing pre-dated cheques	New product	A product designed to manage pre-dated checks for small businesses	Apr 1992	No other bank offers such a comprehensive product
Credit based on pre-dated cheques	New product	A credit line based on accounts receivable (as pre-dated checks) held by small businesses	Apr 1992	No other bank offers such a comprehensive product

BCP main competitors became increasingly good at imitating BCP innovations; however, no one bank could follow all of BCP new product introductions.
BCP also adopted a policy of 'zero-interaction' with other banks in Portugal [for instance on the interbanking network system, SIBS]. This was done in order to reduce the amount of competitive information that was 'leaked-out' from BCP.

Source: Banco Comercial Português, by Pommes, C. de, Taubman, C., Doz, Y. and Herwitch, M. INSEAD case study, 1993.

Exhibit 3.2.4 Organization of the data processing department

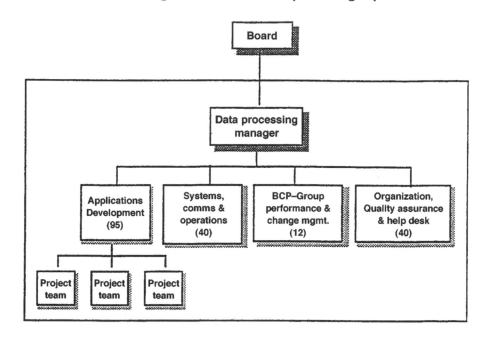

- *Applications development:* This group is responsible for selecting or developing application packages to match user requirements. This group is further subdivided into different project teams. Each project team has a corresponding IT users committee which consists of some technical staff from the project team, business group, senior managers and line staff.

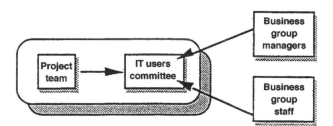

- *Systems, communications and operations:* This group is concerned with the operations of the computer systems, networks and 'production' activities.

- *BCP-group performance and change management:* This is a support group, responsible for coordinating changes in BCP's technology infrastructure.

- *Organization, quality assurance and help desk:* Activities managed by this group include: accounting systems; analysing the performance of systems; quality assurance; change management, i.e. controlling software changes; and a help desk.

Harvard Business School   **9-697-037**
                                          Rev. April 16, 1997

Case 3.3

Case 3.3

*Doctoral Candidate Andrew McAfee prepared this case under the
supervision of Professor David Upton as the basis for class discussion rather
than as an illustration of either effective or ineffective handling of an
administrative situation.*

Vandelay Industries, Inc

On a Monday morning in January 1996, Elaine Kramer was in her Philadelphia office catching up on e-mail. She had been in Minneapolis the previous week for a series of meetings between her firm, Deloitte & Touche Consulting Group/ICS, and the executives and plant managers of Vandelay Industries, Inc., a major producer of industrial process equipment. The week had ended with the two companies signing a contract to work together on a large information systems implementation project.

Her phone rang.

'This is Elaine Kramer.'

'Hi, Elaine, this is George Hall. I manage Vandelay's Dunbarton plant; we met at the project kick-off meetings last week.'

'Oh, yes—how are you, George? I really enjoyed the presentation on your pull-based manufacturing project. You guys have generated some really impressive results. What can I do for you?'

'I was just wondering if I could start getting my people signed up for training on the R/3 system. I know it's early, but we're really eager to dig in here. Your presentation got a lot of people excited about getting rid of our plant's old mainframe. We want to get an R/3 team together so we're ready to work on the system as soon as it's installed at Dunbarton.'

'Well, we haven't started putting training schedules together yet…'

'Well, keep us in mind when you do. Like I said in the presentation, we're a bunch of tinkerers, and that's what has helped us improve so much over the past few years. We think R/3 can really help us past some of our current roadblocks, so we're eager to start experimenting with it.'

'OK, let me get back to you once we're further along with training plans.'

After Kramer hung up the phone she mentally replayed the conversation; it raised an issue that had been in the back of her mind since the meetings. How much should the plants in the network be encouraged to modify their local set-up of the new computer system once it had been installed?

Project background

Vandelay had decided in 1995 to implement a single Enterprise Resource Planning (ERP) information system throughout the corporation. The firm had chose the R/3 system from SAP AG, a German company that was the market leader in ERP products. Vandelay hoped that the R/3 implementation would end the existing fragmentation of its systems, allow process standardization across the corporation, and give it a competitive advantage over its rivals.

Vandelay managers realized that putting R/3 in place would be an enormous effort, of which installation of hardware and software was only a small part. For help with all aspects of the project, from the technical details of an ERP system to widespread business practice changes, Vandelay had engaged the Deloitte & Touche Consulting Group/ICS. ICS assisted clients in managing fundamental changes of the kind entailed by an ERP system, and had significant expertise with SAP's R/3 product.

Kramer, who had been with the firm for over five years, had been chosen to lead the project. At the meetings in Minneapolis she had been impressed with Vandelay's enthusiasm for the project. The plant managers seemed especially excited; many of them had said that they considered their existing information systems an impediment, rather than an aid, to efficient production.

George Hall certainly seemed pleased at the thought of R/3 in his plant, but Kramer was not entirely calmed by their conversation. Hall had evidently assumed that Dunbarton would be free to modify the system at will; Kramer knew this would not be the case. She wondered how to respond to this request for training, and how to let him and the other plant managers know that all decisions about R/3 were not under their control.

VANDELAY INDUSTRIES

Company background

Vandelay Industries, Inc. was an $8 billion corporation that manufactured and distributed industrial process equipment used in the production of rubber and latex. The company was founded in Minnesota during World War II; its initial products proved important on the Home

Front, enabling the much greater productivity required by the war effort. From the beginning, Vandelay's offerings were known for their design quality and innovative engineering; company lore held that this was because wartime rubber shortages necessitated precise, wasteless production.

Markets for Vandelay products were extremely healthy throughout the following decades, and the firm steadily expanded, partly by building new sites and partly by acquiring smaller firms.[1] The company also steadily expanded its product lines, eventually supplying a range of process industries. Vandelay plants were treated as revenue centers and typically were allowed a high degree of independence, provided that they maintained acceptable profit margins. At its peak, the company employed 30 000 people and manufactured on four continents. Employees tended to remain with Vandelay for a long time, taking advantage of generous pay and benefits and a stimulating work environment.

The company began to experience difficult times beginning in the mid-1980s as a result of market shifts and severe competitive pressures. Three strong foreign competitors emerged, offering less expensive alternatives in many of Vandelay's product lines. These machines typically did not include all of the features of the comparable Vandelay product, but were substantially (20–30%) cheaper. As a result, the American company's traditional emphasis on features and customizability became a liability. Its manufacturing operations were never intended to be low cost, but its products could no longer command a large price premium once suitable substitutes were available. The firm's traditionally long lead times also became a problem; the new entrants could fill customer orders much more quickly.

Vandelay fought hard over the next ten years to learn new technologies and new ways of doing business. It adopted lean production methods, rationalized its product lines, and introduced new, simpler and cheaper machines. It also closed three plants, leaving eight in operation[2] and had the first layoffs in its history, reducing its total headcount to 20 000 people. Many of these efforts paid off, and the company returned to profitability in the mid-1990s.

During this decade of realignment Vandelay's executives realized that they would have to accord much higher priority to manufacturing and order fulfilment in order to further drive down costs. They also needed to become quicker; the company's new machines were popular, but still had longer lead times than competitors. Internal investigations had shown that actual manufacturing and material movement times accounted for less than 5% of total lead times experienced. Large parts of the remaining time, Vandelay found, were devoted to information processing and information transfer steps. This was partly because the computer systems in use across the firm to guide order fulfilment and production activities were poorly integrated, and in some cases completely incompatible.

Information systems

Each plant had selected its own system for manufacturing resource planning (MRP), the software that translated customer demands into purchasing and production requirements. In addition, many sites had also installed specialized software to help with forecasting, capacity planning, or

---

[1] Half of all new Vandelay sites from 1945 to 1985 were bought rather than built.
[2] Four in North America, two in Europe, and two in Asia.

scheduling. Information systems for human resources management had also been selected individually. There was a single corporate financial information system, to which each site had built an interface for automatic electronic updates, but this was the only example of corporation-wide systems integration.

As Vandelay reviewed its operations, it uncovered several examples of how this patchwork of information systems added time and expense to the production cycle. Examples included:

- *Scheduling*: The plants' dissimilar manufacturing software often made integration across sites difficult. For example, one of Vandelay's American plants made a variety of machined and stamped metal parts which were used by the other North American assembly sites. This plant used an outdated MRP system which required all data to be entered manually. Requirements from all downstream users were keyed in at the beginning of each week, a task that required almost a full day. No other inputs were allowed during the week, and the plant was deluged with companies about its responsiveness.

- *Forecasting*: Vandelay's European planning group used a forecasting program which grouped all demand for an item into one monthly 'bucket.' Plants were then free to decide when to build the product within that month. Customer orders, however, usually requested delivery within a specific week. If by chance these requests did not line up with the month's production plan, late shipments resulted.

- *Order management*: Customer orders were taken manually by an inside sales organization in each region (North America, Europe, and Asia), then routed via fax to the appropriate plant where they were keyed in to that site's order entry system. Faxes, and therefore orders, were sometimes lost.

- *Human resources*: When a Vandelay employee transferred from one location to another, their complete employee record had to be copied. Because of incompatible human resources software, this data often had to be manually re-entered. In addition to being redundant and time-consuming, this meant that the confidentiality of the information was difficult to guarantee.

- *Financials and accounting*: The manufacturing software used within most Vandelay plants was not integrated with the site's financial package, so information such as labour hours charged to a job, materials purchased, and orders shipped had to be entered into both systems. This introduced potential for errors, and necessitated periodic reconciliations.

Business practices

Vandelay sites' operations practices were as varied as their information systems. There was no uniformly recognized 'best' way to invoice customers, close the accounts at month end, reserve warehouse inventory for a customer's order, or carry out any of the hundreds of other activities in the production process that required computer usage or input. At the kick-off meetings, Kramer had heard 'horror stories' about flawed processes uncovered by plant managers. Some of them were quite vivid:

I walked down to my receiving dock a few days ago and just watched what happened each time a supplier's truck unloaded. First, our receiving guys would verify the quantities. Then they'd leave the boxes on the dock and take the packing lists over to a terminal for our quality system. If it said that the part needed incoming inspection, they'd move the boxes over to quality control. If there was no inspection, they'd take the list over to another terminal and enter the received quantities into the purchasing system, then they'd move the boxes and the list over to stores. Meanwhile, the stores guy is working through a backlog of these boxes, entering the stockroom bin numbers of all the items he's shelved. And if there's a discrepancy in the packing list or a high-priority item hits the dock, things get real complicated.

Teri Buhl, Fort Wayne (IN) plant manager

When I started at the plant last year, I couldn't believe how work got scheduled on the floor. They'd started a system of putting a green tag on high priority workorders to flag that they should be at the head of the queue. That worked for a while, but then someone decided that really high priority jobs should get a red tag. You can guess how it went from there. By the time I got there, no job had a prayer unless it had some kind of tag on it and there were at least a half-dozen colour combinations in play. Our starting queues looked like Christmas trees.

Alain Barsoux, Marseilles plant manager

To alleviate these problems with systems and practices, Vandelay decided to purchase and install a single ERP system, which would incorporate the functions of all the previously fragmented software. The company would also use this effort as an opportunity to standardize practices across sites.

Vandelay saw one other major benefit from an ERP system: gaining visibility, in a common format, over data from anywhere in the company. The company anticipated that once the software was in place, authorized users would be able to instantly see relevant information, no matter where it originated. This would provide the ability to co ordinate and manage Vandelay sites more tightly then ever before; plants could see what their internal customers and suppliers were doing, and network-level managers could directly compare performance across locations.

After a review of leading ERP vendors and implementation support consultants, Vandelay decided to purchase SAP's R/3 software and put it in place with the help of Deloitte & Touche Consulting Group/ICS.

THE SOFTWARE VENDOR: SAP

Company background

SAP AG was founded in 1972 in Walldorf, Germany with the goal of producing integrated application software for corporations. These applications were to include all of the activities of a corporation, from purchasing and manufacturing to order fulfilment and accounting. SAP's first major product was the R/2 system, which ran on mainframe computers. R/2 and its competitors came to be called Enterprise Information Systems (EISs). Within manufacturing firms they were also known as Enterprise Resource Planning (ERP) systems to reflect that they incorporated and expanded on the functions of previous MRP systems.

SAP was one of the first ERP vendors to realize that powerful and flexible client-server computing technologies developed in the 1980s were likely to replace the established mainframe architectures of many large firms. The company began work on a client-server product in 1987 and released the R/3 system in 1992. R/3 capitalized on many of the advantages of client-server computing, including:

- *Ease of use*: Client-server applications often used personal computer-like graphical user interfaces. They also ran on the familiar desktop machines used for spreadsheets and word processing.

- *Ease of integration*: The flexible client-server hardware and operating systems could be more easily linked internally (to process control equipment, for example) and externally, to wide-area networks and the Internet.

- *Scalability* or the ability to add computing power incrementally: Companies could easily expand client-server networks by adding relatively small and cheap machines. With mainframes, computing capacity had to be purchased in large 'chunks.'

- *More open standards*: The operating systems most used for client-server computing were non-proprietary, so hardware from different manufacturers could be combined. In contrast, most mainframe technologies were proprietary, so a mainframe purchase from IBM or Digital locked in the customer to that vendor.

As Figure CS3.3.1 shows, R/3 was extremely successful and fueled rapid growth at SAP. By 1995, the firm was the third largest software company in the world (see Exhibit CS3.3.1). Expansion

Figure CS3.3.1 Growth of SAP and the R/3 system (*Source:* SAP)

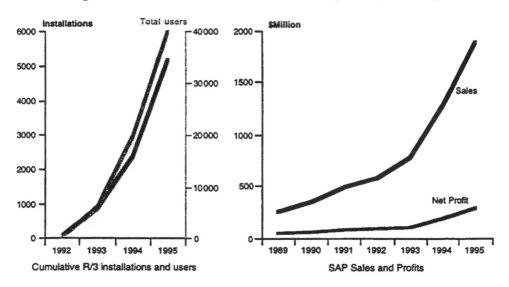

was especially rapid in North America, where SAP went from a very small presence in 1992 to $710 million in sales in 1995. This success was due to several factors, including:

- *Client-server technology*: As large firms moved from mainframe to client-server architectures in the early 1990s, the R/3 system was available to them. Meanwhile, many suppliers of existing 'legacy systems' did not have client-server applications ready for market.

- *Modularity, functionality, and integration*: R/3 functionality included financials, order management, manufacturing, logistics, and human resources, as detailed in Exhibit CS3.3.2. Prior to the arrival of ERP, these functions would be scattered among several systems. R/3 integrated all of these tasks by allowing its modules to share and transfer information freely, and by centralizing all information in a single database which all modules accessed.

- *Marketing strategy*: SAP partnered with most large consulting firms. Together, they sold R/3 to executives as part of the broader business strategy, rather than selling it to information systems managers as a piece of software.

R/3 usage: transaction screens and processes

On a user's machine, R/3 looked and felt like any other modern personal computer application; it had a graphical user interface and used a mouse for pointing and clicking. Users navigated through R/3 by moving from screen to screen; each screen carried out a different transaction. Transactions included everything from checking the in-stock status of a component to changing an assembly's estimated cost; Exhibit CS3.3.3 gives an example of a transaction screen.

A full SAP implementation, including all standard functions, incorporated hundreds of possible transactions. Most common business processes included multiple transactions and cut across

Figure CS3.3.2 SAP modules involved in a single business process. Copyright © ICS, 1995

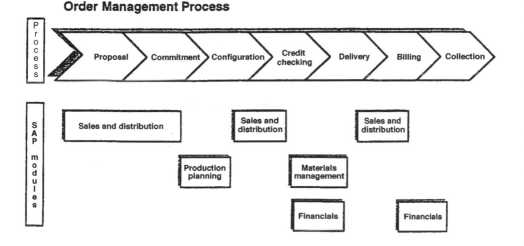

more than one functional area or software module. Figure CS3.3.2 outlines the process of taking a customer order, and shows the SAP modules involved at each step. It shows that without ERP this process could involve three separate information systems—sales and distribution, MRP, and accounting and financials. With R/3, each step would require a different transaction screen, but they would all be part of the same system. They would thus be sharing and updating the same information. This elimination of redundant entry and 'hands-off' between applications was one of the chief advantage of ERP systems.

R/3 usage: system configuration

Configuration tables

Although R/3 was intended as a 'standard' application that did not require significant modification for each customer, it was still necessary to configure the system to meet a company's specific requirements. Configuration was accomplished by changing settings in R/3 configuration tables.

R/3's approximately 8000 tables defined every aspect of how the system functioned and how users interacted with it; in other words, they defined how all transaction screens would look and work. To configure their system, installers typically built models of how a process should work, then turned these into 'scripts,' and finally translated scripts into table settings. For example, after writing a script that defined how a new customer order would be entered, Vandelay would know whether the order taker should have the ability to override the product's 'price' field on the order entry screen.

During an implementation, this configuration activity had to be replicated for all relevant processes and required a great deal of time and expertise. People who were adept at this work, and who understood the impact of each table change, were a feature of every R/3 implementation. Kramer would be relying on several of them at ICS to work closely with the Vandelay project team.

Added functionality

Although the R/3 system was generally recognized to contain more functionality than its competition, it typically could only satisfy 80–95% of a large company's specific business requirements through standard configuration table setting work.[3] The remaining functionality could be obtained in four ways:

- Interfacing R/3 to existing legacy systems

- Interfacing R/3 to other packaged software serving as 'point solutions' for specific tasks

- Developing custom software that extended R/3's functionality, and was accessed through standard application program interfaces

- Modifying the R/3 source code directly (this approach was strongly discouraged by SAP and could lead to a loss of support for the software)

---

[3] According to an SAP estimate.

THE CONSULTANTS: DELOITTE & TOUCHE CONSULTING GROUP/ICS

Company background

Deloitte & Touche Consulting Group (the Consulting Group) was the consulting division of Deloitte & Touche, one of the 'Big 6' audit and tax firms and the product of the 1989 merger of Deloitte, Haskins & Sells and Touche Ross. Consulting had been an important activity for the predecessor firms since the 1950s, and accounted for over 15% of total Deloitte & Touche revenues by the mid-1990s.[4] In 1995, the Consulting Group generated slightly over $1 billion in revenues and employed 8,000 professionals in more than 100 countries.

Deloitte & Touche Consulting Group/ICS was the subsidiary of the Consulting Group which specialized in SAP implementations, offering complementary software products, education and training, and consulting in business process reengineering and change management. ICS was one of the largest worldwide providers of SAP implementation services and one of the most rapidly growing sections of the Consulting Group, employing over 300 professionals on four continents in 1995.

ICS had developed a considerable knowledge base in SAP systems; over 50% of its consultants had more than two years experience with the products. ICS had won SAP's Award of Excellence, which was based on customer satisfaction surveys administered by the software maker, every year since its inception. SAP had also named ICS as an 'R/3 Global Logo Partner.' According to SAP[5],

> The aim of these partnerships is to establish, extend and enhance R/3 expertise. In order to keep these logo partners up to date with the latest developments, SAP maintains very close contact with them, providing an intensive flow of information and offering the following services:
>
> - an R/3 System for internal training;
> - regular R/3 logo partner forums, workshops and training sessions;
> - access to SAP InfoLine, SAP's internal information system;
> - second level support from SAP Consulting, including the consultant hotline.

ICS professionals ranged from general management consultants to SAP specialists. The specialists focused on a functional or technical area of SAP and worked as, for example, experts on the materials management functions or programmers in the system's native ABAP/4 language. Management consultants, meanwhile, had experience with process re-design, systems implementation, change management, or project management. More senior personnel often combined both types of skills.

Technology-enabled change

ICS consultants had adopted a common set of principles for leading large-scale change in a firm. According to Kramer:

---

[4] Professional Service Trends: Deloitte & Touche' Dataquest® report, February 25, 1994.
[5] From SAP Web site. http://www.sap.com. Download 3/1/96.

Change occurs at several levels in an organization: strategy, process, people, and technology. Depending on the particular client situation, there are two approaches which can be taken. The first is 'clean sheet,' where all four dimensions of organizational change are explored without constraints. The second is 'technology-enabled change.' In this situation, the primary technology is selected early in the process and more strongly influences the other three dimensions of strategy, people, and processes, but still enables significant overall business change. The introduction of powerful, flexible, enterprise-wide solutions such as SAP is driving this approach as clients are looking to concurrently replace mainframe legacy systems and achieve significant operating improvement.

The 'right' approach to change is determined based on the client's situation. In Vandelay's case, the latter approach is more appropriate since they have already made the decision to go with SAP. To guide a client through all phases of implementation, ICS uses a structured approach tailored to the client's situation. This methodology captures the collective learning of the practitioners and creates a roadmap for the SAP implementation.

Figure CS3.3.3 illustrates how ICS viewed the difference between the 'technology-enabled' and 'clean-sheet' approaches to process redesign.

THE VANDELAY PROJECT

Vandelay management projected that the implementation would take 18 months and require the full-time efforts of 50 people, including consultants (both process redesigners and SAP specialists) and employees, as well as part-time involvement from many employees at each site. The total budget for the project was $20 million, including hardware, software, consulting fees, and the

Figure CS3.3.3 ICS's view of two models for business process re-design. Copyright © ICS, 1995

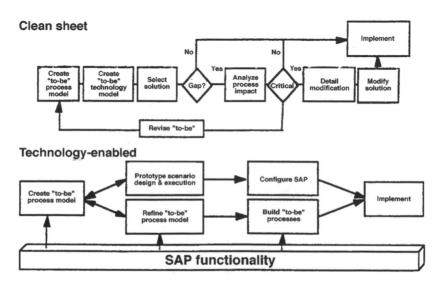

salaries and expenses of involved employees. Based on her prior experiences, Kramer felt that the timeline and budget were very aggressive for the scope of the implementation; she wondered whether all of these elements were in place to achieve the desired change.

R/3 software was to be implemented at Vandelay's eight manufacturing sites and four order entry locations,[6] and at the corporate headquarters in Minnesota. The plant installations would take the longest; each one would require a lengthy preparation period to align its operations with the new business practices. Kramer estimated that two thirds of all Vandelay employees would need training on how to use the new system, with the amount of training required ranging from one day for casual users to two weeks for those who would use R/3 heavily in their jobs.

Initially, the project team would focus about 80% of its effort on designing the 'to be' process model of the organization, and 20% on issues relating to the system implementation. This reflected the fact that the project would begin by establishing the need for business change and setting performance targets, rather than installing software. In the later phases of the implementation the required mix of consulting skills would shift to deeper SAP expertise. During the activities of system configuration, testing and delivery, the emphasis would be reversed; 80% concentration on SAP implementation, 20% on process design.

Team structure

As Kramer put it:

> A project like this one requires a variety of skills. I think the most important are project management ability, SAP expertise, business and industry understanding, systems implementation experience, and change leadership talent. We'll need to field a joint client/consultant team with the right mix of skills at the right time.

Vandelay and ICS had decided to use two teams for managing the project. While senior management on the steering committee would decide strategic issues relating to the implementation, Table CS3.3.1 shows that the project team would be responsible for the bulk of the decisions made, and for the ones which determined how the system would actually work. For this reason, Kramer was eager to structure this team correctly.

Kramer's experience had shown her that there were two basic ways to select participants for the project team. She could simply present a list of the required skills and characteristics for team members to senior-level management and ask them to nominate and approach the people who they felt would be best for the job. Alternatively, she could mandate that the team contain at least one representative from each of Vandelay's implementation sites around the world. While this approach might sacrifice some quality and depth on the team, it could also help to ensure that each site would have a project champion from the outset.

Managing change

Selecting the best development team was only one of Kramer's considerations as she prepared to dig in on the Vandelay engagement. She had led ERP implementations before and was aware

---

[6] Order entry locations: two in North America, one in Europe, and one in Asia.

Table CS3.3.1 Vandelay project team structure

Team name	Team composition and time commitment	Issues addressed by team	% of total issues addressed by team
Steering committee	Division VPs 8 people, meeting monthly	*Business Strategy*, e.g. sequence of site installations, planned changes in mfg. strategy	5%
Project team	Operations employees, e.g. planner/buyers, financial accountants: 20 people, full-time[7]	*Implementation specifics*, e.g. rules for reserving inventory for a customer order, horizons for planning and scheduling	95%

of the challenges involved in assisting a large organization as it attempted to change and standardize its practices. She placed these challenges into a few categories:

Centralization vs. autonomy

There was not way to involve all users in the decisions that would affect them and their jobs. This could create a strong temptation for people to second guess these choices, and to alter the system that was delivered to them. This was especially true at Vandelay, which had a strong tradition of encouraging innovation and autonomy among its employees, as George Hall had demonstrated. Should this tinkering be encouraged, or should systems and processes be 'locked-down' as much as possible? Could Kramer and her teams be confident that their decisions were the right ones throughout Vandelay? If so, should processes be tightly centralized and controlled, and tinkering (and therefore possible innovation) strongly discouraged? If not, what was the point of the long and thorough development and implementation cycle? Kramer had a strong bias toward 'input by many, design by few,' but how could she put this rule into practice?

She knew that this issue was particularly important for global companies like Vandelay. Just as cultures and currencies varied across countries, so did standard business practices, outlooks, and relationships between customers and suppliers. The implementation team would have to be sure that any universal processes did not run foul of local ones.

A closely related question concerned standardization on externally defined 'best practices.' Much of the consultants' expertise came from their previous engagements; they knew what had worked and what hadn't for other clients. In addition, SAP's standard capabilities were the result

---

[7] Includes only client business operations team members; does not include consultants, IT resources, or other staff for project activities such as testing, training, and documentation.

of the firm's accumulated knowledge about the requirements for ERP software. There was thus a set of outside practices involving systems, operations, and processes that could be used at Vandelay. Kramer wondered how they should be incorporated into the project—were they a starting point or the final word?

Kramer could already see one area where plants would have to give up some of their autonomy. R/3 required that each item have a single, unique part number, and Vandelay executives wanted common part numbers across all sites so they could see accurate consolidated information about production, orders, and sales. Each plant, however, had developed its own internal part numbering system over time. Replacing these schemes would be a major effort, involving everything from stockroom storage bins to engineering drawings to part stamping equipment. In addition, plant personnel would have to forget the previous numbers, which they often knew by heart. Kramer saw that part number standardization would be part of the Vandelay R/3 implementation, that the plants would probably resist it, and that they would have no choice in the matter.

Change agents and organizational inertia

Kramer also knew from experience that large implementations went best when a critical mass of early leaders—people who were enthusiastic about the work of change and who were respected within the firm—had been built. She also knew, however, that even with committed change agents in place, most people did not completely accept a new system until they really believed that it was inevitable. She found this paradoxical; companies committed substantial resources up front and stated clearly that the new system was a given, but most employees remained skeptical for a long time. Why was this, and what could she and her early movers do about it?

Software

Although R/3 had broad capability, there would be situations where it would not exactly fit, the desired Vandelay process design. Kramer had observed three primary alternatives to addressing this situation:

1. Change the business process to match the capabilities of the software.

2. Interface R/3 to another package or custom solution.

3. Extend the R/3 system to precisely match the business requirements.

What guidelines should she follow in selecting among these options?

Kramer knew that she would have to get back to George Hall soon about training for his site, but she was unsure what to tell him. If his people weren't allowed to experiment with the system as much as they wanted, would his enthusiasm for the project turn into hostility?

Exhibits

Exhibit CS3.3.1 World's largest software companies. Vertical and cross-industry applications worldwide software revenue, top 10 vendors worldwide, 1993–95 ($ million)

Company	1993	1994	1995	1995 % share	1994 % share	Growth 1994–95 (%)
Microsoft	1,246	1,688	2,484	6.2	4.8	47.1
IBM	1,647	1,607	1,711	4.3	4.6	6.5
SAP AG	414	843	1,322	3.3	2.4	56.9
IBM/Lotus	281	401	540	1.4	1.1	34.6
Computer Associates*	431	425	478	1.2	1.2	12.5
Autodesk	351	392	455	1.1	1.1	16.1
Novell	603	477	443	1.1	1.4	−7.1
Adobe Systems	324	387	438	1.1	1.1	13.3
Cadence Design	336	391	435	1.1	1.1	11.2
Siemens Nixdorf	350	365	361	0.9	1.0	−1.1
All other vendors	24,238	27,895	31,240	78.1	79.9	12.0
Worldwide solutions revenue	30,273	34,930	39,989			14.5

*Includes revenues of Legent Corp. for entire year.

Source: International Data Corporation, 1996.

Exhibit CS3.3.2 Functions included in SAP's R/3 system (diagram trademark of SAP)

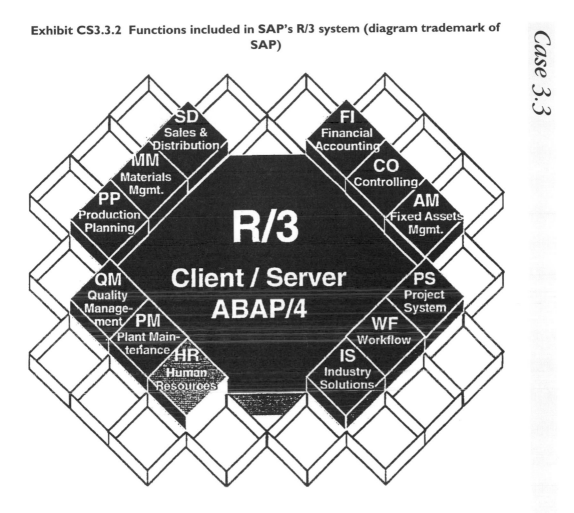

Case 3.3

Exhibit CS3.3.5 Sample SAP screen. Copyright © SAP America, Inc.

```
┌─────────────────────────────────────────────────────────────────────────┐
│ ⊞ Create Standard order: Overview - Single-Line Entry          _ ⌸ ✕    │
│ Sales document  Edit  Overview  Header  Item  Environment  System  Help ⊕│
├─────────────────────────────────────────────────────────────────────────┤
│  ✔  │░░░░░░░░░░░░░░░░░░░░▾│  ☐ ⇦⇧✖ ⌕? ▤▤▤▤                               │
├─────────────────────────────────────────────────────────────────────────┤
│  Business data-Item │ Business data-header │ Schedule lines │ Pricing │ Partners │
└─────────────────────────────────────────────────────────────────────────┘
```

Sold-to party	customer1		
Purch. order no.	test order	Purch. ord. date	05/27/1996
Req. deliv. date	T 05/27/1996	Pricing date	05/27/1996
Sales order		Net value	0.00

All items

Item	Material	Order quantity	UoM	Description
☐	testpart	100		
☐				
☐				
☐				
☐				
☐				
☐				

SDB [1] ¹mko.dec . OVR 06:09PM

Measuring, Evaluating and Rewarding Employee Performance

Chapter

4

Introduction

One of the most often heard management aphorisms focuses on the power of measurement. It takes several forms, some positive, such as 'You get what you measure', or 'What gets measured gets managed' or 'What gets measured gets done', others negative, such as 'What doesn't get measured doesn't get done'. The nuance is subtle but interesting, as the positive formulation puts more emphasis on measurement as a *sufficient* condition for performance, while the negative formulation presents measurement as a *necessary* condition for performance.

Strictly speaking, both formulations are slightly exaggerated. We have all observed examples of individuals engaging in activities that were not being measured and for which they knew they would not receive any explicit organizational reward. Similarly, it is not uncommon to see individuals not paying much attention to areas of performance that *are* being measured by the organization.

Still, measures of performance can be a powerful component of performance enhancement programmes. For example, imagine trying to improve on a performance dimension without the benefit of knowing how well you have been doing and how successful your efforts are. Consider the following examples from daily life.

Imagine you want to improve your basketball shooting skill. You know that practice will help, so you decide to invest the time and effort and go to the basketball court after work. There you start shooting towards the hoop, but as soon as the ball gets close to the rim your vision gets blurry for a second, so that you cannot observe where the ball ended up in relation to the target, whether it is left, right, in front, too far back, inside the hoop. It would be pretty difficult to improve under these conditions.

Or try bowling: you throw the ball into the lane but, as the ball is about to hit the pins, the image disappears and you cannot see how many pins you downed. Of course, for both bowling and basketball, you could hire a coach who knows how to shoot and can help you learn the mechanics. Based on his or her knowledge of 'the perfect shooting motion', the coach would give you suggestions on how to improve your mechanics. But is there such a thing as a 'perfect throwing motion' irrespective of the shooter's characteristics? Note, also, how dependent you would be on that coach. How long do you think you would listen to the coach's instructions in the absence of any observable feedback? In fact, how long would bowling sustain your interest if you couldn't observe the outcome of your efforts?

Or imagine you are engaging in a weight loss programme. A normal step would be to purchase a scale to be able to track your progress: Is this programme working? Am I losing weight? A positive answer would be encouraging and would motivate you to keep up the effort, while a negative answer might lead you to reflect on the process: Am I working on the right diet and exercise programme? Am I doing everything I am supposed to? And so on. Suppose you don't want to set up a sophisticated measurement system and decide to forego the scale. You would still have some idea of how well you are doing from simple methods such as clothes feeling looser, a belt that fastens at a different hole, or simply via observation in a mirror! Now, imagine trying to sustain a weight loss programme without *any* feedback on your progress.

In these three examples, availability of quantitative measures of performance can yield two types of benefits. First, performance feedback can help improve the 'production process' through a better understanding of what works and what doesn't; e.g. shooting this way works better than shooting that way. Secondly, feedback on performance can sustain motivation and effort, because it is encouraging and/or because it suggests that more effort is required for the goal to be met.

Performance measures can be helpful in organizational settings for the same two generic reasons:

1. Quantitative feedback can support *learning* by helping employees better understand the cause–effect relationships involved in their task. This is, of course, one of the basic pillars of the TQM approach; the so called 'seven tools of quality', for example, are different ways of using quantitative data in a learning and problem solving approach. By enhancing learning, quantitative feedback can improve the likelihood that employee effort will translate into superior performance.[1]

2. Quantitative feedback can also contribute to stimulating and/or guiding employee effort. If employees care about the performance feedback, they may invest more effort into improving performance, particularly those aspects of performance captured by the quantitative indicators. Employees may care about the performance indicators because of their link with *extrinsic* rewards, i.e. rewards administered by the firm such as performance appraisal, pay rises/bonuses, career prospects or more simply the quality of the interaction with the boss during the year. Employees may also care for more *intrinsic* reasons, such as their personal pride, the desire to do their job well or to win as a team.

Put together, the two potential benefits of quantitative feedback can have a powerful impact on performance, by contributing to:

■ stimulating more effort,

■ guiding this effort into directions that are (most) valuable for the firm and

■ improving the translation of employee effort into performance.

Clearly, not all performance measures have this positive an impact. That of quantitative

---

[1] Note that the notion of 'effort' includes both quantitative and qualitative dimensions. The quantitative side includes notions like length and intensity of work, the qualitiative side notions like ingenuity, creativity and actions or decisions that benefit the firm, but are personally costly for the employee. Effort thus includes employees working harder, smarter and/or more in the firm's interest.

feedback depends on a number of factors, including the type of indicators used, their properties as performance indicators, and the way they are used by the firm.

Properties of performance indicators

Congruence with organizational objectives

Given our objective of *guiding effort* through the choice of performance indicators, it is obvious that the indicators chosen should be at least consistent with, and ideally should foster the achievement of, the goals of the organization. This is easier said than done. First, some organizations have not explicitly agreed on a clear strategy and thus cannot translate it into a clear set of priorities/performance indicators.

In other cases the firm has developed an overall strategy, but has not been able to decompose it into coherent subsets of priorities and indicators for the lower-level units; the subunits may have developed a set of indicators for their own purposes, but these local indicators have not been explicitly coordinated and linked with the organization's overall strategy and goals. In such cases a subunit may hurt overall firm performance by trying to optimize its own indicator, as such 'local optimization' may destroy more value in other subunits than the value it creates in the first place. A classic example of such global suboptimization would be a service department underspending its budget by 2%, while the business unit serviced by this department misses its revenue target by 20% due to the poor service level it received.[2]

We also have many examples of organizations that, at some point, agreed on a clear strategy and decomposed it into congruent sets of priorities and indicators. Unfortunately, the world changed and the company's performance indicators did not. Congruence of performance indicators is not a static phenomenon; it must be maintained over time through periodic reexamination of the indicators as strategies, structures and objectives evolve in response to (or in anticipation of) changes in the firm's environment.

Completeness

A measure is complete when it captures 'the whole truth' about the subunit's (e.g. the individual's or the team's) performance. Perfect completeness of measurement is hard to achieve because actions and decisions taken today in a given subunit may have consequences today and/or in the future, in this and/or in other subunits.

An incomplete measure captures part of (the impact of) the unit's activities, but fails to reflect other dimensions. For example, it is typically easier to capture the

---

[2] This type of situation occurs very frequently and could be linked with many references. This specific example is reported by a business unit CFO as having occurred in the past at AT&T (Jenson, R.L., Brackner, J.W. and Skousen, C.R., *Management Accounting in Support of Manufacturing Excellence: Profiles of Shingo prize-winning Organizations,* The IMA Foundation for Applied Research, 1996, p.101).

Figure 4.1 Consequences of today's actions

Consequences	*Immediate*	*Delayed*
Felt in the deciding subunit	easier to measure	
Felt in the other subunits		much harder to measure

immediate impact of a decision on the deciding unit (top left quadrant in Figure 4.1), than the future impact of that decision on other administrative units (bottom right quadrant in Figure 4.1). Over time, the unmeasured dimensions are likely to receive less attention from the unit, which may result in suboptimal allocation of unit members' time and energy.[3]

Incomplete measure(s) also result in the unit receiving an incomplete view on the impact of its decisions and actions, which may lead to suboptimal decisions being made. If I am not aware that my actions have a negative impact on some downstream unit, I may hurt them unintentionally.

The lack of completeness of performance indicators can also allow employees to 'game' the system, (that is, to take actions that improve the local indicators to a greater extent than they contribute to real long-term value). Some actions, for example, may improve short-term profitability—which is reasonably easy to measure—at the expense of long-term performance, which is harder to measure and even harder to trace to specific past decisions and subunits. This is a completeness problem.

Controllability

A performance indicator is controllable to the extent that it is only influenced by elements under the unit's control; the indicator tells 'nothing but the truth' about the unit's performance. Most of the time, this ideal is impossible to achieve. For example, the amount of profit generated by a commodity trader is influenced by the trader's decisions and actions, but also by fluctuations of the market over which the trader has no control. Still, some measures can be more controllable than others for a specific manager under specific circumstances. Other things being equal, the more a measure is influenced by 'external' factors, the easier it is for the unit's efforts to be overwhelmed

---

[3] Without going as far as 'What doesn't get measured doesn't get done', the fact that people operate under tight time and energy constraints makes it likely that dimensions that are not measured or captured will receive less attention over time, even if the subunit members know that they are important. When you have more things to do than time available, and when the organization requires you to focus hard on dimensions one to five, it becomes increasingly difficult to set time and energy aside to take full care of dimension six that, by the way, the organization does not really track and discuss.

by the impact of these uncontrollable forces and thus, the weaker the perceived link between the unit's effort and the performance reported by the measure.

Controllability of performance indicators is important because it is often said that in order to maximize motivation and minimize perceptions of unfairness, 'people should only be evaluated based on what they control'.[4] Controllability is also important from a learning point of view; uncontrollable events have a polluting effect on the information contained in the indicator, by making it difficult to untangle the impact of one's efforts from the effect of exogenous factors (e.g. did the measure go up because of my actions or because the overall industry is doing well?).

Completeness and controllability: often a trade-off

While the fairness of the controllability principle is intuitively appealing, its practical implications are more problematic. For example, how do we determine whether a manager has 'control' over a performance dimension? Very few managers 'control' the price of raw materials or the sales price of products; competitive conditions, unions, government regulations and many other factors also play an important role. On the other hand, we still expect managers to try to *anticipate* and/or *adapt* to these exogenous events in order to optimize their impact on the firm's results.

One way to circumvent this problem is to understand the term 'control' in terms of 'administrative control', i.e. a manager has (administrative) 'control' over some element when she or he makes the associated decisions (e.g., purchasing of raw material, setting the sales price). Even in this restricted sense, however, the controllability principle remains problematic in the numerous cases where the unit has much influence over a performance aspect without being in sole charge of it. This type of situation occurs frequently when two or more subunits are interdependent and are supposed to collaborate to produce a joint output.

Take the following example: A group of supply engineers is in charge of ordering electronic components for a factory. The product design engineering group provides our supply engineers with the specifications that must be ordered. Our group's job is to select the suppliers and manage the purchase and delivery of components meeting specifications.

The components are tested on a sampling basis when received at the factory. They are then given to the manufacturing group and inserted into circuit boards. At the end of the manufacturing process the product undergoes a series of tests that can

---

[4] This so-called 'controllability principle' has appeared throughout the accounting and control literature for decades. For example: 'A man should be held accountable for only that which he alone can control' (Dalton, G.W., 'Motivation and control in oganizations', in *Motivation and Control in Organizations,* ed. G. Dalton and P. Lawrence, Irwin 1971, p.27) 'A manager is not normally held accountable for unfavorable outcomes or credited with favorable ones if they are clearly due to causes not under his control' (Arrow, K.J., Control in large organizations' in *Behavioural Aspects of Accounting,* eds. M. Schiff and A.Y. Lewin, Prentice Hall, 1974, p.284) And 'It is generally agreed that it is better to evaluate managers on their *controllable* performance, and to filter out the uncontrollable factors' (Magee, R.P., *Advanced Managerial Accounting,* Harper & Row, 1986, p.265).

identify which specific component is working and which is not. When one of the board's components does not perform, it is taken out of the board and tested to assess whether it meets specifications. If it does, the manufacturing unit pays for the (nonfunctioning) component; if it does not 'meet specs', our supply engineering group gets charged for the part.

The supply engineers were thus measured on the 'percentage of parts that meet the specs'. This sounded reasonable to them; one of their roles was indeed to secure components that meet engineering specifications. When a new departmental manager arrived, however, he changed the main indicator discussed at meetings from 'percentage of parts meeting specs' to 'percentage of parts tested as nonfunctional at the end of the manufacturing process'. The component engineers objected to this change, arguing that they were not in charge of manufacturing and, hence, should not be held accountable for mistakes made by manufacturing personnel. After all, they said, 'if the part meets the specs yet doesn't function in the final product, it must be because manufacturing mishandled the component—you know how these manufacturing people are!'.

The new manager answered that he was less interested in whose fault the problem was than in getting the number of defects down. If a part meets the specs yet does not function in the final product, he said, there must be something wrong somewhere, probably in the assembly process or in the component specifications. Either way, he told his group, 'your skills and knowledge are needed to help the manufacturing people and product design engineers to identify the root causes of these problems and eliminate them. I want you to cooperate with these two groups to decrease the overall unit's problem, rather than simply look inside your subunit at what you're strictly responsible for.'

This case illustrates the frequent dilemma between completeness and controllability of performance measures. The supply engineers wanted to focus and be evaluated on a metric that was 'controllable' by them. Their boss chose instead a measure that was less controllable but more complete, as it captured the impact of the coordination and joint problem solving between the various subunits. The manager found this more complete indicator by measuring *joint performance*, which can involve measuring performance one hierarchical level up.

More generally, performance is harder to measure whenever two or more subunits are *interdependent* because interdependence makes it hard to trace the costs and benefits of each decision to the source of the decision. As a result costs and benefits of activities often end up being recorded in different administrative units (e.g. the cost of rework undertaken by department B to correct mistakes introduced at department A; or lost sales due to quality problems created by manufacturing). One way to deal with this problem is to measure performance at a higher level in the organization (e.g. we can measure performance at the division level rather than for each function). But enlarging the administrative unit also *dilutes* the impact of individual decisions, thus decreasing the extent to which the indicators capture 'nothing but' the activities of a specific group of employees (i.e. decreasing indicator controllability for each member of the unit).

Completeness, controllability and simplicity

Aside from measuring performance of a larger administrative unit, another way of obtaining more completeness in measurement is to use several performance indicators to try to capture more facets of performance. As one manager put it:

> *The biggest problem [. . .] is understanding the behaviour your measurements are going to drive. The problem is that people will scope in on a single metric, whatever it might be; no single metric is appropriate. [For example] there was never a problem with measuring efficiency. Efficiency is a perfectly legitimate measure of performance, [but] if you measure efficiency then you have to measure inventory turns, customer service, and aggregate cost. If you just measure efficiency, you will destroy yourself. When you put a metric in place, you must be careful also to put a countermetric in place to prevent the first metric from causing stupidity.[5]*

Increasing the number of performance indicators probably involves decreasing marginal returns, however. In fact, in some cases the additional indicators may generate negative returns (i.e. they destroy value). First, because measurement typically does not come for free; there are data collection and processing costs. Secondly, because people can only digest and act upon a limited amount of information; they can take very seriously only a limited number of performance indicators. Some companies measure so many dimensions that capture so many trade-offs, that people reach a state of 'information overload' and learn to disregard most of the data they receive.

User-friendly presentation and interpretation

Firms invest significant resources in collecting, processing and distributing quantitative data. One would expect them to try also to make sure that the data presented in reports are (a) understandable and (b) reasonably easy to use by the feedback recipients. Many firms indeed do so. For example, when Taco Bell introduced an information system aimed at helping lower-level staff to make decisions previously made by supervisors, the company made sure the system presented information in graphical and colourful ways; graphical because graphs tend to be easier to grasp quickly than numbers, and colourful both for ease of interpretation and to make the experience more pleasant (and thus more encouraging, hence people will access the system more often and they will use the information to make better decisions).[6]

Surprisingly, however, many firms still produce reports that are tiresome to read and feature key indicators that readers do not understand well. The situation is improving on the presentation front as increasingly powerful hardware allows the use of more user-friendly software, but many firms still reap limited benefits on their data

---

[5] Quote from A.L. Peterson, chief financial officer of AT&T Power Systems (1992 Shingo Prize for Excellence in American Manufacturing), slightly edited from Jenson, R.L., Brackner, J.W. and Skousen, C.R., *op. cit.*, p.107.
[6] See Case 1.1 'Taco Bell (A): Reengineering the Business (1983–95)'.

collection and processing investment because they underinvest in training people to understand the reports.

The link with the organization's *structure*

Consider a 'production process' that has three steps, which we call for now A, B and C. (The following example is set in a manufacturing environment, but it could equally well occur in a service business). The 'product' first goes through step A, then moves to step B and is completed within step C. The three steps are *sequentially interdependent*, meaning that what happens within step A has an impact on the effectiveness and efficiency at step B, which both have an impact on what happens in step C.

It so happens that the company's product range involves five major 'product groups', which are largely processed on the same equipment, but nevertheless involve some specificities. In particular, the product groups do not react the same way to each process step and to interdependencies between the steps. (For example, one of the product groups has a much smaller tolerance to slight imperfections in step A.)

The question is: How should the company be structured? This is a complex issue, which will receive more extensive treatment in Chapter 6 of this book. Intuitively, though, one can feel that the best answer at this point is probably: It depends! Can we split up the 'equipment' into five subunits (one for each product group), or must the product groups share the same equipment? Are there constraining bottlenecks at some stage(s) of the process that require intense coordination across product groups? How large is the workforce for each process step? (If each step involves a very large number of people, it probably is more conceivable to split each step into five units organized along product groups.)

In the real-life situation on which this example is based, the company had chosen, historically, to structure itself along the three process steps. There were three departments, each headed by a management team. Each department monitored a certain set of performance indicators focused on its particular part of the process. In other words, they monitored measures that were largely controllable by them, and then tried their best to optimize these local, incomplete (from the point of view of the firm), performance measures.

As a result, there was no one fully responsible for the overall 'process performance' (such as overall yield). In principle, step C could have been responsible for overall results, but they were so dependent on steps A and B that this would not have been 'fair'. It was also too easy for step C to invoke excuses related to inadequate performance in the first parts of the process: ('Sure, our results are not good, but that's because steps A and B did . . . (or did not do . . .)'.

Similarly, no one was responsible for optimizing the overall process from a product group point of view; each step was trying to optimize its own set of metrics, so the three departments benefited from improving performance on their process step for any given product group, but the impact of a product group on their particular step might be quite different than that group's importance for the firm. (For example, a product group might have a small impact on a department's overall results, but a large one on the overall result of the unit.)

Because no one was explicitly responsible for overall process performance, particularly on a product group by product group basis, limited *attention* was devoted to these issues. People were busy and had more than enough 'things' to take care of within their own areas. As a result, there were *few performance measures available* on overall process performance on a product line basis. (Measurements tend to appear when someone within the organization expresses a request for the specific measure. This is normal and legitimate: we cannot measure everything, otherwise we would spend all our time collecting and processing measurements! Furthermore, we *should not* measure everything; the only measures worth having are those whose benefits to the firm exceed the cost of collecting and processing the data.)

When a new general manager took charge of the unit, he observed a series of 'local optimizations' resulting in 'global suboptimization'. He identified three major alternatives. The first was to make the three department managers report to a dedicated boss (they currently reported to a manager who had a large span of control). The second possibility was to modify the structure more radically by creating five departments defined along product groups. This would allow him to assign responsibility for overall process performance to specific individuals. The manager chose a third alternative, which was to leave basically intact the 'functional' structure, but complement it with additional structural mechanisms and enlarged accountability. More specifically, he made the three department managers *jointly responsible* for overall process performance (for all five groups of products), and to help them in their task he formed cross-departmental teams that were assigned specific responsibilities regarding *one* product group. The number of performance measures available on overall process performance has drastically increased since these changes were put in place, as more attention is now focused in that direction.

This general manager's decision may not be most appropriate in every case. He eliminated the second solution (reorganizing along product lines) because he felt that each production step required high technical expertise and breaking up the three departments would weaken the technical specialization and expertise development within each step. He was tempted by the first solution (adding one hierarchical level between himself and the department managers), but he first wanted to try a 'self-managed management team' approach. The three department managers agreed to give it a try, with the understanding that they would revisit the decision if it proved unworkable.

More generally, the point illustrated by this example is the interdependence between organizational structure and performance measurement. On one hand, structure defines boundaries which will guide managerial attention, allocate *decision rights* (what do I have authority on?) and *accountability* (what am I responsible for?) across managers. As a result, managers will request and pay more attention to some measures than others. To modify the way managers allocate their attention and efforts, firms can change the structure to reshape intraorganizational boundaries and reallocate responsibilities, and/or modify the performance measures that managers are required to monitor to draw more attention in specific directions. To take the example of 'attention to overall process performance', an initially functional structure might be turned around and restructured along process lines; process owners might be named (in addition to 'functional managers'); and/or, performance measures capturing overall performance could be defined, reported, and included in some way in managers' objectives. We will come back to these questions in Chapter 6.

Types of indicators

There are many typologies of performance indicators, but the following two distinctions are particularly useful.

Financial vs. nonfinancial measures

This distinction is simple to define. Basically, financial measures are indicators expressed in currency terms. They also include ratios involving one or more financial measure. For example, ROI (return on investment = net profit/investment) is a financial measure even though it is not expressed in currency as such. The other indicators are known as 'nonfinancial', or 'operational' indicators.

Leading vs. lagging, means vs. ends

Decisions and actions taken today can have many implications, some observable fairly quickly, others that only become apparent after some time.

Take, for example, an insurance company's claim processing unit, organized in a variety of cross-functional cells each supervised by a first level manager. Assume that one of these managers decides to implement a performance measurement system for the unit they manage. Say that, partly because of the content of the innovation and partly because of the way the manager presented it, the unit's staff reacts very negatively and their work is perturbed. Some clients' claims will be processed as well as usual, others will experience problems (e.g. with accuracy and/or speed of processing, courtesy over the phone, etc.). Among the clients who experience problems, a proportion will complain to the company, another proportion will complain to their friends and some will decide to take their business elsewhere.[7] The clients who take their policy to the competition will not do so immediately; it will take a few weeks, a few months, in some cases. Similarly, some of the friends who came to know about our client's problem will move to the competition. Some will move now, others will move after one more negative incident, etc.. As a result, the impact of our unit manager's

---

[7] Some clients will do all three, of course. As an aside, British Airways estimates that they receive one official complaint for 20 dissatisfied customers (*Financial Times*, 24 February 1994, p.6).

innovation on the firm's revenue will take some time to materialize and it will probably never be traced to the unit manager.

This example illustrates why financial measures are typically considered to be 'lagging indicators of performance'. They record the *effect* of decisions *not* when the decisions are made, but rather as the *financial impact* of these decisions materializes, which can be long after the decision was made. The financial indicator thus captures performance with a time lag, which is a problem for both learning and performance evaluation.

In contrast, it would have been possible to capture some of the impact of the unit manager's actions by measuring the performance of the unit, its 'well being', and customers' reactions. For example, frequent measurement of speed and accuracy of processing, courtesy over the phone, of the unit's morale/attitude and of customer satisfaction/complaints, would have allowed the identification that 'something was wrong', which then could become the basis for an analysis and a discussion.

In this (typical) case, nonfinancial performance indicators capture performance more quickly than financial measures. To the extent that there will be negative financial consequences in the future, the nonfinancial measures are also 'leading indicators' of financial performance. In fact, one could organize the performance measures along a continuum where, for example, staff morale might be a leading indicator of unit performance (e.g. speed, accuracy and courtesy), which is itself a leading indicator of customer satisfaction, itself a leading indicator of financial performance. In this sense, the notion of 'leading' and 'lagging' indicator is *relative*. Customer satisfaction is a leading indicator of financial performance, but it is probably a lagging indicator compared to unit performance. Or again, unit performance is a leading indicator of customer satisfaction, but a lagging indicator of the way the team is managed.

A similar ordering of performance measures is sometimes made in terms of the indicator being a *means* or an *end*. In the previous example, each indicator is a means on a chain leading to one end, which is itself a means to another end, and so on. Still, the distinction is relevant because, while a measure is a means to an end at the company level, a subunit might consider it the end as far as it is concerned. Such partitioning of responsibilities is healthy in principle, as it allows each subunit to select performance indicators that are somewhat 'controllable' by its members. It is also important, however, to remind subunits of the context in which they operate, i.e. that there are other 'ends' further down the line. Being mindful of the means–end chain can foster integration and cooperation between subunits by helping strike an appropriate balance between completeness and controllability of performance indicators.

Financial vs. nonfinancial measures: pros and cons

We mentioned above the major weakness of financial indicators: they capture the impact of decisions with a significant time lag. As a result, they tend to be less proactive indicators of potential problems than operational (nonfinancial) indicators. Many articles published in the 1980s following the TQM movement emphasized the need to complement financial indicators with nonfinancial ones. Several articles went a step further and recommended reducing the focus on financial indicators, which were said to encourage managers to make decisions that were not in the best interests of the company.

We now know that financial and nonfinancial indicators should not be viewed as substitutes. While financial measures tend to be lagging indicators of performance (they tend to capture the impact of decisions only when their financial consequences materialize, which can be long after the decision was made), they also have two important benefits. First, they represent the impact of decisions in a comparable measurement unit, money, which allows aggregation of results across units; and secondly, they capture the cost of trade-offs between resources as well as the cost of spare capacity. Let's examine both aspects in turn.

First, financial measures capture everything with a single measurement unit, money. Aside from the fact that money happens to be a significant preoccupation of for-profit organizations, this single measurement unit has the advantage of capturing trade-offs between resources. Quality of products and services, high customer and employee satisfaction, high flexibility of processes and development of innovative products are all important parameters, but they ultimately must lead to profitable operations. While some parties question whether shareholder value maximization should be the only goal of a firm, very few dispute that adequate profitability is at least a condition for survival. Improving operations and redesigning processes for enhanced quality and flexibility often requires investment money. At some point, this investment must pay, which some companies overlook in their zeal to improve operations.[8]

Secondly, it is important to remember that the first impact of process improvement is often the creation of spare capacity. Consider the simple, but powerful equation that follows:

$$\text{Resources available} = \text{Resources consumed} + \text{Spare capacity}^9$$

Question: Where is 'money' in this equation? Under which term would you place the salaries that the firm pays, the payments for space, etc. Resources available or resources consumed? Clearly, when money is spent, money is consumed. But paying salaries and rents does not consume the *resources* thus created, it makes the resources *available* for the firm (e.g. eight hours a day of clerk time, twenty-four hours of machine availability or one floor of office space). These resources can then be used (productively or unproductively), or not used.

---

[8] A particularly vivid illustration of this phenomenon was provided by Wallace Company, a 1990 Malcolm Baldridge National Quality Award winner which filed for bankruptcy in early 1992. In a *Business Week* magazine article (The Ecstasy and the Agony, 21 October 1991, p.40), a Wallace manager explained that while customers valued the improvement in on-time delivery from 75% in 1987 to 92% in 1990, overhead increases over that period of time led to price increases that customers did not see as a sufficiently good deal (cited in Shapiro, E.C., Eccles, R.G. and Soske, T.L., 'Consulting: Has the Solution Become Part of the Problem', *Sloan Management Review,* Summer 1993, pp. 89–95.
[9] This distinction was first made and discussed by Robin Cooper and Robert S. Kaplan in 'Activity-based systems: Measuring the cost of resource usage', *Accounting Horizons,* September 1992, pp. 1–12.

Consider the previous example of the insurance claim processing department that became more effective and more efficient. A very important department is now performing better, but is the firm making more money? In time, improved claim processing performance may lead to increased premium income, which would improve profit. In the short run, better processing may result in increasing customer satisfaction while reducing claim payments, which would increase the firm's profit immediately.[10] From an overhead point of view, however, the firm still employs as many clerks and uses as much space. Some of these clerks may work fewer hours, which might save some overtime pay depending on the pay arrangement, but that does not take away their basic salary. The same reasoning applies to a bank restructuring its back-office operation to centralize it into service centres. The centres will be more efficient than the branches' back offices, but removing the work from back offices does not, by itself, make the branches space and employees go away.

The punchline is that, in the short run, process improvements often generate more spare capacity than real profit increases. Spare capacity offers an *opportunity* to improve financial performance, but management must take actions to capitalize on this opportunity. Actions can go along either, or both of the following lines: *increasing sales* (through increased net price and/or increased volume), and/or *managing the spare capacity out of the system* (which means laying off people, selling equipment, etc.). It could be one, it could be the other, it could be a little bit of both, but it's got to be something!

This reality is easy to miss when one looks at nonfinancial measures only. Nonfinancial measures are very helpful in guiding operational improvements, but they do not capture how much it has cost to improve operations, nor whether the spare capacity liberated by the operational improvements has been used and/or redeployed. Financial measures can help answer both questions because they capture trade-offs between resources and they capture the cost of capacity. Furthermore, business organizations exist in large part to create value for shareholders; financial performance thus remains an essential parameter. Ultimately, improvement on nonfinancial measures should translate into superior financial performance.

The notion of the Balanced Scorecard[11]

The concept of the Balanced Scorecard came out of this realization that no single performance indicator can capture the full complexity of an organization's performance. First proposed five years ago, in the first of three *Harvard Business Review* articles on the subject by Robert Kaplan and David Norton, the *Balanced Scorecard* is a short document summarizing succinctly a set of leading and lagging performance indicators

---

[10] This win–win situation can occur when improved operations allow more clarity for customers about what risk is covered by the policy and what is not, and when the burden of reduced payments is borne mainly by third parties such as garages and contractors.

[11] This section draws on Epstein, M.J. and Manzoni, J.-F., 'The Balanced Scorecard and Tableau de Bord: Translating strategy into action, *Management Accounting,* August 1997, pp. 28–36.

Figure 4.2 Translating vision and strategy: four perspectives. Adapted from 'Linking the Balanced Scorecard to strategy', Kaplan, R. and Norton, D. P., *California Management Review*, July 1996

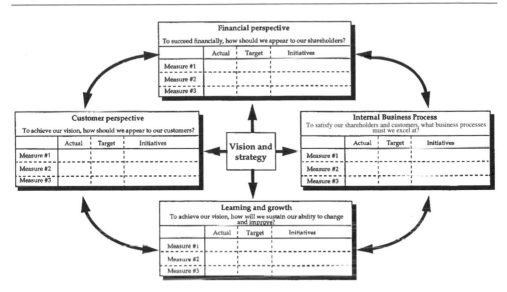

grouped into four different perspectives: financial, customer, internal processes, and learning and growth (see Figure 4.2).[12]

The idea of having some form of balanced picture of company performance was not new in itself; many companies had for years tracked and reported multiple indicators, and many countries had specific traditions in this respect. In France, for example, companies have been using a related tool called 'Tableau de Bord' for over fifty years. Still, Kaplan and Norton's Balanced Scorecard goes one step further because of four important characteristics:

1. It presents *on a single document*, a series of indicators providing a more complete view of the company's performance.

2. This document is supposed to be *short* and connected to the company's information system for further detail (rather than the monthly 'book' that many organizations still produce and which requires enormous managerial time and skill to digest).

3. Instead of listing indicators in an ad hoc manner, the Balanced Scorecard groups the indicators into four 'boxes', each capturing a distinct perspective on the company's performance (see Figure 4.2).

---

[12] See 'The Balanced Scorecard—Measures that drive performance', *Harvard Business Review*, January–February 1992, 'Putting the Balanced Scorecard to work, *Harvard Business Review*, September–October 1993, and 'Using the Balanced Scorecard as a strategic management system, *Harvard Business Review*, January–February 1996. Kaplan and Norton recently expanded on their ideas in a book called *The Balanced Scorecard: Translating Strategy into Action*, Harvard Business School Press, 1996.

4. Last, but maybe most important, performance indicators presented in the Balanced Scorecard must be chosen on the basis of their link with the vision and strategy of the firm/unit. Rather than starting from the set of performance measures already available within the firm, the selection process should be a conscious, deductive effort starting from the objectives the firm is trying to achieve and the critical means that will get it there. This process often results in the selection of performance indicators that are not currently available, and for which a data collection process must be developed.

Why four perspectives? The financial perspective focuses on the shareholders' interests: Is the company generating satisfactory return on investment and creating shareholder value? The three other perspectives can be explained through the following reasoning. How does a company succeed financially? Through a combination of two elements: one is creating value for customers; we thus need to know how customers perceive our performance. But a firm can delight customers all the way into bankruptcy, so it also needs to make sure that it performs well on key internal dimensions. For example, we can improve customer service by having massive numbers of employees servicing customers, or by having fewer employees whose time we utilize more efficiently and whom we support with excellent information technology. Creating value for customers only translates into shareholder value if it is based on effective and efficient key internal processes.

The next step is to make this value creation sustainable over time. The company may create value for customers and make excellent use of its resources *today*, but the world does not stand still and performance requirements keep ratcheting up over time. To make sure that the company will still be appreciated by tomorrow's customers and will keep making excellent use of its resources, the organization and its employees must keep learning and developing. This perspective should thus group indicators capturing the company's performance with respect to innovation, learning and growth.

From this reasoning comes a set of four perspectives, each characterized by a small set of performance measures. The specific content of these four 'boxes' must be adapted to the circumstances of each organization. In particular, the four sets of indicators should reflect and operationalize the organization's mission and strategy. A company following a low-cost strategy will have different key success factors than one creating value through very innovative products targeted at a subset of the overall market. These two organizations should track different indicators to assess how well they are doing and guide performance improvement programmes.

The concept of Balanced Scorecard can be cascaded down through the organization to achieve a dual purpose: (1) customization of the Balanced Scorecard to the (sub)unit by identifying its own set of actionable performance indicators, and (2) alignment of the subunits within the company's overall vision and strategy. Figures 4.3 and 4.4 provide examples of these processes. (See also Case 4.4 'Mobil USM&R (B): New England sales and distribution').

Figure 4.3 Translating corporate objectives into functional objectives

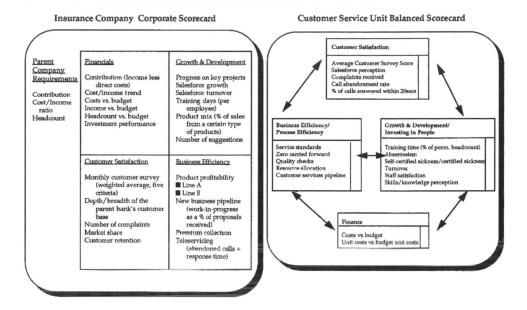

Figure 4.4 Cascading the strategy through nested scorecards

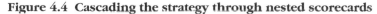

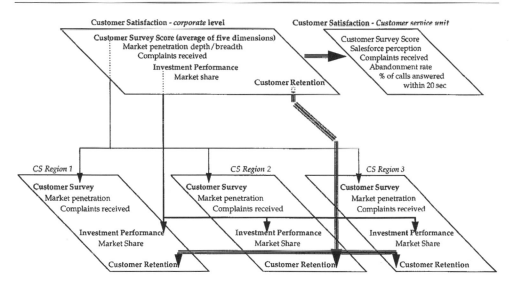

Figure 4.3 shows the Balanced Scorecard for an insurance company affiliated with a bank, itself part of a large banking group. The insurance company developed a scorecard for the company as a whole, featuring the four 'traditional' perspectives. It also asked its various subunits to develop scorecards at their own level. The complete scorecard for the 'customer services' subunit is also shown in Figure 4.3, while Figure 4.4 describes in greater detail how one perspective (in this case, the 'customer perspective'), can be 'cascaded down' the organization.

As shown in Figure 4.4, all the measures selected by the company as a whole can be tracked at the regional level. The regions' customer box is thus identical to that used by the overall company. The customer service unit, on the other hand, needed to adapt some of the indicators to its specific conditions (e.g. salesforce perception); its customer box thus includes a mix of corporate and local indicators (see Figure 4.3). Note also how indicators can sometimes be placed in different perspectives depending on the unit. Teleservicing is part of the customer perspective for the customer service unit, but it is considered part of business efficiency for the company as a whole.

More generally, cascading the Balanced Scorecard involves two interrelated processes: taking the part of the overall strategy and indicators that are applicable to the subunit, and designing other indicators that reflect local needs. While the Balanced Scorecard concept is meant to translate a unit's strategy and is thus best applied to 'units' that *have* a strategy, it can be cascaded all the way down to individual managers, who could use the scorecard's four perspectives to organize their personal goals and anchor them in the larger unit's strategic framework.

A Balanced Scorecard contains a set of performance metrics, some considered 'lagging indicators', others considered 'leading indicators'. In practice, the notion of leading vs. lagging indicator should really be thought of as a *continuum*. Consider the following example, this time taken from a manufacturing environment. Customer satisfaction is a *leading* indicator of financial performance, but—assuming on-time-delivery is an important factor for the firm's customers—it is also a *lagging* indicator of on-time-delivery. On-time-delivery is a *leading* indicator of customer satisfaction, but it is determined in part by, and thus is a *lagging* indicator of, production cycle time and quality of both product and process. It is possible to reflect such relationships between means and ends, or between causes and effects, through the Balanced Scorecard (see Figure 4.5).

Management use of measures: diagnostic vs. interactive

When managers are asked which performance measures they track regularly, they typically list a relatively large number of indicators. Yet we know that managers are busy and tend to have limited time for systematic study of long reports, which raises an interesting question: How can managers pay equal attention to 10, 15 or 20 indicators? The answer, of course, is that managers do *not* give equal attention to all the indicators they follow. In particular, it is helpful to distinguish two basic ways of using performance measures.[13]

At one end of the continuum is 'diagnostic control', also known as 'management-

---

[13] This section is based on Robert Simons' work. For more information see Simons, R., *Levers of Control: How Managers Use Innovative Control Systems to Drive Strategic Renewal,* Harvard Business School Press, 1994, and Simons, R., 'Control in an age of empowerment', *Harvard Business Review,* March–April 1995.

Figure 4.5 Organizing performance indicators in a causal chain. Adapted from Kaplan, R.S. and Norton, D.P. 'Using the Balanced Scorecard as a strategic management system' *Harvard Business Review* **(January–February 1996).**

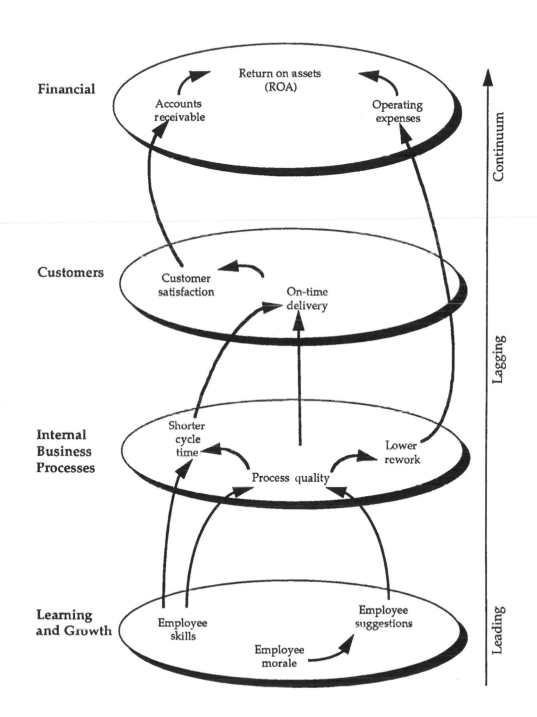

by-exception'. Indicators tracked on a diagnostic basis are followed regularly, but they do not trigger specific action or discussion as long as they remain within preset limits. Diagnostic control is meant to protect management attention and concentrate it on areas that might benefit most from it, i.e. those areas that triggered variances from expectations, particularly unfavourable ones.

At the other extreme is 'interactive control', which is characterized by the following three dimensions: 'The process demands *frequent and regular attention* from *operating managers* at all levels of the organization; Data are interpreted in face-to-face *meetings of superiors, subordinates and peers*; and the process relies on the *continual challenge and debate* of underlying data, assumptions and action plans'.[14] Interactive control is not limited to unfavourable variances, it is systematic; we *always* discuss where we stand on this metric.

Simons' notion of diagnostic vs. interactive control helps us to understand how managers can focus on so many measures at the same time: They use some measures 'interactively' and the others on a more diagnostic basis. Some managers are even quite articulate about this distinction; they may not use Simons' terms but the principle remains. One manager, for example, clearly distinguished two sets of indicators; those he called the 'traditional measures of manufacturing performance', such as performance to schedule, which he only followed on an exception basis. He explained:

> *I get reports on these measures, I talk to people about them, but I don't focus on them in meetings. Typically, I don't have to get involved at all in these measures: We have now reached the point where we rarely miss a commitment. But there is an escalation process, whereby if any traditional metric is not going to be made, I expect the people to let me know about it early enough for me to get some involvement.*

> *I don't focus on these measures during staff meetings because we're trying to continuously improve the business, not review things that are running well constantly; we need to look forward and to spend our time on what we need to improve on. But we also need to run well with what we've got and that's where the exception basis comes in. It's sort of like your oil pressure gauge in your car; you expect it to be up there, but if there is going to be a problem, you want to know about it.*

In contrast, 'process and nontraditional measures' were reviewed *systematically* during staff meetings to foster continuous improvement. They were also reviewed during quarterly off-site meetings, where the objective was to 'change the way we do business'.

The purpose of interactive control is to 'act as a catalyst for an ongoing debate about underlying data, assumptions and actions plans'.[15] In this context, quantitative

---

[14] Simons, R. 'Strategic orientation and top management attention to control systems', *Strategic Management Journal*, 1991, p.50; italics in original.
[15] Simons, R. 'Control in an Age of Empowerment, *Harvard Business Review*, March–April 1995, p.87.

data are used not as ends in themselves, but rather as a *means* to *understand* and *improve* the underlying *activities*. Diagnostic control remains necessary because it is both unnecessary and impractical to use all performance indicators on an interactive basis; impractical because interactive control is a time-consuming process, and unnecessary because if the measure(s) followed interactively are well chosen, the important points will already be raised. Managers should thus think carefully which measure(s) and system(s) they want to operate on an interactive basis.

Linking target achievement with rewards

The question of whether or not to link performance feedback with rewards has been one of the most hotly contested issues in management research over the last 20 years. Both extremes are advocated, with strongly held views on both sides.[16] Before examining these views, it is important to understand that the link between performance measures and rewards/punishments can be direct or indirect, and more or less explicit (see Figure 4.6). A *direct* link between performance measures and rewards would typically take the form of an incentive plan, which clearly associates predetermined performance levels with some reward, typically financial.

Performance on quantitative indicators (POQI) can also have an *indirect* impact on the rewards received by employees. This indirect impact occurs mainly through the boss(es)'s influence on employees' rewards (e.g. pay rise, career prospects, quality of their relationship with their boss), and the key question becomes: How strongly does POQI influence the boss(es)' evaluation of employee performance?

Figure 4.6 Potential links between extrinsic rewards and quantitative indicators of performance

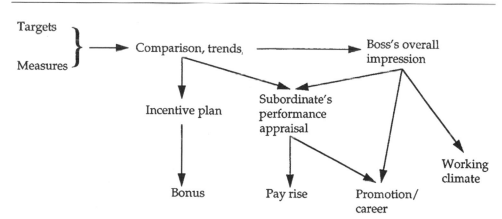

---

[16] For example, when an article by A. Kohn, entitled 'Why incentive plans cannot work', appeared in the September–October 1993 issue of *Harvard Business Review*, several experts wrote to *HBR* to comment on, mostly to attack, its author's views. One of the nine experts whose comments were published in *HBR*'s next issue (November–December 1993) associated the author's views with communism and another suggested that he was advocating outrageous views in order to develop a consulting practice!

There are two major reasons to link rewards with performance on quantitative indicators:

1. The prospect of *inducing more effort*, as individuals work harder and smarter to achieve the goals in order to receive the associated rewards.

2. The possibility of *guiding this effort* in specific directions captured by the performance indicators.

For example, it is not unusual for salespeople to have some form of variable compensation that ties pay level to sales performance. These arrangements, meant to 'motivate' salespeople to sell more, can also be tailored to push some products and services more than others.

There are, also, three major sets of arguments *against* creating a strong link between rewards and performance on quantitative indicators.

1. The first set of arguments builds on the previously discussed limitations of the performance measurement process, particularly with respect to completeness and controllability of the performance indicators. Because performance is too multidimensional and complex to be captured by quantitative performance indicators, the performance picture painted by the indicators is likely to fail to capture some performance dimensions (i.e. to be incomplete), and/or to be affected by exogenous events that cannot be influenced by the actors being measured (i.e. to be partly uncontrollable). As a result, the firm may not observe the positive motivational impact they are looking for.

When measures are severely uncontrollable, linking them with rewards may have limited impact on subordinate effort; the impact of exogenous factors on performance measures weakens the link between *effort* and rewards, which decreases the motivational power of the rewards associated with measured performance and may lead to perceptions of unfairness on the part of the members of the subunit.

Measures that are controllable but incomplete may trigger the opposite effect: the firm may get exactly what it asks for (e.g. better local performance on dimensions X and Y), but as a result may not get enough of important unmeasured performance dimensions (e.g. cooperation with other subunits and preoccupation with long-term performance). In some cases, the trade-offs are simply a function of reallocating limited attention and energy, resulting in more effort invested into rewarded areas and less into unrewarded areas. In other cases, the trade-off may reflect gaming of the performance measures, where the subunit takes actions that clearly benefit the performance measures more than they do real long-term performance (i.e., the subunit takes advantage of the incompleteness of the performance measures to secure excessive rewards). Some subunits may try to resist the temptation of gaming the system and keep investing in the unmeasured dimensions, but these attempts are often hampered by two realities: 'Rowing against the current' tends to get tiring and frustrating over time, and managers who do not meet performance standards on the measured dimensions may not last very long.

2. A number of authors have also argued that heavy emphasis on quantitative indicators for performance evaluation purposes can have a severe negative impact on the extent to which these data can be used to support learning and problem-solving. This is an important issue, particularly given the growing recognition that 'the only competitive advantage the company of the future will have is its managers' ability to learn faster than their competitors'.[17]

An intense focus on performance measures could inhibit learning in two ways. First, it has been argued that learning requires experimentation. The problem is that modifications to existing processes tend to generate short-term performance dips. A strict emphasis on short-term quantitative targets could then discourage subordinates from engaging in such experimentation, to avoid being penalized for unstable results. Secondly, some evidence suggests that external pressure—such as that created by an intense focus on meeting very challenging short-term targets—may increase employee stress to levels that have a detrimental effect on their creativity, their ability to innovate and their willingness to collaborate.[18]

3. Finally, several authors have argued that, while linking rewards and performance measures may increase employees' *extrinsic* motivation to improve the measures, it also decreases employees' *intrinsic* motivation to perform. It is as if, these authors argue, the additional extrinsic motivation came at the expense of intrinsic motivation. The danger is to start paying for something that people might have done anyway, and once we start paying them to do it, they lose their internal drive for this task and only perform it because they are paid to do so.

The evidence on the issue is mixed. On one hand, some experiments have indeed shown that subjects paid to complete puzzles stopped playing with the puzzles when they were not paid anymore, while unpaid subjects continued to play during break periods. This phenomenon suggests that while people would play puzzles (or be nice to customers) 'just for the fun of it', they may start doing it to secure the rewards or avoid punishments—rather than for the fun of it—if we make some reinforcement structure contingent on performing the task.

It should be said, however, that this effect, called the 'over-justification effect', does not occur in all circumstances. In particular, the impact on intrinsic motivation is much less clear when the task is *not* intrinsically interesting to the subject, when the subjects are college-age students rather than younger children, and when the reward offered is contingent on performance rather than on performing the task.[19]

Still, this is an important question. If extrinsic motivation can displace or reduce intrinsic motivation, we may be setting up conditions that will force the organization to start 'paying' (or punishing) for more and more specific acts or performance dimensions. Making the link between rewards and performance measures more indirect

---

[17] 'Planning as learning', by Geus, A. de, *Harvard Business Review*, March–April 1988, p.74.
[18] See, for example, Apter, M.J. *Reversal Theory: Motivation, Emotion and Personality,* Routledge, 1989.
[19] For a review, see Tang, S.-H. and Hall, V.C., 'The overjustification effect: A meta-analysis', *Applied Cognitive Psychology*, 1995, pp. 365–404.

(i.e. going through the boss's personal judgment) does not necessarily make the situation much easier, as it creates the potential for subjectivity and perceived unfairness in the attribution of rewards, and it places a significant time and emotional burden on the boss.

Concluding remarks

W.E. Deming was a famous Quality expert who helped Japan develop its manufacturing practices after World War II. In recognition of his influence, Japan named its equivalent to the American Baldridge Quality Award the Deming Award. One of the main pillars of W.E. Deming's philosophy was: 'In God we trust, everyone else bring data'. Data are not necessarily quantitative performance measures, but quantitative measures can certainly provide data.

More generally, it is very hard to improve any process without some form of feedback that allows us to understand better the cause–effect relationships involved in this process. Quantitative performance measures can also contribute to stimulating and/or guiding employee effort.

It is probably excessive to say that 'things only get done when they get measured and rewarded'; there are lots of examples of activities people engage in without much external reinforcement. Still, it is important for companies to align their performance measurement, evaluation and reward system with their strategic and behavioural objectives. In spite of the common sense of this remark, it is quite remarkable to observe how often it is disregarded in practice.

This problem is not exactly new. One of the most frequently cited management articles, entitled 'On the folly of rewarding A, while hoping for B', was written more than 20 years ago by Steve Kerr, now vice-president and chief learning officer of General Electric.[20] In the article Kerr reviewed several cases of organizations that said they wanted their employees to behave a certain way, but actually rewarded them for behaving in ways that were different than, sometimes contrary to, what they said they were hoping for (see Table 4.1 for a summary list).

Twenty years later, *Academy of Management Executive* reprinted Kerr's article and conducted an informal poll of its executive advisory panel, a group of over 50 top executives representing companies from all over the world. Of the respondents, 90% believed that Kerr's folly is still prevalent in corporate America today, and over half assessed that the folly is widespread in their companies (see Table 4.2). When asked what they believed were 'the most formidable obstacles in dealing with the folly', the panel identified three themes, all related to the way the firms measure, evaluate and reward performance: inability to break out of the old ways of thinking about reward

---

[20]Kerr's article was originally published in *Academy of Management Journal* in 1975 (**18**, pp. 769–783).

Table 4.1 Kerr's common management reward follies (1995 version)[21]

We hope for . . .	But we often reward . . .
Long-term growth; environmental responsibility	Quarterly earnings
Teamwork	Individual staff
Setting challenging 'stretch' objectives	Achieving goals; 'making the numbers'
Downsizing; rightsizing; delayering; restructuring	Adding staff; adding budget; adding Hay points
Commitment to total quality	Shipping on schedule, even with defects
Candor; surfacing bad news early	Reporting good news, whether it's true or not; agreeing with the boss, whether or not he or she is right

Table 4.2 Contemporary management reward follies, as seen by the *Academy of Management Executive*[22]

We hope for . . .	But we often reward . . .
Teamwork and collaboration	The best team members
Innovative thinking and risk-taking	Proven methods and not making mistakes
Development of people skills	Technical achievements and accomplishments
Employee involvement and empowerment	Tight control over operations and resources
High achievement	Another year's effort

and recognition practices, lack of overall system view of performance factors and results (too much emphasis on subunit performance), and continuing focus on short-term results by management and shareholders.

Shall we get a more favourable assessment, 20 years from now?

---

[21]Kerr, S., 'An academy classic: On the folly of rewarding A, while hoping for B', *Academy of Management Executive,* February 1995.
[22]*Academy of Management Executive* is the practitioner journal of the Academy of Management. See 'More on the folly', *Academy of Management Executive*, February 1995, pp. 15–16.

Case 4.1

Case 4.1

*This case was prepared by Professors Soumitra Dutta and James Téboul in the Technology Management Area at INSEAD. It is intended to be used as a basis for class discussion rather than to illustrate either effective or ineffective handling of an administrative situation.**

Friends Provident: Reengineering Customer Services (condensed)

INTRODUCTION

In 1993, Friends Provident (FP), a major UK insurance company, was in the midst of a complete reorganization of its administrative structures to support its strategy of providing the highest level of customer service. In 1989, the administrative activities within FP were organized by product and function, and split between the branches and Head Office. Such an organization was causing several problems in the provision of a high level of customer service. Starting in 1990, FP embarked on a quick-paced reorganization which redefined the relationship between branch and Head Office activities and focused on the gradual reorganization of administrative processes to support the company's customer focus. Functional demarcations within Head Office were blurred with the creation of service centres to handle all needs of a particular set of customers. Emphasis was increasingly made on cross-functional teams and multiskilled individuals. Friends Provident was also actively investigating whether further improvements in important processes were possible with the use of work-flow and image technologies.

While most of the changes associated with the reorganization had been fairly successful, several questions on the scope and depth of the change effort remained unanswered. Had the reorganization gone far enough? Was the reorganization focused too much on Head Office? Were the current structures and processes best suited for ensuring the highest levels of customer service? Should FP consider a more radical redesign of its customer facing processes?

COMPANY AND INDUSTRY BACKGROUND

The UK insurance market is large, sophisticated and profitable. It ranks only behind the US, Japan and Germany in terms of total gross premiums. The market is very fragmented with nearly 200 authorized insurance companies (the largest company, Prudential, has under 10% of the market share). Mutuals[1] control about 50% of the UK market. The market is witnessing a gradual

*Copyright © 1995 INSEAD, Fontainebleau, France.
Financial support from the INSEAD Alumni Fund European Case Writing Programme is gratefully acknowledged.
[1] A mutual is wholly owned by its policy owners (similar to a cooperative).

consolidation with new entrants such as large non-UK insurance companies, banks, building societies and even retailers (such as Marks and Spencer) from outside the financial services. Exhibit CS4.1.1 provides some general information about the UK insurance market.

Friends Provident is one of the United Kingdom's leading mutual life insurance and investment companies with a little less than 4% of the market share. It has major business activities in the areas of life insurance, personal pensions, company pensions and health insurance. In the United Kingdom, FP has almost 2 million individual policyholders and also writes substantial volumes of group business. In 1992 total assets under management at FP exceeded £8 billion and group premium income was £1.2 billion.

Established by the Society of Friends in 1832, FP has a network of branches in major towns and cities throughout the United Kingdom. Friends Provident was one of the first insurance companies in the UK to adopt a multidistribution strategy comprising independent financial advisers, appointed representatives,[2] direct (field) sales and direct (telephone and mail) marketing. Friends Provident expects its new business for individual policies in 1994 to be distributed among these four distribution channels in the following proportion: independent financial advisers (50%), appointed representatives (14%), direct sales (25%) and direct marketing (11%).

Exhibit CS4.1.2 gives the position of FP relative to other major insurance companies along different product lines. It also provides a summary of the image of FP relative to its competitors as perceived by independent financial advisers. A senior FP manager commented on the perception of FP among the brokers:

> 'We are among the top four or five choices of the brokers, but quite often not their first or second choice.'

CUSTOMER SERVICES ADMINISTRATION

The Customer Services Division (CSD), also known as the Head Office, provides the back-office administrative support function for FP. Headed by Roger Hallett (since 1989), the CSD is organized in to two areas: Individual Business and Group Business distributed over three different locations, Dorking, Salisbury and Manchester. Exhibit CS4.1.3 describes the overall organization of FP and provides details about the structure of the CSD (in 1993).

There are three broad categories of 'customers' served by CSD: FP branches, independent financial advisers and appointed representatives, and individual policyholders and groups (i.e. companies). FP branches are distributed all over the UK and report to the Sales Division of FP. Individual and group customers contact FP branches directly or more commonly through independent financial advisers to purchase FP products. Both individual and group customers usually contact CSD directly for servicing and claims processing on existing policies. To maintain a high level of customer service, FP encourages independent financial advisers to allow their customers to contact CSD directly for servicing and claims. Exhibit CS4.1.4 lists the most important aspects of 'service' to the independent financial advisers.

---

[2] While independent financial advisers (or brokers) are free to sell the products of any insurance company, appointed representatives are agents who are independent, but have a special agreement to sell only FP products.

Case 4.1

Each FP branch has an administration manager and a sales manager reporting to the branch manager. The sales support staff under the administration manager is responsible for functions such as new business quotes, the processing of new business proposals and selective 'one-touch'[3] servicing of policies. The new business inspectors under the branch sales manager focus on obtaining new sales by visiting brokers and agents. The Head Office provides help on complex underwriting cases, completes the processing of new business proposals and services policies. Exhibit CS4.1.5 illustrates how a proposal for a new policy is currently processed at both a branch and the Head Office.

STIMULI FOR CHANGE

Until the end of 1989, the functions performed by CSD were distributed over two separate divisions: Marketing and Administration as illustrated in Figure A of Exhibit CS4.1.6. Pensions reported to the marketing general manager for primarily historical reasons. A reorganization in January 1990 aggregated all departments responsible for servicing customers (shown shaded in Figure A of Exhibit CS4.1.6) into a new division called the Customer Services Division. The name 'Customer Services Division' was deliberately chosen to emphasize a customer orientation in the activities of the concerned departments. Figure B in Exhibit CS4.1.6 illustrates the partial structure of the CSD after the reorganization of 1990.

After the reorganization, CSD retained its organization around product and functional lines such as new business, claims, premium administration and servicing (see Figure B of Exhibit CS4.1.6). Such a structure often caused problems and inefficiencies within CSD. A senior CSD manager described some of these problems:

> *The functional demarcation lines within CSD were quite dysfunctional. No one had a clear responsibility for the customer and this led to a degradation in the level of customer service we could provide. For example, assume that a customer sent us a claims letter with a request for premium adjustments. This letter would first be processed by the claims group, and then passed to the premium administration group where it would wait to be processed. The delays added up within CSD—leading to a degradation of the overall level of customer service we could provide.*

Several customers were also visibly dissatisfied with the level of service they were receiving from FP. For example, Endsleigh[4] had communicated the need for better customer service to FP management. Roger Hallett summarized the service offered by FP in 1989: 'Our speed was variable, quality was indifferent and cost was worrying!'

In 1987, when Abbey National joined FP on a five-year contract, Roger Hallett was given a free hand in quickly setting up a servicing facility for Abbey National. He decided to pilot a new customer service model for servicing Abbey National in Manchester (a new site about 200 miles away from Dorking and Salisbury). That was the origin of the concept of service centres within CSD.

---

[3] New business policies which can be underwritten and processed to completion at the branches without the necessity of referral to the customer services division (head office).
[4] Endsleigh is an independent sales company associated with Friends Provident as an appointed representative.

The Manchester site was operational in 1988, and by mid-1990 it was clear that the Manchester site was incurring lower expenses and generating fewer complaints from its customer (Abbey National). In December 1990, a four-member group consisting of managers from CSD and Sales (branch) Administration was formed to determine ways to improve the level of customer service provided to Endsleigh. The group looked at the Manchester site in detail and recommended the creation of a new service centre at Manchester to service all Endsleigh needs (which were previously serviced from the departments at Salisbury).

The Endsleigh Service Centre was operational in Manchester by 1st April 1991. The notion of a service center was relatively new for CSD. A service center implied a reorganization across functional lines to focus the responsibility for customers at one point and allow a single centre to satisfy all customer needs.

ORGANIZING FOR CHANGE

In May 1991, a branch/head Office project was started with the aim to 'improve administration of office insurance products at branch and Head Office locations'. The project team consisted of four managers—two from the Head Office and two administration managers from branches—and was given the charter to focus on improving the quality of service and determine ways to deliver that service cost-effectively. The team visited several branches, Head Office departments, brokers, appointed representatives and Endsleigh insurance offices and conducted interviews over a period of five weeks. Their study lead to a better understanding of the needs of branches and agreed service levels between the Head Office and branches. The team arrived at the following conclusions:

- Branches should concentrate on the acquisition of new business
- CSD should be reorganized into service centers
- Policy servicing work should be transferred from branches to the Head Office
- Staff should be transferred from branches to Head Office to increase their customer awareness
- The number and roles of branches should be reviewed.

In response to the conclusions of the branch/head Office project an external consultant was brought in on 20 June 1991 to organize a one-day 'Change Day' for sixteen CSD managers. At the end of the Change Day, the sixteen managers were asked about what should be done next. All of them said there was a need to spread awareness about the need for change, and 15 of them recommended that a dedicated task force should be formed to devise a new coordinated customer service approach. Interestingly, no one said 'do nothing'. Jane Stevens commented on the results of the Change Day vote:

'If you had asked the managers two days earlier about change, most of them would not have even thought about change, let alone agree that change was necessary.'

In response to the recommendations of the Change Day, a group of seven managers from Head Office was set up on 1 July 1991 and given the task of coming up with the practical details for reorganizing CSD. Any proposed solutions had to address the following needs:

- To clearly identify customers and thus gain a greater understanding of their requirements

- To provide flexibility for changing needs in the future

- To gain the commitment of the CSD management and staff

- To maintain technical knowledge/skills

- To ensure service provision did not deteriorate during any change.

Roger Hallett was not a member of the group, but interacted periodically with them.

PLAN FOR CHANGE

The group of seven managers presented their view of the change process to Roger Hallett on 11 July 1991. The recommendations of the group outlined how CSD should be reorganized (see Exhibit CS4.1.7) into nine service centers, each focused on a known customer, i.e. a group of geographically co-located FP branches or a particular distribution agent (such as Endsleigh). Each service centre would contain the expertise necessary to provide a complete service for all products for its assigned customers. Each team within a service centre would remain focused on one particular function, but would have to work alongside other functional teams. Staff were not expected to become 'multifunctional' overnight, but it was envisaged that the proximity to other functional teams would make them more aware of the customer service process as a whole.

To ensure the maintenance of high technical standards, the group also recommended the formation of a few specialist units to support the service centres. The distinction between the service center and specialist units would be invisible to the customer and the service centre would always be responsible for all customer needs. Agreed service levels would be measured from the date of receipt of post to date of dispatch to the customer inclusive of all referrals to specialist units.

Of particular importance was a new specialist group, under the heading Operating Standards, which would be set up at each site and led by a small number of senior people. This group was to be responsible for systems, processes and training and for ensuring consistency of approach and standards over all sites and for building on best practice.

Another important initiative in the reorganization was the creation of a Customer Information Centre at each site staffed by experts in each function to provide an interface between Head Office and its customers and give a 'one-stop' answering service for simple queries. The Customer Information Centre would forward a customer call to the relevant service center only if it needed specialized assistance.[5] Jane Stevens explained the importance of the Customer Information Centre:

> *Previously when a policyholder called Friends Provident, the receptionist decided the functional division to which the call would be forwarded. Receptionists frequently made errors and as a result, customers were passed around from one function to another.*

---

[5] In 1993, the Customer Information Centre had a staff of 16 employees. It received an average of 5000 calls each week. Less than 10% of incoming calls needed to be forwarded to a service centre.

Answering customer queries quickly and efficiently is important as 80% of our policyholders contact us directly for service.

The group also considered some key implementation issues related to work transfers (across locations), personnel requirements, training, technology changes and communication. Technology changes were kept to the minimum (such as entering new function codes, changing addresses and modifying sign-on authorities) and no new systems were planned. A detailed implementation plan was drawn up and the group concluded that the earliest possible date for implementation was 1 October 1991.

IMPLEMENTING CHANGE

On 19 July 1991, a 'Customer Day' was organized for all managers (about 50 in number) from within CSD. To develop awareness about customer needs, the managers first went through exercises pretending to be customers and trying to ascertain problems faced by customers while dealing with CSD. Next, the group of 7 managers presented their view of change and asked the managers to brainstorm on issues related to the proposed changes.

Having obtained the backing of CSD managers, a presentation of the proposed changes was made to the general management of FP on 2 August. With the support of FP general management behind them, Roger Hallett and the group started matching new job descriptions with manager competencies. It was clearly understood from the start of the process that there would be no redundancies.

Two days later, the management team communicated the changes to all supervisors (the next level down from managers). Specially prepared documents were given to them to aid the presentation and emphasize the fact that the customer was the focus of the changes. The fact that there would be no redundancies in the new organization was also stressed. The supervisors were assigned the task of distributing their staff within the service centres and ensuring that the skills were correctly distributed within each centre. Whenever possible, entire (functional) teams were relocated together.

A day later, all lower level staff were informed about their new positions. Jane Stevens commented on the reactions of the staff:

The staff knew that change was coming. They were on the front line and were living through the kind of problems which were now being voiced by the managers. There was little resistance to the new assignments. We also tried to be as flexible as possible in accommodating requests for changes.

The staff were not expected to become multifunctional from day one, but it was clear that there would be increased amounts of cross-training in the future with the aim of breaking boundaries between teams within a service centre. However, some training was necessary prior to implementation as there were only a few experts in some areas, and it was not possible to split these people evenly among the service centres. All necessary preimplementation changes were completed within the next five weeks and the service centres were up and running as scheduled on 1 October 1991.

Case 4.1

CHANGE IN WORK PROCESSES

Prior to the creation of service centres, product and functional divisions were organized into teams. Team members worked individually on cases, but often interacted with others while handling a complex assignment. Team leaders were responsible for allocating work among team members according to the complexity of tasks and skills of team members. Customer queries requiring processing by multiple functional departments flowed from one functional team to another with arbitrary hands-off delays across functions. As no one function was responsible for servicing a customer, the level of service provided was often inadequate. There was poor continuity in customer relationship as service requests were allocated to different teams on an *ad hoc* basis.

The creation of service centres changed the work flow dramatically. Each service centre contained different functional teams and handled all aspects of service related to a particular set of customers. The overall work-flow changed from an *ad hoc,* hands-off across teams to a focused flow across colocated teams. Responsibility for servicing customers was localized in the managers of the concerned service centres. There was increased continuity in customer relationship as the same service centre satisfied all requests from a particular customer. These differences in the overall work-flows are represented in Exhibit CS4.1.8.

CONSEQUENCES OF THE REORGANIZATION

After the reorganization, an overall customer service plan was formed and each individual service centre was asked to make its own plan in accordance with the overall plan. Due to the mutually agreed service level objectives, each service centre had incentives to follow up on different aspects of service to a customer. A manager commented on the service centre plans:

> *It all seems simple, but it was the first time we had ever done it. Why had we not done it before? I think that it was the fact that now people had a clear customer they were trying to achieve for.*

Staff within service centres also seemed to be more motivated by knowing 'who they were working for' and by 'seeing the whole picture'. Pay for performance was introduced and salaries could vary by as much as 60% of the base salary. A number of nonmonetary rewards and incentive plans were also introduced to reward excellent ideas and/or performance.

The service centres changed the relation between CSD and branches as each service centre was responsible for a fixed set of branches. This increased the continuity of the relationship between them and resulted in increased satisfaction on both sides. A branch sales manager noted the following:

> *If I went a long way to visit a broker and spent the first 10 minutes listening to him complain about our service, then there is little use in me discussing our products any further with him. The provision of a higher quality customer service from CSD has helped to alleviate such concerns.*

> *Also, the branches had for a certain period lost their focus on sales. These changes (within CSD) have helped to increase awareness about the importance of sales to the company's future.*

Case 4.1

Cost savings were also realized from the reorganization. Administrative work (such as group business processing) was moved from the branches to CSD with no increase in staff. Further cost savings were obtained with a reduction in branch administrative staff by 12% to 400.

QUALITY INITIATIVES

In December 1992, two service centres (in Dorking and Salisbury) started on a special quality programme called 'Pursuit of Excellence' (POE). Specific objectives were set for the programme:

- To reduce start to finish processing time by 25% for six key functions
- To achieve 5% productivity gain
- To maintain 100% agreed service levels
- To reduce 'rework'[6] by 50%
- To create an environment of continuous improvement

The initiative started with a brainstorming session with all service centre staff on 125 different issues. Next, teams were set up to produce solutions. One team, the 'Quick Fix' team looked at minor problems which could be fixed easily for quick gains. The other team, the 'Top 10' team focused on the top 10 issues identified in the brainstorming session. Process groups were created to redesign three key processes: maturities, new business and personal pension quotations. A project group was also set up to coordinate all POE activities and to communicate all plans, actions and achievements.

Jane Stevens commented on the POE initiatives:

The most important benefit of the quality initiatives is the change it has produced in the attitude of staff. Now they always question: 'Is this the best way to do it'? Before, they would not do so. The bigger challenge for us is to decide how to spread best practices to other centres.

FUTURE CHALLENGES

When Roger Hallett took charge of the CSD in late 1989, he set himself two short-term (two-year) and two long-term (five-year) goals. The two short-term objectives were: (a) standardizing business results and (b) restructuring CSD. The two five-year objectives were (a) winning a major UK Service Award and (b) reducing servicing costs to 1985 levels in absolute terms (without accounting for inflation).

In late 1993, Roger Hallett could take some pride in having achieved his initial objectives to a fair degree. Business results for most aspects of CSD had been standardized and important aspects of the restructuring of CSD had been completed successfully. More important was the fact that FP won a major Service Award for the first time from the National Federation of Independent Financial Advisers in October 1993 who rated FP as number 1 in overall service. Unit servicing

---

[6] Rework is defined as work returned by customer due to error or incompleteness.

Case 4.1

costs were also headed downwards. Taking the unit servicing cost in 1990 as 100, unit servicing costs in 1992 had fallen to 77 (in absolute terms). The goal was to achieve the unit servicing cost of 1985 (equal to 66). Exhibit CS4.1.9 summarizes the changes in the perception of FP by independent financial advisers during the period September 1991 through March 1993.

FP had gone through major changes in its administrative processes over the last couple of years. The benefits of the reorganization were becoming evident. Customer satisfaction had increased significantly and costs were down. A clustering of branches had reduced the number of branches from 40 in 1991 to 28 in late 1993. This had enabled the administrative staff in branches to be reduced further by 26% to about 320.[7] Roger Hallett commented on the changes:

> *The top few insurance companies now have very similar products with comparable prices. Systems and investment returns are not that different either. So the quality of service provided becomes a key differentiator from a competitive perspective.*

However, some important concerns remained related to the management of relationships with customers. While the degree of satisfaction among the brokers and branches had increased, could CSD do more to understand its end customers (individual policyholders)? Most individual policyholders contacted CSD directly for service. Could FP exploit these interactions with the end customer further? Were the current systems and procedures adequately focused on the individual customer?

---

[7] The headcount reduction was occurring largely through 'managed attrition'. With only a few exceptions, there were no forced layoffs.

Exhibits

Exhibit CS4.1.1   Major UK Insurance Companies (1993)

Company	Assets £ million	Liabilities £ million
AXA Equity & Law	5 847	4 598
Clerical Medical and General	5 329	4 757
Cooperative Insurance Society	7 871	5 829
Eagle Star	6 211	5 977
Equitable Life	9 565	8 721
Friends Provident	**8 747**	**7 972**
Legal & General	14 821	12 401
London Life	12 312	11 112
National Provident Institution	5 439	4 909
Norwich Union	20 822	18 104
Pearl Assurance	9 184	6,536
Prudential	29 233	24 708
Royal Life	5 243	4 487
Scottish Amicable	7 141	6 377
Scottish Widows	9 884	7 903
Standard Life	24 461	20 316
Sun Alliance	5 458	4 773
Sun Life	6 857	6 093

Source: Money Management, pp. 8–10, November 1993.

Value index ratings (performance of low-cost endowment policies)

Company	Value index
Equitable Life	253
Friends Provident	**235**
Eagle Star	231
Nat West	*231*
Standard Life	229
Gen Accident Life	225
TSB	*222*
Clerical Medical	220
Midland	*215*
Lloyds (Black Horse)	211
Barclays Life	*210*

Note: (1) Banks are italicized in above table. This highlights the threat of new entrants into the traditional insurance market. (2) The value index is obtained by dividing the Maturity value by the total outlay.
Source: Savings market as quoted in the *Sunday Times*, 3 April 1994.

Case 4.1

Exhibit CS4.1.2   Business placed with independent financial advisers (1993)
(*continues*)

The following table gives the companies with whom the surveyed independent financial advisers (IFAs) have placed the *most* business during the past 6 months. The data comes from a face-to-face interviews of more than 500 IFAs in the period 8 March–2 April, 1993. The survey was conducted by an industry market research association consisting of Friends Provident and eighteen other major UK insurance companies. The names of other companies participating in the survey have been disguised due to the confidentiality of the data.

Product	Company	IFAs using company most (%)	Position of company in overall ranking (out of 19)
Mortgages	Company 1	28	1
	Company 2	21	2
	Friends Provident	10	5
Savings and investment	Company 1	25	1
	Company 3	24	2
	Friends Provicent	12	7
Personal protection	Company 4	30	1
	Company 5	26	2
	Friends Provident	8	7
Corporate protection	Company 4	17	1
	Company 6	11	2
	Friends Provident	4	7
Individual pensions	Company 1	29	1
	Company 7	26	2
	Friends Provident	18	8
Group pensions	Company 7	15	1
	Company 1	13	2
	Friends Provident	4	12

Source: Internal Friends Provident documents.

Exhibit CS4.1.2 *(continued)* **Image Ratings of Friends Provident (1993)**

The following table gives the percentage of independent financial advisers (IFAs) stating that Friends Provident was excellent/good along different dimensions of a company's image. As before, the data comes from a face-to-face interviews of more than 500 IFAs in the period 8 March–2 April, 1993. The survey was conducted by an industry market research association consisting of Friends Provident and eighteen other major UK insurance companies.

Image dimension	IFA's rating FP as excellent/good along dimension (%)	Rank of FP in ranking along dimension (out of 19 companies)	IFA's rating company ranked No. 1 as excellent/ good (%)
Financial strength	73	5	87
Past performance	79	3	90
Price of products	68	5	68
Technical literature	56	9	72
Overall service	72	1	72
Past performance (with profit)	82	2	91
Past performance (unit linked)	59	9	82
Quotation service	80	1	80
Product range	66	3	71
Consultant support	67	2	71
Client literature	61	6	69

Source: Internal Friends Provident documents.

Exhibit CS4.1.3 Structure of Friends Provident and Customer Services Division (1993)

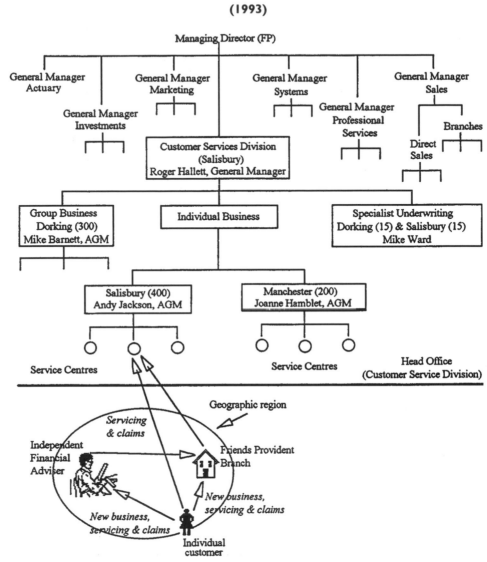

Note: Each service centre serves a fixed set of customers from a certain geographical region. Customers included FP branches, independent financial advisers and individual customers.
Source: Internal Friends Provident documents.

Exhibit CS4.1.4   Important aspects of service

Case 4.1

The following table gives the most important aspects of 'service' as perceived by IFAs. Similar to the tables of Exhibit 4.1.2, the data comes from face-to-face interviews of more than 500 IFAs in the period 8 March–2 April, 1993. The survey was conducted by an industry market research association consisting of Friends Provident and eighteen other major UK insurance companies.

Aspect of service	IFAs saying 'very important' (%)	Relative rank of Friends Provident
Fast rectification of errors	79	1
Efficient premium collection	71	1
Well-trained branch staff	63	1
Inform of failure to collect direct debit mandate (DDM)	67	2
Fast, flexible underwriting	48	3
Inform of delays	61	1
Staff on top of administration	60	2
Speed of response	65	1

Source: Internal Friends Provident documents.

Exhibit CS4.1.5   Processing of a proposal for a new insurance policy (1993)

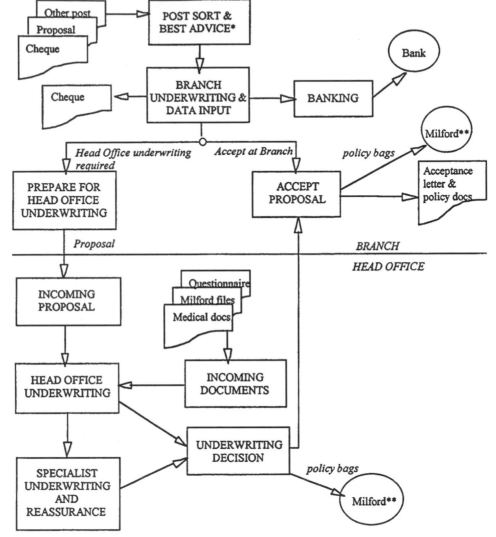

Note: * Best advice refers to checking compliance with the UK Financial Services Act requiring all financial service institutions to offer the right product for the customers' needs. ** Originals of all documents are stored in Milford.
Source: Internal Friends Provident documents.

Exhibit CS4.1.6   Structure of Friends Provident and the Administration Division (1989)

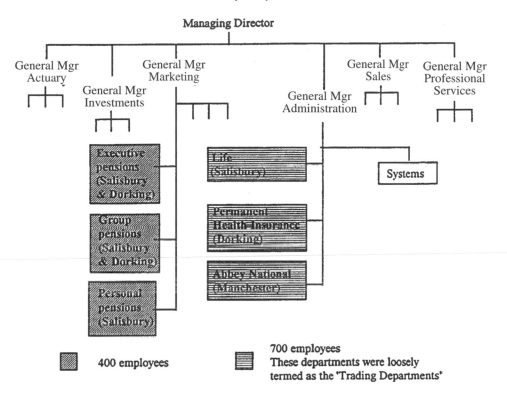

Note: (1) The shaded departments were consolidated into the Customer Services Division. The 'Systems' department under the Administration Division was moved to the Information Systems Division during the reorganization of January 1990. (2) The total number of employees in the shaded departments is as mentioned above.

Source: Internal Friends Provident documents.

Case 4.1

Exhibit CS4.1.6 *(Continued)* **Traditional (partial) structure of Customer Services Division (1990)**

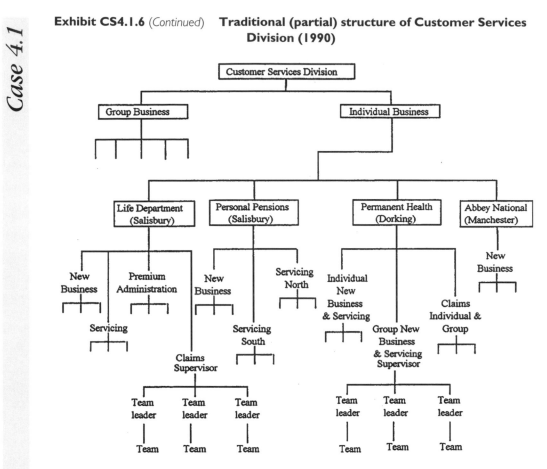

Note: Within each function, there were several teams. Members of the team worked mainly individually on projects, but helped each other out on specific tasks when required. The team leader was responsible for allocating work among the team members.

Source: Internal Friends Provident documents.

Exhibit CS4.1.7   New (partial) structure of Customer Services Division (1991)

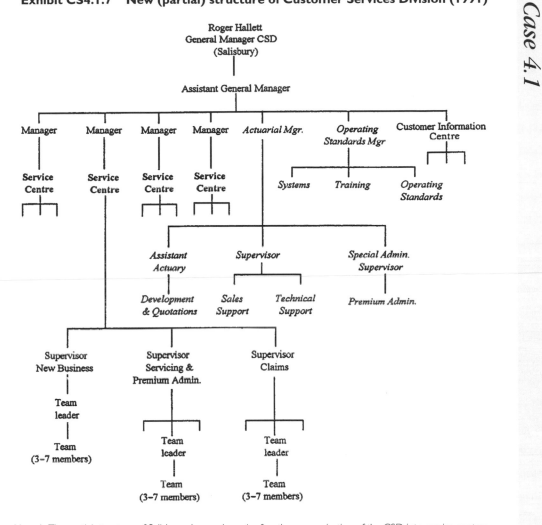

Note: 1. The partial structure of Salisbury shown above is after the reorganization of the CSD into service centres in August 1991. 2. Service centres are shown in bold and specialist units are in italics.
Source: Internal Friends Provident documents.

Case 4.1

Exhibit CS4.1.8   Work-flows before and after creation of service centres

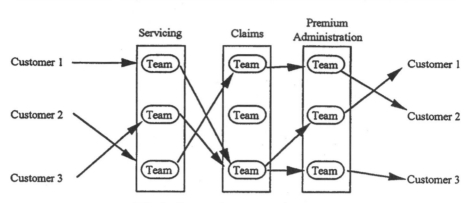

Work-flow prior to service centres

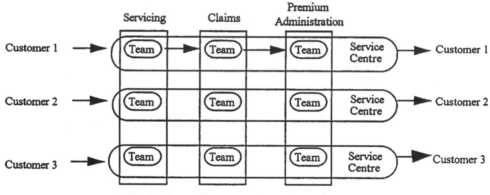

Work-flow after creation of service centres

Note: Before the creation of service centres, work was handed off across functional departments in an *ad hoc* manner. No one was responsible for the customer. After the creation of service centres, work flowed through different teams within a service centre serving the same customer.
Source: Internal Friends Provident documents.

Case 4.1

Exhibit CS4.1.9   The perception of FP along key dimensions

The following graph depicts the changes in the perception of Friends Provident along key dimensions during the period from September 1991 through March 1993. The base comprises all independent financial advisers using Friends Provident.

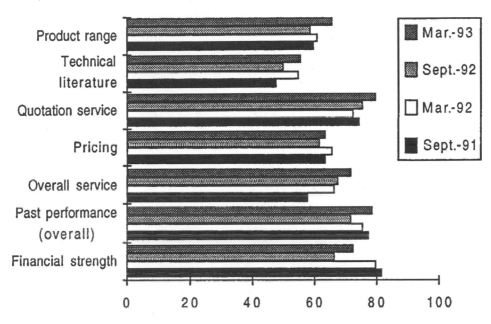

Source: Internal Friends Provident documents.

Harvard Business School

9-694-047
Rev. May 19, 1994

Case 4.2

This case was written by Research Associate Susan Rosegrant as part of a joint effort between the Kennedy School of Government and the Harvard Business School. The case was prepared as the basis for class discussion rather than to illustrate either effective or ineffective handling of an administrative situation.

A measure of delight: The pursuit of Quality at AT&T Universal Card Services (A)

In the lobby of the headquarters building of AT&T Universal Card Services Corp., a crystal Malcolm Baldrige National Quality Award rotates silently on a pedestal within a glass case. On one marble wall, above a sheet of flowing water, are the words, 'Customers are the center of our universe.' The inscription on the opposite wall reads, 'One world, one card.' Despite a steady stream of visitors and employees, the lobby has the hushed and serene atmosphere of a shrine.

But on the upper floors of the building—in the heart of the operation that brought home the nation's top quality award—it is never silent. In a honeycomb of open and brightly lit cubicles, about 300 men and women are speaking intently but pleasantly into telephone headsets while deftly keying information and instructions into the computer terminals before them. In all directions, one phrase is repeated so often that it seems to hang in the air: 'AT&T Universal Card, how may I help you?'

The workers appear private and autonomous, connected only to their customers on the other end of the line. Yet at any moment, day or night, there may be someone else listening. It may be a co-worker, monitoring the call in order to suggest how a request might be handled differently. It may be a team leader, gathering information that will figure in next year's raises. It may be a senior manager, putting on a headset to listen to calls while working out in the company gym. Or it may be a quality monitor in another building, scribbling ratings and comments on a one-page sheet that will help determine whether everyone in the company gets a bonus for that day. And all the while, as the pleasant voices talk on, a computer tracks every call that comes in, continually measuring how long it takes to answer each call, how many seconds are spent on each conversation, and whether any customers hang up before their calls are answered. In the eyes of the executives who designed the company's operating philosophy and strategic plan, the monitored calls were an indispensable component in boosting AT&T Universal Card over its competitors, and making it a true quality company.

THE CHALLENGE

In the summer of 1993, Universal Card Services (UCS) was, by most standards, in an enviable position. The wholly owned AT&T subsidiary had broken into the highly competitive credit card business in 1990, determined to build on the AT&T name with a philosophy of 'delighting' customers with unparalleled service. To help do that, the company had created an innovative measurement and compensation system to drive the pursuit of quality and customer satisfaction. Now just three years later, UCS, with nearly 12 million accounts, was the number two credit card issuer in the industry. Not only that; in 1992 UCS was the youngest company ever—and one of just three service firms—to win the coveted Malcolm Baldrige National Quality Award.

But despite these successes, there was a sense within the Jacksonville, Florida-based company that some fundamental changes were in order. In particular, Rob Davis, vice president of Quality, was searching for ways to push UCS's quest for quality one step further. A number of factors triggered this critical self-examination. Competitors had begun to close the gap opened by UCS when it pioneered its innovative policies and practices three years earlier. The departure of two key architects of company policy underscored the fact that now, with more than 2,700 employees and three new sites across the country, UCS was no longer in a startup, entrepreneurial phase. Finally, Davis and other senior managers were questioning many of the basic concepts underlying the measurement system that had helped the company achieve so much. Nearly everyone agreed that changes were needed in what the company measured, how it measured, and what it did with the information. There was no consensus, however, on exactly what to do.

THE FOUNDING

When AT&T recruited Paul Kahn in 1989 to lead its foray into the credit card business as UCS's president and CEO, the information technology giant had two main goals. It wanted to offer a combined credit card and calling card that would bolster its long-distance calling revenues. Perhaps more important, it wanted to regain the direct link to the customer that it had lost in 1984 when a court decision forced the spin-off of its regional Bell operating companies. With the backing of AT&T, Kahn, a 10-year veteran of First National Bank of Chicago and Wells Fargo Bank, developed a bold plan for breaking into the market, where unchallenged pricing practices and highly profitable operations were the norm. First, the new company did away with the annual membership fee, saving cardholders who signed up during the first year $25 or more typically charged by issuing banks. Second, UCS set its interest rate on unpaid balances below what most bank issuers were charging, pegging it to the banking industry's prime lending rate.

In addition to such pricing strategies, AT&T and Kahn shared a vision of the kind of company they wanted to create: an organization where motivated and empowered employees would set new standards for quality in customer service. To achieve this ambitious goal, the new company set out to measure almost every process in sight. 'We decided that we had to create an environment where the net takeaway to both the parent company and to the consumer was an experience superior to anything they'd had before,' explained Kahn.

On March 26, 1990, during the Academy Awards, UCS aired its first ad. The combination of the AT&T name and the waived annual fee proved more potent than anyone had imagined. In the first 24 hours, UCS's 185 employees received 270,000 requests for applications or information. The company opened its one millionth account 78 days after launch.

PILLARS OF QUALITY

In contrast to many established companies that have struggled to superimpose 'quality' on an existing corporate culture, UCS had the luxury of establishing quality as an overarching goal from the start. In fact, quality was less a goal than an obsession. The seven core company values— customer delight, continuous improvement, sense of urgency, commitment, trust and integrity, mutual respect, and team work—were emblazoned on everything from wall plaques to T-shirts and coasters. Senior management was convinced that quality processes—with the end result of superior customer service and efficiency—would give UCS a key competitive advantage in the crowded credit card marketplace. As a company brochure noted:

> *Each time a customer contacts UCS, it's a moment of truth that can either strengthen our relationship with them or destroy it. Each call or letter is an opportunity to create a person-to-person contact that makes the Universal Card, and AT&T, something more than another anonymous piece of plastic lost in a billfold.*

In order to provide such unprecedented customer service, the Business Team, an executive committee of a dozen top vice presidents headed by Kahn, took a number of steps (see organization chart, Exhibit CS4.2.1). They made sure that the telephone associates—Universal's designation for its customer service representatives—were carefully selected, and then trained to 'delight' customers. They set up benchmarking studies, comparing UCS both to direct competitors and to other high-performing service companies. They conducted a Baldrige-based quality assessment in the very first year, as well as each successive year, and used the results as the basis for a companywide strategic improvement process.

But the most unusual mechanism built into the organizational pursuit of quality was a unique and multi-faceted measurement system, designed to measure performance on a number of levels both within the company and without. While it was not unusual for credit card issuers to monitor certain aspects of customer service, UCS's efforts went far beyond industry standards. Nor were most measurement systems designed to achieve so many purposes: to locate problem processes; to promptly address any problems discovered; to constantly assess how well customers were being served; and to reward exceptional performance. 'We had an expression here, 'If you don't measure it, you can't move it,' recalled Mary Kay Gilbert, a senior vice president who helped develop the original business plan for AT&T. 'If you're not measuring a key process, you don't even know if you have a problem.' UCS was determined not to let that happen.

THE QUALITY ORGANIZATION

As one of the first initiatives, Rob Davis and his Quality team developed two extensive surveys. The Customer Satisfier Survey was a questionnaire to gather market research data on what the company termed 'customer satisfiers,' the products, services, and treatment—including price and customer service—that cardholders cared about most. An outside market research firm conducted the survey, talking to 400 competitors' customers and 200 UCS customers each month. More unusual were the Contactor Surveys, for which an internal team each month polled more than 3,000 randomly selected customers who had contacted the company, querying them within two or three days of their contact. UCS's survey team administered 10 to 15 different Contactor Surveys, depending on whether a customer had called or written, and on the customer's particular reason for contact, such as to get account information or to challenge a bill. Survey questions such as, 'Did the associate answer the phone promptly?' and 'Was the associate courteous?' were designed to gauge overall satisfaction as well as the quality of specific services.

But the effort most visible to telephone associates and other employees, and the one that had the most profound effect on the company's day-to-day operations, was the gathering of the daily process performance measures. Senior managers had debated every aspect of this so-called 'bucket of measures' at the company's formation, and it was at the heart of how UCS operated.

The Business Team had agreed that the best way to drive quality service and continuous improvement was to measure the key process that went into satisfying the consumer—*every single day*. Building on the experience of credit card industry veterans recruited at startup, such as Fred Winkler, executive vice president for Customer Services, and adding information gleaned from the Customer Satisfier and Contactor Surveys as well as additional benchmarking studies, the Business Team assembled a list of more than 100 internal and supplier measures it felt had a critical impact on performance (see Exhibit CS4.2.2 for an example of how UCS linked internal process measures with key satisfiers).

The original list was top-heavy with actions directly affecting cardholders—such as how soon customers received their credit cards after applying, and whether billing statements were accurate. But the list gradually expanded to include key production, service, and support processes from every functional area of the company, many of which were invisible to customers but which ultimately impacted them (see Exhibit CS4.2.3 for a list of such processes). By the middle of 1991, vice president Jean Collins and her Relationship Excellence team, the independent monitoring group within UCS charged with collecting the measures, were tracking about 120 process measures, many considered confidential. Indicators ranged from the quality of the plastic used in the credit cards to how quickly Human Resource responded to job resumés and issued employee paychecks, and how often the computer system went down.

UCS did more than measure, though; it set specific standards for each measure and rewarded every employee in the entire company when those standards were met on a daily basis. To make clear the importance of quality, the bucket of measures was linked directly to the company's compensation system: If the company as a whole achieved the quality standards on 95% of the indicators on a particular day, all the associates—or non-managerial employees— 'earned quality' for the day, and each 'quality day' meant a cash bonus, paid out on a quarterly basis.[1] Although some top managers questioned the compensation/quality link, arguing that, in essence, the achievement of quality should be its own reward, Kahn felt the tie to compensation was essential. 'I think we ought to put our money where our mouth is,' he declared. 'We wanted quality, and we ought to pay for it.' The financial incentives were not insignificant: The bonus system gave associates the ability to add more than $500 to their paycheck every quarter, and managers could earn 20% above base salary.

The daily push to earn quality—and to earn a bonus—was an omnipresent goal. Video monitors scattered around the building declared the previous day's quality results. Every morning at 8:00, Fred Winkler, in charge of operations, presided over a one-hour meeting of about a dozen senior managers to discuss the latest measures, identifying possible problems and proposing solutions. A summary of the 'Fred meeting,' as one manager dubbed it, could be dialed up on the phone later that morning. In each functional area, managers convened a similar quality meeting during the day, examining the measures for which they were responsible and, if they had failed to meet a particular indicator, trying to figure out what went wrong (see Exhibit CS4.2.4 for a sample report showing telephone associate performance). Furthermore, the bucket of measures figured prominently in monthly business meetings, the Baldrige assessments, focus groups, and other regular process improvement meetings. According to Deb Holton, manager of quality, the daily measures were on everyone's minds: 'It is virtually impossible to be in this building for 10 minutes without knowing how you did the day before.'

THE EMPOWERED EMPLOYEE

At UCS, customers were referred to as 'the center of our universe.' At the center of the business, however, were the telephone associates who, although entry-level workers, had the highest pay and status among non-managerial employees. They, after all, were the front-line representatives who determined what impression customers took away from their dealings with UCS. Indeed, telephone associates were responsible for almost all customer contact—answering phones, taking applications, handling correspondence, and even collecting from overspenders and trying to intercept fraudulent card users.

To make sure that it had the right people for the job, UCS put applicants through a grueling hiring process: Only one in 10 applicants won an offer of employment after the two-part aptitude test, customer service role-playing, handling of simulated incoming and outgoing calls, credit check, and drug testing. Once hired, telephone associates received training for six weeks and two more weeks on the job. Instruction began with a two-day cultural indoctrination dubbed 'Passport to Excellence,' introducing concepts such as mission, vision, quality objectives, and empowerment. But the main purpose of the lengthy training was to give associates detailed coaching in telephone skills and the management of all phases of a customer inquiry, from initiation to conclusion.

---

[1] For managers to earn quality, they also had to meet standards on a separate set of indicators tied to vendors' products and services. Managers' bonuses were then based on three components: quality days, individual performance, and the company's financial performance.

UCS did not expect to get commitment and excellent customer service from the telephone associates, however, without giving them something in return. In fact, the company's vision of 'delighting' customers rested on having 'delighted' associates. Much of what the rest of the organization—from Human Resources management to information support systems design and the measurement system itself—revolved around ensuring that telephone associates were able and motivated to provide the quality service that was the company's stated goal.

The Information Services group, for example, developed and continually upgraded U-WIN, an information management system tailored to the specific needs of the telephone associates. Drawing in part on the company's U-KNOW system which gave managers on-line access to the customer, operational, and financial information in UCS's database, known as UNIVERSE— the U-WIN system allowed associates to pull up on their workstation screens information ranging from cardmember files, to form letters, to special product offers (see Exhibit CS4.2.5 for an overview of the information management system). U-WIN even gave associates a head start on serving customers by automatically calling up cardmembers' accounts as their calls were being connected. 'We're high touch, high tech,' explained Marian Browne, vice president of Customer Relationships, the service area in which telephone associates handled general correspondence and responded to customer calls. 'That means we work with our people and focus on our customers, but we can't do either unless we have leading-edge technology.'

UCS top management was also determined to involve associates, listen to their ideas and concerns, and draw them into most facets of the business. Associates served side by side with senior managers on teams decided issues ranging from what awards the company should bestow to how computer screens should be designed for maximum efficiency. They were encouraged to ask questions at monthly business reviews and at 'Lakeside Chats'—quarterly question-and-answer sessions with Winkler held in the company cafeteria. And the UCS employee suggestion program, 'Your Ideas... Your Universe,' was broadly publicized, with impressive results: in 1991, more than half the workforce participated, and management accepted and acted on almost half of the more than 5,000 suggestions.

In addition to these 'empowerment' oriented activities, the company looked for concrete ways to please associates. UCS provided generous fringe benefits, for example, including a free on-site fitness center for employees and their spouses, and reimbursement for undergraduates and graduate courses. The company supported a substantial reward and recognition program, sponsoring 6 companywide awards, 3 companywide recognition programs, and more than 30 departmental awards. And the Business Team encouraged managers to look for reasons to celebrate. Indeed, boisterous ceremonies in the cafeteria marking such events as all-company achievements or the bestowal of specific awards were a regular occurrence. 'The culture we've developed is very focused around rewards and celebration and success,' said Melinda Stickley, compensation/recognition manager. 'We've got more recognition programs here than any company I've heard of.'

The far-ranging programs and activities appeared to be paying off. According to annual employee opinion surveys, associates rated the company significantly higher in such categories as job satisfaction, management leadership, and communication than the norm for employees at high-performing companies. Not only that, absenteeism was low, and employee attrition was far below the average for financial services companies (see Exhibit CS4.2.6 for selected employee opinion survey results and attrition and absenteeism rates).

Despite the efforts of senior managers to create a positive environment, however, the telephone associate's job was not easy. Many stresses arose simply from working for a 24-hour customer service operation—stresses that may have been particularly trying for UCS's well educated employees.[2] Telephone associates, organized in teams of about 20, spent long days and nights—as well as periodic weekends and holidays—on the phone, performing a largely repetitive task. There was often mandatory overtime, particularly during unexpectedly successful card promotions, and associates knew their schedules only two weeks in advance.

Along with these largely unavoidable downsides, the particular culture of UCS imposed its own stresses. The pressure to achieve quality every day was an ever-present goad. Furthermore, the company's determination to continuously improve—captured in an oft-used phrase of Fred Winkler, 'pleased, but never satisfied'—frequently translated into increased performance expectations for the associates. As the telephone technology systems got better, for example, managers expected associates to take advantage of the increased efficiencies by lowering their 'talk time,' the average amount of time they spent on the phone with each customer.

Finally, there was the monitoring. About 17 process measures were gathered in Customer Relationships, the general customer service area. To begin with, the information technology system tracked the average speed of answer, the number of calls each associate handled, and how long each associate spent on the phone. As a result of their exposure to the daily printouts detailing these statistics, most associates could rattle off with deadly accuracy how many calls they handled in a day—typically about 120—as well as how many seconds they spent on an average call—in the range of 140 to 160.

Perhaps more daunting, telephone associates were directly monitored by a number of people both inside and outside of Customer Relationships. As part of the gathering of the daily measures, specially trained monitored in both the Relationship Excellence group and an internal quality group listened in on a total of 100 customer calls a day.[3] The monitors—or quality associates—rated telephone associates on accuracy, efficiency, and professionalism, recording their comments on a one-page observation sheet (see Exhibit CS4.2.7 for a description of these measures and how they were gathered).

Any 'impacts'—UCS's term for a negative effect on a customer or the business—were reported at Customer Relationship's daily quality meeting, attended by representatives from both Relationship Excellence and the internal quality group.[4] Negative reports were then passed on to the team leaders of the associates involved to discuss and keep on file for performance reviews.

Other parts of the organization monitored calls as well, each with a slightly different purpose. Team leaders listened to 10 calls a month for each of the approximately 20 associates in their groups, using the observations to review and 'develop' the associates. And *all* managers at UCS, regardless of their function, were encouraged to monitor at least two hours of calls a month to

---

[2] Because of underemployment in the Jacksonville area, and the desirability of working for AT&T, UCS had been able to recruit a highly qualified workforce. Sixty-five percent of telephone associates had college degrees.
[3] Relationship Excellence originally did the entire 100-call sample, but Customer Relationships began co-sampling when it created its internal quality department in November 1990.
[4] The ten areas in which impacts could occur had been identified as (1) telephone contact, (2) correspondence contact, (3) application contact, (4) change of address, (5) claims, (6) credit line increase, (7) payment receipt, (8) statements, (9) plastic card production accuracy/timeliness, and (10) authorization availability/accuracy.

stay in touch with services and practices. Rob Davis, vice president of Quality, for example, held a regular monthly listening session with all his staff, followed by a discussion period to analyze the quality implications of what they had heard. Finally, the results of the Customer Contactor Surveys, including verbatim remarks from cardholders about how associates treated them, were turned over to managers in Customer Relationships who could easily identify which associates handled a particular call if there was an 'impact' or other problem to resolve.

The combination of high corporate expectations and these multiple forms of monitoring and feedback created considerable pressure at UCS not only to perform well, but to do so under intense scrutiny, at least for telephone associates. Some managers felt this took a toll. 'The quality process, daily sampling, and feedback were not without pain,' claimed Mary Kay Gilbert, who as senior vice president of Cardmember Services oversaw the Customer Relationships operation. 'I had to stop people and say, Wait we're here to make sure we're delivering the right service to customers. This isn't personal.'

But others argued that the way the associates were monitored, and the way team leaders and managers delivered feedback, kept it from being a negative or stressful experience. Company policy dictated that all supervisors and managers were to treat associates with respect and to view mistakes as a learning opportunity. If an associate were overheard giving inaccurate information to a customer, for example, the team leader was not to rebuke the associate, but to explain the error and provide additional training, if necessary, so that the mistake would not occur again. 'The positive stress for workers here is high risk, high demand, high reward,' asserted Deb Holton, manager of Quality. 'It is not the stress of coming in in the morning and checking their brains at the door.'

RAISING THE BAR

Thanks in large part to the customer-pleasing work of the telephone associates, by the close of 1991 financial analysts had declared UCS a major success for AT&T. During that year, holders of the UCS card had dramatically increased AT&T calling-card usage. And after less than two years in business, UCS ranked a stunning third in the dollar volume of charges on its card, with $3.8 billion in receivables, $17.2 billion in total sales volume, and 7.6 million accounts. Industry kudos included a 'Top Banking Innovation' award from *American Banker* and 'Best Product of 1990' from *Business Week*.

Despite this stellar performance, the Business Team was convinced that it was time to shake things up—that everyone could do better. Although some executives initially balked to the prospect of a change, after a series of debates the Business Team agreed to 'raise the bar' on the number of indicators the company had to achieve to earn quality. A compelling argument for the increase was the fact that associates were meeting or exceeding standards so consistently. During 1991, associates had made quality at least 25 days out of every month, and in August they had earned quality every day, often achieving 97 percent or more of the indicators. Managers, too, were doing well. 'We wanted to take it up,' explained Davis, 'because of our strong commitment to continuous improvement.'[5] Added Marian Browne, vice president of

---

[5] In fact, the threshold for managers had already changed: Since January 1991, managers had to achieve 96 percent of their quality indicators for full compensation, receiving only three-quarters of the bonus for 95 percent.

Customer Relationships, 'Everything was going fine, but if you look at perfect service every day, we weren't giving perfect service every day.'

With the Business Team's blessing, Kahn sent the following letter to all employees on December 26, just five days before the change was to take effect:

> *Dear UCS Colleague,*
>
> *In the spirit of continuous improvement, UCS will take another step in our never-ending commitment to customer delight. Beginning Jan, 1, 1992, the quality objective for associates will move from 95 percent to 96 percent. The quality objective for managers will move to 96 percent for the target goal and 97 percent to 100 percent for the maximum goal. UCS's Excellence Award program will continue to reward quality as it has in the past—the only difference will be that the objective will be moved up for both managers and associates.*
>
> *UCS people have demonstrated our value of customer delight since 'day one.' As we continue to improve our ability to delight customers, we'll also continue to evaluate and revise our quality standards and measurements. I'm extremely proud of the work each of your performs. Your dedication to our seven values continues to make UCS a leader in the industry.*

What the letter didn't mention was that the raising of the bar was actually a double challenge: Not only did employees have to achieve a higher percentage of measures, but individual standards had been raised on 47 of the indicators, making each of them harder to earn. In addition, Collins and her Relationship Excellence team took advantage of the start of the calendar year and the relative lull after the holiday season to retire and replace a substantial chunk of the measures. While only 15 indicators had been dropped in all of 1991 and 26 added, the monitoring group abruptly cut out 48 indicators, many of them among the most consistently achieved, and replaced them with 46 new ones. In effect, this meant that close to half of the measures by which associates judged their daily performance—and were judged—were now different.

The reaction to the change was immediate. Associates earned only 13 quality days in January and 16 in February, and managers fared even worse. Not only was the company failing to make the new goal of 96%, it was missing quality by as much as six percentage points on a given day, well below the worst daily performance of the previous month. 'We fell flat on our faces as far as the number of days we were paying out as a business,' Davis recalled. Added Collins, 'For most of the days we were well below even the old standard.'

The abrupt dropoff took management by surprise. According to Robert Inks, who started as a telephone associate in May, 1990, associates were not so much mad as they were concerned—concerned that higher standards of efficiency might make it harder to deliver quality service, and concerned that regular bonuses might be a thing of the past. 'The associates looked at it as, well, this is my money,' explained Inks, 'I'm not going to be getting my money.' Added Pam Vosmik, vice president of Human Resources: 'There was probably some grousing in the hallways.'

It was no consolation that UCS was on the verge of logging its first profit. In fact, at a business meeting open to all employees, associates accused management of having raised the bar as a cost-cutting measure to avoid paying compensation. Nor was the timing of the slump propitious. UCS was ready to make an all-out push to win the Baldrige award, and although the site examiners would not arrive until September, it was critical that employees be motivated and on board. 'I went to the Business Team,' recalled Gilbert, and I said, 'Look, we raised all these indicators and measures and I don't think the people around this table understand the impact. But if we start beating people up as a result of this, you can kiss the Baldrige good-bye.'

Senior managers took the performance plunge to heart. In fact, according to Davis, some managers were so concerned by the apparent associate disaffection that they were ready to lower the bar to its previous level. Instead of backing down, however, the Business Team concocted an alternate scheme to reignite associate enthusiasm. In March, the same month that UCS submitted its Baldrige application, the company announced the 'Triple Quality Team Challenge.' The special incentive program allowed associates and managers to earn triple bonuses that month for each quality day they achieved beyond a base of 20 quality days. If employees earned 22 quality days, for example, they would get credit for 26. A four-foot by 16-foot calendar board mounted in the cafeteria and small boards in each functional area displayed daily progress toward the goal. In explaining the incentive program, *HOTnews*, an internal publication reserved for important communiqués, noted:

> ...*quality results in January and through February 26 shows UCS not doing as well as it did even before we raised our quality standards in 1992. Many of the current problems have nothing to do with our new standards or indicators, but are failures of basic courtesy and accuracy. 'I know we can do better,' says Kahn. 'The results concern me and I know they concern you. It's important that we work together to meet our quality goals and delight our customers. The 'Triple Quality Team Challenge' must be a team effort—we need to help each other achieve our indicators, not look around for who's not making theirs and punish them.'*

SOFTENING THE SYSTEM

The Triple Quality Challenge was a rousing success. Associates' quality days spiked back up to 25 in March, and managers earned 19 days (see Exhibit CS4.2.8 for an overview of quality days achieved over time). But the organizational upset engendered by the raising of the bar, along with fears that telephone associates—on whose dedication the company's success depended—could become disillusioned, prompted a harder look at making both measures and feedback more participatory and more palatable. In the months that followed, UCS even abandoned the 'pleased but never satisfied' expression because it gave associates a sense of inadequacy and futility.

Efforts to reach out to associates took a number of forms. Managers in Customer Relationships continued to coach team leaders, one-third of whom had been promoted from the associate level, to make sure they were comfortable and skilled at giving feedback. 'We've got a lot of young, inexperienced team leaders, and what you have to teach your team leaders is that you can't use feedback as a club,' noted Marian Browne. 'You use it as a development tool. You don't do it to beat people up, or to catch people.' Customer Relationships also began to experiment with peer monitoring, having telephone associates critique each other rather than relying solely on team leaders for development review.

Relationship Excellence, which had already been sharing the gathering of the daily measures with Customer Relationships since the end of 1990, helped other functional areas set up internal quality departments to co-sample, with the plan that they might eventually take over the measures entirely. Although some executives were concerned that this shift might hurt the integrity of the sample, Ron Shinall, a Relationship Excellence team leader, insisted it was a necessary evolution. 'There's going to be a natural aversion to someone telling you how to make your process better if that person hasn't worked with you or been in that process,' he declared.

Relationship Excellence also changed what it did with call observations. The daily Customer Relationships quality meeting, which had served largely as a chance for quality associates to report the mistakes they had caught, became, instead, a forum for discussion and learning. Telephone associates from the floor were invited to join the internal and external quality representatives, and the entire group debated whether negative impacts had occurred without even identifying those who had handled the questionable calls. 'It's helped get a lot of buy-in from the associates,' remarked Darrin Graham, who had led Customer Relationships' internal quality department. 'Back at the beginning, when you would hear that there is this group out there listening to my calls, you just naturally started to get an us/them mentality, and they're out to get us. Now that mentality is going away.'

As part of this overhaul, Relationship Excellence experimented with no longer giving associates— or their team leaders—feedback on calls monitored for the daily measures. But although the experiment had been urged by an associate focus group, the so-called Nameless/Blameless program lasted only a few weeks. 'The majority of the people wanted to know if they'd made a mistake,' Browne explained. Feedback resumed, but with two important differences: negative impacts no longer went into associates' files, and team leaders received, and handed on, both good news and bad. The internal quality group also worked harder to stress the positive. 'We used to walk up to people's desks and we'd have a piece of paper in our hand, and they'd be like, 'Oh no, here they come,' recalled Paul Ferrando, team leader of Customer Relationship's first internal quality group. 'And I'd say, "Someone on your team had an excellent call." When you bring good news, they don't grimace when you walk up to them anymore. People aren't afraid of quality, and they aren't afraid of this monitoring anymore.'

The steady evolution of the system appeared to have increased associate acceptance of the measures. There would always, of course, be some employees who balked at being measured, as the following response to the June 1992 employee survey indicated:

> A big handicap is being monitored constantly. The people are not relaxed. They are under so much stress that they will get a variance, that they don't do their job as well as they could. Monitoring should be used as a learning tool—we're all human and sometimes forget things.

But most telephone associates professed their support. 'The reason that we're measuring is to find out what we're capable of, and what we're doing right, and what we can improve on, and what we don't need to improve,' declared Cheryl Bowie, who took a large paycut from her former managerial position to become a telephone associate in 1992. 'There is no problem here with the feedback. You're not branded or anything. It's just a learning experience.'

On October 14, 1992, near the end of a challenging year of growth and change. Universal Card was awarded a Malcolm Baldrige National Quality Award. At a black-tie celebration party recognizing employees' part in the companywide effort, associates received a $250 after-tax bonus and a Tiffany pin, and a small group of associates, selected by lottery, travelled to Washington, D.C., for the actual Baldrige presentation. But the award did not lessen the sense of urgency at UCS. 'When we learned we had won the Baldrige,' recalled Quality manager Deb Holton, 'our second breath was, "But we will not be complacent."'

In truth, UCS would have had to change, whether it sought to or not. Paul Kahn announced his resignation in February 1993 over differences within the company as to whether UCS should expand into new financial products, and Fred Winkler defected for archrival First Union Corp. in April. Although David Hunt, the banking industry executive who replaced Kahn and Winkler's successor, AT&T veteran Gerald Hines, quickly won widespread acceptance, the departure of these two critical and charismatic leaders created anxiety about the company's future direction.

The competitive landscape within which UCS operated was also changing. Although by early 1993 the company had captured the number two ranking among the 6,000 issuers of credit cards, with almost 12 million accounts and 18 million cardholders, it was becoming increasingly difficult for UCS to make its product stand out. Competitors such as General Motors Corp. had introduced their own no-fee cards, and the variable interest rates pioneered by UCS had become common. 'The sad part is, our competition is catching up with us,' lamented Mark Queen, manager of Customer Listening, and overseer of the Customer Contractor Surveys. 'Where we need to continue to distinguish ourselves is in service.'

But continuous improvement—finding ways to motivate associates beyond what they had already accomplished—was not an easy task. For one thing, with the company's growth slowing, it would no longer be possible for as many associates to quickly ascend the corporate ladder to team leader and other managerial positions. Moreover, the current measurements no longer seemed to be driving the quest for improvement, and Davis and others had become convinced that it was time to retool a system that no longer fit the needs of the company. Ironically, considering how much Universal Card had already done to create meaningful and effective measures, among the Business Team's top 10 goal for 1993 was the development of a world-class measurement system.

WEIGHING THE OPTIONS

By the summer of 1993, Davis's Quality organization was assessing a range of new approaches to measuring. In particular, a specially convened Measures Review Committee under Thedas Dukes, a senior manager now responsible for the daily measures, was taking a hard look at what to change.

Customer-centered measures

A project of particular interest to Davis was the company's early experimentation with customer-centered measures (CCM). While CCM might not change what UCS was measuring advocates argued it would more concretely and powerfully express how the company was serving cardholders by stating this performance in terms of customer impacts.

Instead of reporting that 98% of cardholder bills were accurate on a given day, for example, a CCM report might state that 613 customers did *not* get a correct bill. 'We are trying to change the language away from percentages and indexes to a language of customers,' explained Davis. Added Ron Shinall, Quality Team leader, 'It's hard to tell the difference between 99.8% and 99.9%, but in some of the high-volume areas, that can mean a tremendous number of people are actually impacted. Fractions of a percent mean a lot when you're talking about 40,000 daily calls.'

UCS had been considering customer-centered measures since visiting early Baldrige winner Federal Express Corp. in the summer of 1991. Unlike UCS, with its 100-plus measures, Federal Express had selected just 12 processes it deemed critical to serving customers, and had based its reward system on that 12-component CCM index. In January 1993, Universal began a six-month test of CCM, reporting customer impacts on 13 existing process indicators that measured different aspects of accuracy and professionalism. The now 30-member Relationship Excellence group, which had changed its name to Quality Applications in December, sent out its first CCM report in March.

But the jury was still out on what impact CCM would have. Linda Plummer, a senior manager in Customer Relationships, applauded the idea of expressing error in human terms. Yet she found the initial reports, which simply listed the number of customers impacted in each category along with the effects per thousand contacts, to be meaningless. 'Someone needs to tell me at what point I have a concern,' she complained. 'Is it when 100 customers are impacted, or 2,000 customers are impacted? I don't even look at them anymore because I don't know how to interpret them.' Jean Wentzel, another senior manager in Customer Relationships, agreed: 'Until we've really communicated it effectively and tied it back to the compensation system, it's not going to have the same buy-in or impact.'

But increasing the relevance of CCM by tying it to the compensation system would not be easy. In fact, the cross-functional CCM group responsible for the pilot project had recently agreed to shelve temporarily the issue of whether to create a compensation link, concluding that the points raised were too complicated to tackle all at once. Unresolved questions included how to set standards for customer impacts; whether the compensation system should include both business-centered measures reported the old way and customer-centered measures reported the new way; and whether UCS should retire its bucket of measures and move instead to a system more similar to that at Federal Express, with compensation based on just a dozen or so service measures, rather than on a broad range of company functions. This last possibility, which would result in many people and processes no longer being measured, fundamentally challenged the company's founding philosophy of having all employees work together, be measured together, and earn quality together.

Statistical process control

Statistical process control was another tool Quality Applications was examining. There was a growing conviction within UCS that the company needed to adopt a more long-term outlook in quality measurement. This belief was further fueled by feedback, late in 1992, from a committee that had evaluated UCS for AT&T's prestigious Chairman's Quality Award, noting that 'there is no evidence of a statistical approach to data analysis, including determining out-of-control processes, identifying special and/or common causes, and the approach to prioritizing improvement opportunities.'

In fact, the gathering of measures on a daily basis, as well as UCS's commitment to a 'sense of urgency'—one of its seven values—had contributed to the focus on the short term. Only recently had Universal switched from monthly to quarterly business reviews, and the group that met every morning to discuss the daily measures, now headed by Fred Winkler's replacement, Jerry Hines, was for the first time adding a quarterly quality review. Remarked Davis, 'With our daily focus on measurements and our fix-it-today mentality, the thing that sometimes suffers is looking at the long-term trends in the data.'

Statistical process control (SPC) seemed to provide at least a partial answer to this shortcoming. The quality improvement methodology, developed at Bell Laboratories in the 1920s to chart manufacturing processes and identify events that affect product output, had been broadly defined in recent years to include such tools as cause and effect diagrams and Pareto charts, as well as control charts to statistically examine process capability and variation. But SPC had only rarely been applied in a service environment. The challenge at UCS, therefore, was to adapt the manufacturing tool to its customer service business.

Pete Ward, a process engineer within Quality Applications, was confident this could be done. He had already begun to prepare individualized reports for associates, allowing them to use SPC to chart and trend such daily productivity measures as talk time and number of calls handled. In contrast to mere daily statistics, Ward explained, the SPC charts would help telephone associates see the impact that one action—such as spending too much time on the phone with customers—had on another, as well as aid them in spotting cyclical patterns in their own performances.

But SPC, like CCM, raised questions about the existing measurement system. It was unclear, for example, whether it was valuable to apply statistical tools to something as ambiguous and subjective as deciding whether an associate had been courteous enough or had spent too much time with a customer. In addition, SPC charts, which allowed a more meaningful and long-term look at performance than the daily measures, presented ammunition for the argument that it was time for Universal Card to switch from its obsession with daily goals and rewards to a reliance on more statistically significant trends.

A link to external results

These and other questions had revived old complaints that the measures did not accurately reflect how customers actually viewed Universal, nor how the company was performing. Mark Queen, manager of Customer Listening and overseer of the Customer Contactor Surveys, acknowledged that although the internal measures were designed to measure processes important to customers, missing quality days internally didn't necessarily show up in dissatisfied customers. When internal quality results took a nosedive after the bar was raised in early 1992, for example, the Customer Contactor Surveys indicated only a slight blip in customer satisfaction—in fact, Queen says, that 'was driving everybody crazy.'

Similarly, Queen noted that although recent customer feedback indicated that cardholders viewed associates as somewhat less courteous than before, the internal quality monitors listening in on phone calls had not logged an increase in negative impacts. 'There is not a clear enough linkage,' Davis admitted. 'What people would really like would be for me to say, "OK, if you can take this internal customer measure and raise it from 96% to 99%, I guarantee it will take customer satisfaction up by X amount." But we can't say that, yet.'

Linkage aside, on occasion, the internal measures seemed to be at cross purposes with the company's financial goals. Greg Swindell, who in late 1992 became vice president of Customer Focused Quality Improvement, for example, described an unexpectedly successful marketing promotion for a new credit card product that left understaffed telephone associates unable to keep up with the rush of calls. Although the surge of new business was good for the company, the telephone associates were, in effect, doubly punished: first by having to frantically field additional calls, and second by missing their quality indicators and losing compensation. 'The question is, is that high response rate a bad thing?' Swindell asked. 'And my answer is no. We're here to bring on more customers, to become more profitable. So how do we balance this focus on these metrics and our business and strategic objectives? For me, this offers a very perplexing problem.'

A NEW LOOK AT THE MEASURES

Spurred by these and other questions, there was talk at UCS of a radical rearrangement of the bucket of measures. Although it was not clear what would take the bucket's place, more and more managers were beginning to feel that UCS's drive for continuous improvement was being held hostage by the relentless and short-term push to bring home the daily bonus. What had originally been designed as a means for identifying and improving processes and as a motivational tool, critics charged, was now holding the company back rather than driving it forward.

Greg Swindell was one who questioned the status quo: 'Perhaps it is a very good tool to help us *maintain* our performance, but I'm not sure it's the kind of tool that will help take us into the next century and really get a lot better at what we're doing.' Swindell was particularly concerned about how inflexible the system had become in the wake of associates' intense reaction to the raising of the bar. Managers rarely suggested adding new measures, even when they spotted an area in need of improvement, he remarked, because they did not want to make the goals to challenging and jeopardize the all-important bonus. Mary Kay Gilbert agreed: 'The more focus and pressure you put on your quality standards, the less people are willing to raise their hand and say, "I think this process should be measured," she declared 'Tying compensation to it just kind of throws that out the window.'

Similarly, associates had grown to resist having measures retired, not only because that usually meant the loss of an 'easy win,' but also because it required workers to realign their priorities and goals (see Exhibit CS4.2.9 for charts illustrating the decline in measurement system changes after 1992). In part to address the issue of stagnation in the system, Quality Applications, in a just-released draft on measurement methodology, urged managers to regularly review old measures and create new ones, noting particularly that 'danger lies when the primary reason for a measurement is to adapt to the [compensation program] rather than to improve the performance of the team or process... Our measures should be used to aid in our continuous improvement programs.'

To keep the measures flexible, Davis was considering a 'sunset law' on measures that required all indicators to be retired and replaced after one year. But although he had heard the compensation plan referred to as 'an entitlement,' he remained a supporter of the basic concept. 'Some people in our business believe that if we didn't have measurements tied to compensation, then people would be more willing to measure the right things,' he mused. 'My feeling, though, is that I'll take all the negatives that go with it any day in order to get the attention.' Telephone associate Robert Inks agreed: 'I don't think we would have gotten as far as we have today without it, because people can look at our monitors and say, "We didn't do too good yesterday, we're not getting that money." And then they look at the future and say, "Well, we have eight more days in the quarter. We're going to really focus on quality and make it, because if we don't get those eight days, that's $100 I lose."'

Although Davis was well aware of the measurement debates, he doubted that Universal would abandon its daily measures any time soon. In fact, he had more down-to-earth concerns: In January 1994, UCS was planning to raise the bar again, and Davis was already planning how to make the transition smoother this time around. Although he anticipated some resistance, Davis was convinced that the ongoing quest for continuous improvement was necessary. 'We'll have to hit hard on the fact that we're going to keep raising the standards, it's not going to stop,' he declared. 'And if we think it is, we're just fooling ourselves.'

But Pam Vosmik, vice president of Human Resources, voiced a separate concern. Recalling Winkler's 'pleased but never satisfied' expression, she made a plea for balance. 'You need to keep people focused,' Vosmik asserted, 'but by the same token, in the worst case scenario, you can make an organization dysfunctional if there is never a hope that you're going to be satisfied.'

Case 4.2

Exhibits

Exhibit CS4.2.1 UCS organization chart, August 1992

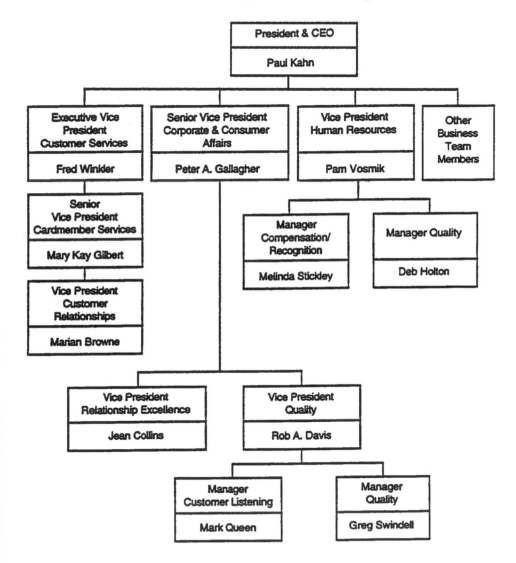

Exhibit CS4.2.2 Internal process measurement linkages to customer satisfiers

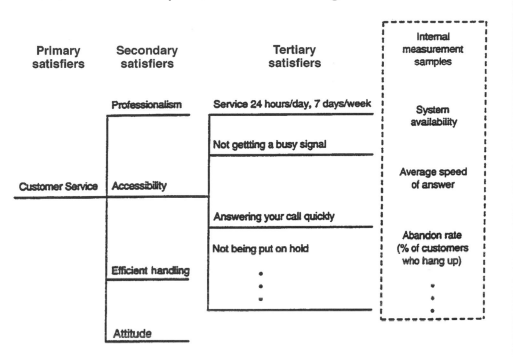

Source: Universal Card Services.

259

Case 4.2

Exhibit CS4.2.3 Key UCS and supplier processes

Key Processes	UCS or Supplier
Business Processes	
Strategic and Business Planning	UCS
Total Quality Management	UCS
Support Services Processes	
Collections	UCS
Management of Key Constituencies	UCS
Customer Acquisition Management	UCS
Financial Management	UCS
Human Resource Management	UCS
Information and Technology Management	UCS
Product and Service Production and Delivery Processes	
Application Processing	Supplier
Authorizations Management	Supplier
Billing and Statement Processing	UCS
Credit Card Production	UCS
Credit Screening	Supplier
Customer Acquisition Process Management (Prospective Customer List Development and Management)	Supplier
Customer Inquiry Management	UCS
Payment Processing	UCS
Relationship Management (Service Management, Communications Management, Programs and Promotions, Brand Management)	UCS
Transaction Processing	Supplier

Source: Universal Card Services.

Exhibit CS4.2.4 Sample Daily Reliability Report—Telephone Associate Performance

| Measure | Wednesday 06/30/93 | | | Month-to-Date | |
	Standard	Sampled	Performance	Sampled	Performance
Average Speed of Answer (ASA)	20 seconds	39,278	12.42 seconds	1,114,722	11.70 sec
Abandoned Rate	3%	39,278	1.24%	1,114,722	1.25%
Accuracy	96%	100	100%	2,400	98.58%
Professionalism	100%	100	100%	2,350	99.91%

Source: Universal Card Services.

Case 4.2

Exhibit CS4.2.5 UCS's integrated data and information systems

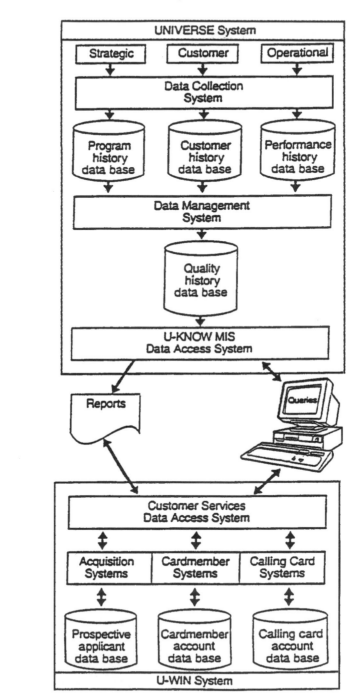

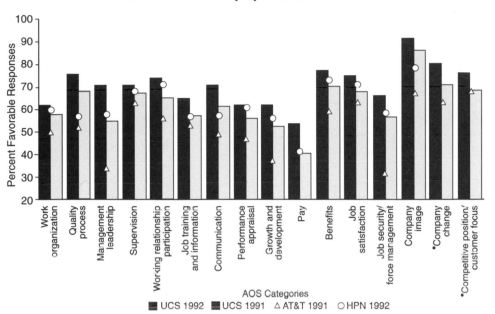

Exhibit CS4.2.6 Employee satisfaction data

AOS Categories
■ UCS 1992 ■ UCS 1991 △ AT&T 1991 ○ HPN 1992

Note: 1992 data for AT&T unavailable; AT&T conducts its AOS biannually. *HPN not available.

Note:

AOS = Annual Opinion Survey.

HPN = High-Performing Norm (average response for a group of high performing organizations that use the same survey).

Adverse Indicators	1990	1991	1992	Benchmark
Employee Turnover				
UCS Total	9.7%	10.1%	12.3%	N/A
Managers	8.7%	9.0%	7.2%	14%
Associates	10.1%	10.5%	14.1%	23%
Customer Contact Associates	10.2%	10.7%	13.5%	23%
*Absenteeism Rate**				
Managers	N/A	1.3%	1.1%	1.3%
Associates	N/A	2.2%	3.3%	1.9%

*includes pregnancy and disability

Source: Universal Card Services.

Case 4.2

Exhibit CS4.2.7 Telephone associates measurement regime

Measure	Description	Sampling and Scoring Regime	Performance Standard (1Q93)
Average Speed of Answer (ASA)	Average time between completion of customer connection and answer by telephone associate	100% sample by automated call management system (CMS)	20 seconds
Abandon Rate	Percentage of calls initiated by customers, but abandoned prior to being answered by telephone associate	100% sample by automated call management system (CMS)	3% of incoming calls
Accuracy	A qualitative measure of the level of accuracy of information given by associates to customers	Random sample of 100 calls per day evaluated by quality monitor Scoring system includes predefined criteria for evaluating customer impacting errors, business impacting errors, and non-impacting errors	96%
Professionalism	Professionalism (courtesy, responsiveness) shown by telephone associate	Random sample of 100 calls per day evaluated by quality monitor Scoring system includes predefined criteria for evaluating customer impacting errors, business impacting errors, and non-impacting errors	100%

Source: Universal Card Services.

Exhibit CS4.2.8 Quality days performance and bonuses

Associate quality days and bonus performance

Quarter	# Quality Days as % of Total	Bonus as % of Salary
4Q90	76.1%	6.4%
1Q91	87.8%	11.4%
2Q91	92.3%	9.9%
3Q91	96.7%	12.0%
4Q91	95.7%	11.6%
1Q92	70.3%	10.6%
2Q92	75.8%	7.5%
3Q92	76.1%	7.9%
4Q92	95.7%	10.8%
1Q93	84.4%	9.4%

Management quality days and bonus performance

Period	# Quality Days as % of Total	Bonus as % of Salary
1991	87.9%	5.6%
1992	66.1%	4.7%
1Q93	76.7%	5.6%

Source: Universal Card Services.

Case 4.2

Exhibit CS4.2.9 Changes in standards and measures

Number of increases in standards for existing measures

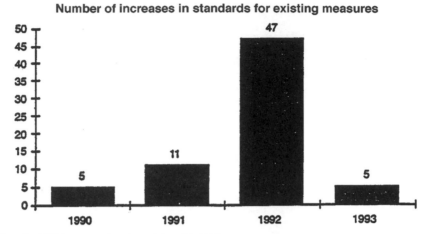

Note: Data for 1993 is year-to-date through June 30, 1993.

Number of additions and deletions of measures

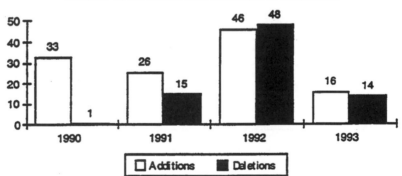

Note: Data for 1993 is year-to-date through June 30, 1993.

Source: Universal Card Services.

266

Harvard Business School

N9-197-025
September 5, 1996

Case 4.3

Robert S. Kaplan prepared this case as the basis for class discussion rather than to illustrate either effective or ineffective handling of an administrative situation. Mr. Ed Lewis of Mobil's Business and Performance Analysis group provided invaluable assistance.

Mobil USM&R (A): Linking the Balanced Scorecard

From what I can see, we had a good quarter even though financial results were disappointing. The poor results were caused by unusually warm winter weather that depressed sales of natural gas and home heating oil. But market shares in our key customer segments were up. Refinery operating expenses were down. And the results from our employee-satisfaction survey were high. In all the areas we could control, we moved the needle in the right direction.

Bob McCool, Executive Vice President of Mobil Corporation's US Marketing and Refining (USM&R) Division, had just commented on first quarter 1995 results.

One executive thought to himself:

This is a total departure from the past. Here was a senior Mobil executive publicly saying, 'Hey, we didn't make any money this quarter but I feel good about where the business is going.'

Case 4.3

MOBIL US MARKETING & REFINING

Mobil Corporation, headquartered in Fairfax, Virginia, and with operations in more than 100 countries is among the top three, with Exxon and Shell, of the world's integrated oil, gas, and petrochemicals companies. Mobil's 1995 return-on-capital-employed of 12.8% ranked it fourth among the 14 major integrated oil companies; its 19.1% average annual return to shareholders from 1991 to 1995 was the highest among the 14 major oil companies and exceeded the average annual return on the S&P 500 by more than 2 percentage points. Summary sales and earnings information are shown in Exhibit CS4.3.1.

The corporation consists of five major divisions: Exploration & Producing (the 'upstream' business), Marketing & Refining (the 'downstream' business), Chemical, Mining & Minerals, and Real Estate. The Marketing & Refining (M&R) division processes crude oil into fuels, lubricants, petrochemical feedstocks and other products at 20 refineries in twelve countries. M&R also distributes Mobil products to 19,000 service stations and other outlets in more than 100 countries. Total product sales had grown more than 5% per year over the past five years.

The United States Marketing and Refining (USM&R) division was the fifth largest US refiner. It operated five state-of-the art refineries and its more than 7,700 Mobil-branded service stations sold about 23 million gallons per day of gasoline. This represented a 7% national share (number four in the US). Mobil's retail network was highly concentrated. In the eighteen states where it sold nearly 95% of its gasoline, Mobil had a 12% market share. Mobil was also the largest marketer of finished lubricants in the US, with a 12% market share, and recent growth rates of about 3%, especially in premium quality blends.

In 1992, USM&R had reported an operating loss from its refining and marketing operations, and ranked 12 out of 13 oil companies in profitability from US marketing and refining operations.[1] A profit turnaround started in 1993 and earnings and return-on-assets, which had been depressed in 1991 and 1992, soon exceeded industry averages. Summary financial data of the USM&R division are presented in Exhibit CS4.3.2.

Until 1994, USM&R was organized functionally. The supply group obtained crude oil and transported it to one of Mobil's refineries. The manufacturing function operated refineries that processed crude oil into products like gasoline, kerosene, heating oil, diesel fuel, jet fuel, lubricants, and petrochemical feedstocks. The product supply organization transported refined petroleum products, through pipelines, barges, and trucks, to regional terminals around the country. The terminal managers received, stored, and managed the extensive inventories of petroleum products, and distributed the products to retailers and distributors. The marketing function determined how USM&R would package, distribute and sell Mobil products through wholesalers and retailers to end-use consumers.

REORGANIZATION: 1994

In the early, 1990s, USM&R faced an environment with flat demand for gasoline and other petroleum products, increased competition, and limited capital to invest in a highly capital intense business. McCool recalled;

---

[1] *Source*: 'Benchmarking The Integrated Oils, 1995,' U.S. Research (Goldman Sachs, July 15, 1996), pp. 83, 85.

> *In 1990 we weren't making any money, in fact there was a half-billion dollar cash drain. Expenses had doubled, capital had doubled, margins had flattened, and volumes were heading down. You didn't need an MBA to know we were in trouble.*

McCool spent the next couple of years attempting to stabilize the business to stop the bleeding.

> *We succeeded, but then we had to confront how we could generate further growth.*

A climate survey in 1993 revealed that employees felt internal reporting requirements, administrative processes, and top-down policies were stifling creativity and innovation. Relationships with customers were adversarial, and people were working narrowly to enhance the reported results of their individual, functional units. McCool, with the assistance of external consultants, initiated major studies of business processes and organizational effectiveness. Based on the studies, McCool concluded that if USM&R were to grow, it had to make the most of its existing assets, and to focus more intensively on customers, giving motorists what they want, not what the functional specialists in the organization thought motorists should want.

In 1994, McCool decided to decentralize decision-making to managers and employees who would be closer to customers. He reorganized USM&R into 17 Natural Business Units (NBUs) and 14 Service Companies (see Exhibit CS4.4.3). The NBUs included sales and distribution units, integrated refining, sales and distribution units, or specialized products (e.g., distillates, lubricants, gas liquids) and process (stand-alone refinery) units. McCool commented on the need for the reorganization.

> *We had grown up as a highly functional organization. We had a huge staff, and they ran the business. We needed to get our staff costs under control. But more important, we had to learn to focus on the customer. We had to get everyone in the organization thinking not how to do their individual job a little bit better, but how to focus all of their energies to enhancing Mobil products and services for customers.*

Brian Baker, Vice President of USM&R, concurred:

> *We were a big central organization that had become a bit cumbersome and perhaps had lost touch with the customer. We didn't have the ability to move quickly with new marketing programs in various parts of the country.*

USM&R's reorganization occurred simultaneously with a newly-developed strategy on customer segmentation. Historically, Mobil, like other oil companies, attempted to maintain volume and growth by marketing a full range of products and services to all consumer segments. The gasoline marketing group had conducted a recent study that revealed five distinct consumer segments among the gasoline-buying public (see Exhibit CS4.3.4 for descriptions of the five segments):

- Road Warriors   (18%)
- True Blues   (16%)
- Generation F3   (27%)
- Homebodies   (21%)
- Price Shoppers   (20%)

Case 4.3

USM&R decided that its efforts should be focused on the first three of these segments (61% of gasoline buyers), and not attempted to attract the price-sensitive but low-loyalty Price Shopper segment that accounted for only 20% of consumers. The new strategy required a commitment to upgrade all service stations so that they could offer fast, friendly, safe service to the three targeted customer segments. It also required a major shift in the role for Mobil's on-site convenience stores (C-stores). Currently, C-stores were snack shops that catered to gasoline purchasers' impulse buying. USM&R wanted to redesign and reorient its C-stores so that they would become a destination stop, offering consumers one-stop, convenient shopping for frequently purchased food and snack items.

USM&R BALANCED SCORECARD

The newly appointed business unit managers had all grown up within a structured, top-down, functional organization. Some had been district sales managers, others had managed a pipeline or a regional distribution network. McCool anticipated problems with the transition:

> We were taking people who had spent their whole professional life as manages in a big functional organization, and we were asking them to become the leaders of more entrepreneurial profit-making businesses, some with up to a $1 billion in assets. How were we going to get them out of their historic area of functional expertise to think strategically, as general managers of profit-oriented businesses?

McCool realized that the new organization and strategy required a new measurement system. Historically, USM&R relied on local functional measures; low cost for manufacturing and distribution operations, availability for dealer-based operations, margins and volume for marketing operations, and environmental and safety indicators for the staff group in charge of environment, health and safety. McCool was unhappy with these metrics:

> We were still in a controller's mentality, reviewing the past, not guiding the future. The functional metrics didn't communicate what we were about. I didn't want metrics that reinforced our historic control mentality. I wanted them to be part of a communication process by which everyone in the organization could understand and implement our strategy. We needed better metrics so that our planning process could be linked to actions, to encourage people to do the things that the organization was now committed to.

Baker also noted the need for new metrics:

> Our people were fixated on volume and margins at the dealer level. Marketing didn't want to lose gasoline dealers. But we didn't have any focus or measurement on dealer quality so we often franchised dealers who didn't sustain our brand image. Also, we drove so hard for short-term profit that when volumes declined, our marketing people attempted to achieve their profit figure by raising prices. You can do that for a while if you have a strong brand, which we have, but you can't sustain this type of action for the long term.

In mid-1993, Ed Lewis, formerly the Financial Manager for US marketing, was on a special assignment with Dan Riordan, Deputy Controller of USM&R, to examine the effectiveness of financial analysis for the entire division. They concluded that a lot of excellent financial analysis was being done—plenty of measures, plenty of analysis—but none of it was linked to the division's strategy. In late 1993, Lewis saw an article on the Balanced Scorecard[2] and thought:

> This could be what we are looking for. We were viewed as a flavour-of-the-month operation. Our focus shifted frequently so that if you didn't like what we were doing today, just wait; next month we will be doing something different. Nothing we did tied to any mission. The Balanced Scorecard seemed different. It was a process that tied measurement to the organization's mission and strategy. It could start us on the journey to implement USM&R's new organization and strategy by keeping us focused on where we were heading.

Lewis and Riordan recommended to McCool that USM&R develop a Balanced Scorecard. McCool was receptive since he had heard of the concept in a briefing he had received earlier that year. USM&R's senior management team launched a BSC project in early 1994. They hired Renaissance Solutions, the consulting company founded by David Norton, a co-author of the Balanced Scorecard article, to assist in the process.

A senior-level Steering Committee, consisting of McCool, Baker, the vice presidents of all staff functions, the division controller, and the manager of financial analysis of downstream operations, provided oversight and guidance for the BSC project. The actual project team was led by Lewis and Riordan, assisted by Renaissance consultants.

Starting in January 1994, Lewis and his project team conducted two-hour individual interviews with all members of the leadership team to understand each person's thoughts on the new strategy. The team synthesized the information received from the interviews, and, with David Norton facilitating, led several workshops to develop specific objectives and measures for the four Balanced Scorecard perspectives: financial, customer, internal business process, and learning and growth. The workshops always involved active dialogues and debates about the implications of the new strategy. Lewis noted:

> Forcing the managers, during the workshops, to narrow the strategy statements into strategic objectives in the four perspectives really developed alignment to the new strategy. You could just see a consensus develop during the three month period.

Among the new aspects of the USM&R scorecard was a recognition that the division had two types of customers. The immediate customer was, of course, the extensive network of franchised dealers who purchased gasoline and petroleum products from Mobil. The other customer was the millions of consumers who purchased Mobil products from independent dealers and retailers. The project team wanted the customer perspective on the scorecard to incorporate strategic objectives and measures for both types of customers.

By May 1994, the project team had developed a tentative formulation of the USM&R scorecard. At that point, they brought in more managers and split into eight sub-teams to enhance and refine strategic objectives and measures: a Financial team (headed by the VP of Strategic

---

[2] Kaplan, R. S. and Norton, D. P., 'Putting the Balanced Scorecard to Work,' *Harvard Business Review*, September–October, 1993.

Case 4.3

Planning); two Customer teams—one focused on dealers, the other on consumers; a Manufacturing team, focused on measures for refineries and manufacturing cost; a Supply team, focused on inventory management and laid-down delivered cost; an Environmental, Health and Safety team; a Human Resources team; and an Information Technology team. Each sub-team identified objectives, measures, and targets for its assigned area.

The teams also identified when new mechanisms were needed to supply some of the desired measures. For example, the strategy to delight consumers in the three targeted market segments required that all Mobil gasoline stations deliver a speedy purchase, have friendly, helpful employees, and recognize consumer loyalty. At the time, however, no measures existed to evaluate dealer performance on these now critical processes. The consumer-focused customer sub-team developed a Mystery Shopper program in which a third party vendor purchased gasoline and snacks at each Mobil station on a monthly basis. The shopper scored dealer performance on 23 specific items related to exterior station appearance, service islands, sales area, personnel and rest rooms. The mystery shopper performance score would be a measure in the customer perspective of USM&R's Balanced Scorecard.

The dealer-focused customer sub-team launched an initiative to support the dealer development strategy. The team developed a tool kit that would help marketing representatives evaluate and work with dealers to improve performance in seven business areas: financial management, service bays, personnel management, car wash, convenience stores, gasoline purchasing, and a better buying experience for customers. The marketing representatives would give a rating to dealers to identify existing strengths and opportunities for improvement. The goal was to increase the profit performance of dealers and wholesale marketers of Mobil products, as measured by Total Gross Profit of dealers and the monthly gross margin from Alternative Profit Centers (APC)—convenience stores and service bays.

By August 1994, the eight sub-teams had developed specific strategic objectives for the four Balanced Scorecard perspectives (see Exhibit CS4.3.5) and selected the initial set of measures for these objectives (see Exhibit CS4.3.6). The process had consumed two-to-three full-time equivalent weeks from all members of the Executive Leadership Team (McCool and all his direct reports, including the managers of the business units).

Between June and August, 1994, while the sub-teams had been refining the strategic objectives and measures, the Steering Committee went through each perspective to identify one or two critical themes. The project team produced a brochure to communicate these strategic themes to all of USM&R's 11,000 employees. In August 1994, USM&R announced its initial Balanced Scorecard and distributed the brochure (see Exhibit CS4.3.7).

LINKING THE BALANCED SCORECARD TO NBUs AND SERVCOS

While the USM&R scorecard was still being developed in April 1994, the project team launched pilots to develop business unit scorecards (in the West Coast and Midwest NBUs). Senior management wanted the NBUs to work from the strategic themes established at the USM&R division level and to translate the division strategy into local, NBU objectives and measures that would reflect the particular opportunities and competitive environment encountered by each NBU. This was part of McCool's belief that NBU managers had to learn to take responsibility for the strategy of their business units.

Ed Lewis, with consultant support, went to the NBUs and replicated the scorecard development process with their personnel:

> We did the interviews, conducted the workshops, and, over a six week period, developed a local scorecard. We used the USM&R scorecard as a guiding light, but that's all it was, a light. When an NBU developed a scorecard, it was their scorecard and they would live by it.

McCool concurred:

> Mobil in the Midwest is not the same as Mobil in New England, or on the West Coast. In each market, the consumer looks at us differently, our competition in each region is different, and the economics of operating in each market are different. I don't want to dictate a solution from Fairfax. We have a basic strategy and set of support programs that we can roll out to each NBU. We do have a few constraints: we want our dealers to operate under a sign that says 'Mobil,' there's a basic design for the station and for the C-store that we want to share across regions, and we think we have a winning segmentation strategy with fast and friendly service. But if an NBU thinks it has a better driver for success, I'm willing to hear it. I want the NBU head to tell me, here's my business, this is my vision and strategy, and this is how I am going to get there from here. Our job in Fairfax is to approve (or disapprove) the strategy and ask what additional resources they might need to get the job done.

Lewis recalled:

> When I first started speaking to NBU managers in the workshops about the C-store focus and measurement, 90% of them thought I was off my rocker. They told me that their business was selling gasoline, not snacks. We told them that the USM&R scorecard was assessing C-store progress through a measure of APC revenues per square foot. Eventually, the NBUs agreed to include this measure on their scorecards so that they would be aligned with the Fairfax scorecard.

> A year later, however, all the NBUs had brought into the C-store strategy. They concurred that developing a new C-store design, and making it a destination stop for consumers, was now a differentiator for the company. They even forced us to change the way we measure C-store sales, from aggregate revenues per square foot to revenues per month (same store sales per period) and gross profit percentage because that's the way the best retailers like 7–11 and Walmart evaluated themselves.

The NBU scorecards, in general, mirrored the USM&R scorecard, though with slightly fewer measures, particularly in the internal perspective since the NBUs were focused on particular functions—such as regional marketing and sales, refining, and distribution – so the full range of internal measures were not relevant to each NBU.[3]

---

[3] See USM&R (B) and (C) for description of the Balanced Scorecards developed by the New England Sales & Distribution and Lubes NBUs.

Several of the NBUs devoted a section of their monthly newsletters to Balanced Scorecard information. In the first few issues, the section reviewed a single scorecard perspective, explaining the importance of the perspective, articulating the reasoning behind the specific objectives that had been selected, and describing the measures that would be used to motivate and monitor performance for that perspective. After communicating the purpose and content of the scorecard in the first few issues, the content of the newsletter section shifted from education to feedback. Each issue reported recent results on the measures for one of the perspectives. Raw numbers and trends were supplemented with the human stories of how a department or an individual was contributing to the reported performance. The vignettes communicated to the workforce how individuals and teams were taking local initiatives to help the organization implement its strategy. The stories created role models of individual employees contributing to strategy implementation through their day-to-day activities.

SERVCO SERVICE AGREEMENTS AND BALANCED SCORECARDS

The Steering Committee also wanted the servcos to be accountable for their performance. Previously, each staff function operated from the Fairfax headquarters, providing strategy, direction, and services to the field organization. After the reorganization, staff functions were not free-standing service units that had to sell services to the NBUs and get agreement from them on prices and level of service provided. USM&R established buyers committees, consisting of 3–5 representatives from the NBUs, to work with each servco. In this way the offerings from every servco would be linked to the mission and strategy of NBUs and to USM&R. Eventually, each servco and its buyers committee agreed on the priorities and prices for the offerings it would provide. Dan Zivny, Manager of Finance and Information Services, endorsed the new process:

> Discussions with the buyers committee helped us to communicate to the NBUs about what we do and what our deliverables will be. Previously, the NBUs would complain about the costs and charges for information services. Now the NBUs are part of the process that specifies the outputs we will produce and the prices we will charge.

Several of the servcos began to develop their own Balanced Scorecards.[4] Marty Di Mezza, Manager of the Gasoline Marketing servco, noted:

> The service agreement with the buyers committee and our Balanced Scorecard have enabled my organization to become more customer focused. People now realize they have to sell their services and that we have to fit into the entire picture of USM&R.

In addition to developing their own BSC's, key servco people were assigned to collect the data and report on each measure on the USM&R scorecard. Each measure was assigned to a 'metric owner' (see Exhibit CS4.3.8 for list of metrics and metric owners). For example, DiMezza's Gasoline Marketing organization owned the Mystery Shopper and Dealer Gross Profit metrics. The metric owners verified that the measures appropriately reflected the strategic objectives, and could, based on feedback from the field, make recommendations to the Executive Leadership Team for modified or new measures. People within the metric owner's servco

---

[4] See USM&R (D) for description of the Balanced Scorecard developed for the Gasoline Marketing servco.

collected the actual data from operations and reported current values of the measures to the metric owner.

LINKING THE BALANCED SCORECARD TO COMPENSATION

All salaried employees of USM&R were tied to the Mobil corporate award program. This program was based on performance relative to Mobil's top 7 competitors on two financial measures: return-on-capital employed and earnings-per-share growth (see Exhibit CS4.3.9). This program awarded up to 10% bonus if Mobil ranked #1 on ROCE and EPS growth.

McCool initiated an additional program within USM&R that awarded bonuses up to 20% to managers in each business unit. NBU employees got 30% of the award based on USM&R performance, and 70% based on their NBU performance. Servco employees also got 30% on USM&R performance, 20% on the linkage to other business units, and 50% onto their servco BSC. The linkage measures for servcos represented the objectives and results they could influence either in the NBUs or at USM&R.

The bonus plan was part of the new variable pay compensation programme. Employees' base pay reference point had been reduced to 90% of competitive market wages. The remaining 10% of compensation could be achieved with average performance on three factors:

- A component based on the two corporate financial performance competitive rankings

- A division component based on the USM&R Balanced Scorecard metrics

- A business unit component based on key performance indicators, from the NBU or servco Balanced Scorecard metrics

An additional 20% of compensation could be received for exceptional performance along these three components. The theory for the variable pay plan was simple: award below average compensation for below average performance, average pay for average performance, and above average pay for above average performance.[5]

McCool wanted each business unit to work with the metric owners to develop its own targets for the scorecard measures. In addition, the BUs assigned a percentage weight associated with achieving this target. This percentage, which summed to 100 across all targeted measures, would determine the relative contribution of each scorecard measures to the bonus pool. Most business units chose to weight all measures on their scorecards; the remaining one still weighted most of their scorecard measures. Only one business unit put more than a 50% weight on its financial measures.

The business units, beyond establishing targets for each scorecard measures, also assigned a performance factor that represented the perceived degree of difficulty of target achievement (see Exhibit CS4.3.10). The performance factor would be multiplied by the weight assigned to

---

[5] In addition, the plan included individual awards, administered within a narrow range, to adjust for performance not captured by the metrics. Business unit managers were awarded a fixed 'pot of money' for such individual awards, but this allowance could not be overspent.

the measure to arrive at a total performance amount, much the way a diving competition is scored (absolute performance on a dive gets weighted by the dive's degree of difficulty). The maximum index score of 1.25, occurred when the target would put the Mobil unit as best-in-class. An average target received a performance factor of 1.00 and a factor score as low as 0.7 would be applied when the target represented poor performance, or was deemed very easy to achieve. The individual business units proposed the performance factors for each measure, but these had to be explained and defended in a review with the Executive Leadership Team and metric owners. Business unit managers also were able to see (and comment on) the targets, weights, and performance factors proposed by the other BUs.

Brian Baker was a strong advocate for the indexed targets:

> *Historically, people were rewarded for meeting targets and penalized when they missed a target. So sandbagging targets became an art form around here. I prefer the current system where I can give a better rating to a manager who stretches for a target and falls a little short than to someone who sandbags with an easy target and then beats it.*

REVIEWING THE BALANCED SCORECARDS

McCool reflected on the experience to date with the Balanced Scorecard:

> *It's enabled us to teach the NBU manages about strategy; about lead and lag indicators, and to think across the organization, not just in functional silos. It's exposed the managers to issues outside their expertise and to understand the linkages they have with other parts of the organization. People now talk about things that are outside their immediate responsibility, like safety, environment, and C-stores. The scorecard has provided a common language, a good basis for communication.*

> *We were also fortunate that when Mobil asked us to go to a pay-for-performance plan we could use our scorecard measures. Variable pay plans only work if you have a good set of metrics. Managers accepted the compensation plan based on the scorecard since they believed the measures represented well what they were trying to achieve.*

> *The learning and growth perspectives has been the biggest problem. Ultimately, that perspective will be the differentiator of the company, our people's ability to learn and to apply that learning. The good news is at least we now talk about learning, as much as we talk about gross margin. But we are struggling to get good output measures for the learning objective.*

McCool commented on the changes in the meetings he conducts with NBU managers:

> *For a meeting with an NBU manager, like West Coast, I have the manager plus representatives from various servcos, like supply, marketing, and C-stores. And we have a conversation. In the past we were a bunch of controllers sitting around talking about*

variances. Now we discuss what's going right, what's gone wrong. What should we keep doing, what should we stop doing? What resources do we need to get back on track, not explaining a negative variance due to some volume mix.

The process enables me to see how the NBU managers think, plan, and execute. I can see the gaps, and by understanding the manager's culture and mentality, I can develop customized programs to make him or her a better manager.

Baker commented on the reviews he recently conducted with the managers of nine NBUs and four Servcos that reported to him:

I went into these reviews thinking they would be long and arduous. I was pleasantly surprised how simple they were. Managers came in prepared. They were paying attention to their scorecards and using them in a very productive way to drive their organization hard to achieve the targets. How they weighted their measures spoke clearly about their priorities of relative importance up and down the four perspectives.

Basically, there's no way I can understand and supervise all the activities that report to me. I need a device like the scorecard where the business unit managers are measuring their own performance. My job is to keep adjusting the light I shine on their strategy and implementation, to monitor and guide their journeys, and see whether there are any potential storms on the horizon that we should address.

Baker felt that relying on only a single financial measure, like earning or return-on-capital-employed was dangerous.

A big shareholder may not care about local business conditions or competitive environments. Just achieve a 12% ROCE, produce the money, and don't tell me about your problems. That's his right as the shareholder, and some people would say, 'Those are the rules, and let's set strict earnings objectives for each of our business units and that's it.'

But there's another side of me that says to motivate people there are things managers can influence and things they cannot. In a strong market, you can do a bloody bad job and have a great year. And you can do a superb job and fall way short of earnings because the market was so weak. The scorecard had several elements that help me understand how well a manager performs against the market. Without the understanding we now have from the scorecard, we would force people to do some pretty bizzare things to make short-term earnings targets, and they could be gone before the problems fall in.

Managers do seem to be using the scorecard for their management processes. They're not just doing it because McCool and I have imposed it on them. It's a system they know that everyone is using; all the other business units are living by the same set of rules. That's incredibly important. Also, the degree of difficulty index allowed them to be more ambitious and aggressive in setting their targets.

Case 4.3

McCool concluded:

> In 3–4 years, we have come from an operation that was worst in its peer group, draining a half billion dollar a year, to a company that ranks #1 in its peer group, and generates hundreds of millions of dollars of positive cash flow.
>
> The Balanced Scorecard has been a major contributor. It's helped us to focus our initiatives and to keep them aligned with our strategic objectives. It's been a great communication tool for telling the story of the business and a great learning tool as well. People now see how their daily job contributes to USM&R performance. Our challenge is how can we sustain this performance. We have just seen the tip of the iceberg. I want people to use the scorecard to focus attention on the great opportunities for growth.[6]

---

[6] USM&R's 1995 income per barrel of $1.02 greatly exceeded the industry average of $0.65. Global operating return from refining, marketing and transportation operations of 10.1% per dollar of assets was the highest in the industry (up from 8.6% and 5th place in 1994). Source: 'Benchmarking The Integrated Oils, 1995,' U.S. Research (Goldman Sachs, July 15, 1996) pp. 7,9.

Exhibits

Exhibit CS4.3.1 Mobil summary financial information, 1991–95 (000,000)

	1991	1992	1993	1994	1995
Revenues	$63,311	$61,156	$63,975	$67,383	$75,370
Operating earnings	1,894	1,488	2,224	2,231	2,846
Capital and Exploration Expenditures	5,053	4,470	3,656	3,825	4,268
Capital Employed at year-end	25,804	25,088	25,333	24,946	24,802
Debt-to-Capital Ratio	32%	34%	32%	31%	27%
Rates of Return Based on:					
Average S/H equity	10.9%	8.8%	13.2%	13.2%	16.2%
Industry average				10.0%	14.0%
Average capital employed	9.4%	7.5%	10.2%	10.3%	12.8%
Industry average				8.1%	10.0%

Case 4.3

Exhibit CS4.3.2 US Marketing and Refining: financial summary, 1991–95 (000,000)

	1991	1992	1993	1994	1995
Sales and services					
Refined petroleum products	$10,134	$10,504	$10,560	$10,920	$12,403
Other sales and services	3,879	3,702	3,481	3,522	3,698
Total sales and services	$14,013	$14,206	$14,041	$14,442	$16,101
Excise and state gasoline taxes	2,421	2,606	2,957	3,663	3,965
Other revenues	80	118	90	88	108
Total revenues	$16,514	$16,930	$17,088	$18,193	$20,174
Operating costs and expenses	16,304	17,125	16,822	17,792	19,796
Pretax operating profit	$210	$(195)	$266	$401	$378
Income taxes	94	(50)	115	160	152
Total US M&R Earnings	$116	$(145)	$151	$241	$226
Special Items	(96)	(128)	(145)	(32)	(104)
US M&R Operating Earnings	$212	$(17)	$296	$273	$330
Assets at year-end	$6,653	$7,281	$7,248	$7,460	$7,492
Capital Employed at year-end	4,705	5,286	5,071	5,155	5,128
Earnings: Gasoline & Distillate (cents/gallon)	3.6	0.2	3.7	4.1	4.6
(industry average)	3.5	2.2	4.0	3.6	2.6
Return on Assets	4.2%	(0.2%)	5.2%	4.8%	5.9%
(industry average)	7.0	4.5	7.6	6.8	4.9
Gasoline Market Share (top 18 states)			11.4%	11.6%	11.9%

Exhibit CS4.3.3 Natural business units (NBUs) and service companies (SERVCOs) USM&R

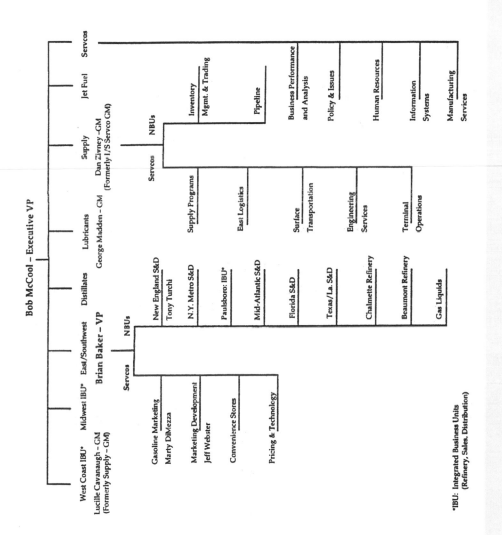

Bob McCool – Executive VP

West Coast IBU*   Midwest IBU*   East/Southwest   Distillates   Lubricants   Supply   Jet Fuel   Servcos

Lucille Cavanaugh – GM
(Formerly Supply – GM)

Brian Baker – VP

George Madden – GM

Dan Zivney – GM
(Formerly I/S Servco GM)

Servcos

Gasoline Marketing
Marty DiMezza

Marketing Development
Jeff Webster

Convenience Stores

Pricing & Technology

NBUs

New England S&D
Tony Turchi

N.Y. Metro S&D

Paulsboro: IBU*

Mid-Atlantic S&D

Florida S&D

Texas/La. S&D

Chalmette Refinery

Beaumont Refinery

Gas Liquids

Supply Programs

East Logistics

Surface
Transportation

Engineering
Services

Terminal
Operations

Servcos

NBUs

Inventory
Mgmt. & Trading

Pipeline

Business Performance
and Analysis

Policy & Issues

Human Resources

Information
Systems

Manufacturing
Services

*IBU: Integrated Business Units
(Refinery, Sales, Distribution)

Exhibit CS4.3.4 Five gasoline buyer segments

Road Warriors (18%)   Generally higher-income middle-men who drive 25,000 to 50,000 miles a year, buy premium gasoline with a credit card, purchase sandwiches and drinks from the convenience store, will sometimes wash their cars at the carwash.

True Blues (16%)   Usually men and women with moderate to high incomes who are loyal to a brand and sometimes to a particular station; frequently buy premium gasoline and pay in cash.

Generation F3 (27%)   (F3—fuel, food, and fast) Upwardly mobile men and women—half under 25 years of age—who are constantly on the go; drive a lot and snack heavily from the convenience store.

Homebodies (21%)   Usually housewives who shuttle their children around during the day and use whatever gasoline station is based in town or along their route of travel.

Price Shoppers (20%)   Generally aren't loyal to either a brand or a particular station, and rarely buy the premium line; frequently on tight budgets; the focus of attention of marketing efforts of gasoline companies for years.

Exhibit CS4.3.5 USM&R Balanced Scorecard: objective statements *(continues)*
Strategic Objectives

FINANCIAL

Return on Capital Employed – Earn a sustained rate of return on capital employed (ROCE) that is consistently among the best performers in the US downstream industry, but no less than the agreed corporate target ROCE of 12%

Cash Flow – Manage operations to generate sufficient cash to cover at least USM&R's capital spending, net financing cost, and pro rata share of the Corporate shareholder dividend

Profitability – Continually improve profitability by generating an integrated net margin (cents per gallon) that consistently places us as one of the top two performers among the US downstream industry

Lowest Cost – Achieve sustainable competitive advantage by integrating the various portions of the value chain to achieve the lowest fully-allocated total cost consistent with the value proposition delivered

Meet Profitable Growth Target – Grow the business by increasing volume faster than the industry average, and by identifying and aggressively pursuing profitable fuels and lubes revenue opportunities that are consistent with the overall division strategy

CUSTOMER

Continually Delight and Targeted Customer – Identify and fulfill the value propositions of our target consumers (speed, smile, stroke) while maintaining and improving the 'price of entry' items

Improve the Profitability of Our Dealer/Wholesale Marketers – Improve Dealer/Wholesale Marketer profitability by providing consumer-driven services and products, and by helping develop their business competencies

Exhibit CS4.3.5 *(continued)*

INTERNAL

Marketing

Product, Service and Alternate Profit Center (APC) Development – Develop innovative and mutually profitable services and products

Dealer and Wholesale Marketer Quality – Improve the franchise team to a level equal to best-in-class retailers outside the oil industry

Manufacturing

Lower costs of Manufacturing Faster Than the Competition – Create a competitive advantage by continuing to increase gross margins and reduce manufacturing expenses faster than the competition

Improve Hardware Performance – Optimize the functioning of our refinery assets through improved yields and decreased downtime

Safety – Strive to eliminate work-related injuries by constantly focusing efforts on improving the safety of our refinery work environment through continued employee education and prevention of workplace hazards

Supply, Trading, and Logistics

Reducing Laid Down Costs – Continue to lower supply acquisitions and transportation costs to reduce light products laid-down costs, such that we strive to supply products to our terminal at a cost equal to or better than the competitive market maker

Trading Optimization – Maximize spot market sales realization from refinery-finished and unfinished light products laid down costs, such that we strive to supply products to our terminals as a cost equal to better than the competitive market maker

Inventory Management – Optimize light products inventories while maintaining satisfactory customer service levels

Improve Health, Safety, and Environmental Performance – Be a good employer and neighbour by demonstrating commitment to the safety of all of our facilities and active concern about our impact on the community and the environment

Quality – Manage the operations to provide the consumers with quality products supported by quality business processes that are timely and performed correctly the first time throughout the value chain

LEARNING & GROWTH

Organizational Involvement – Enable the achievement of our vision by promoting an understanding of our organizational strategy and by creating a climate in which our employees are motivated and empowered to strive toward that vision

Core Competencies and Skills – (a) *Integrated view* – Encourage and facilitate our people to gain a broader understanding of the marketing and refining business from end to end (b) *Functional Excellence* – Build the level of skills and competencies necessary to execute our vision (c) *Leadership* – Develop the leadership skills required to articulate the vision, promote integrated business thinking and develop our people

Access to Strategic Information – Develop the strategic information support required to execute our strategies

Case 4.3

Objective	Measure	Frequency
FINANCIAL		
Return on Capital Employed	ROCE (%)	S
Cash Flow	Cash Flow Excl. Div. ($MM)	M
	Cash Flow Incl. Div. ($MM)	M
Profitability	P&L ($MM after tax)	M
	Net Margin (cents per gallon before tax)	M
	Net Margin, Ranking out of 6	Q
Lowest Cost	Total Operating Expenses (cents per gallon)	M
Meet Profitable Growth Targets	Volume Growth, Gasoline Retail Sales (%)	M
	Volume Growth, Distillates Sales to Trade	M
	Volume Growth, Lubes (%)	M
CUSTOMER		
Continually Delight the Targeted Consumer	Share of Segment (%)	Q
	– % of Road Warriors	Q
	– % of True Blues	Q
	– % of Generation F3's	Q
	Mystery Shopper (%)	M
INTERNAL		
Improve the Profitability of Our Partners	Total Gross Profit, Split	Q
Improve EHS Performance	Safety Incidents (Days Away From Work)	Q
	Environmental Incidents	Q

Exhibit CS4.3.6 *(continued)*

Case 4.3

Objective	Measure	Frequency
Product, Service and APC Development	APC Gross Margin/Store/Month ($M)	Q
Lower Costs of Manufacturing Vs Competition	Refinery ROCE (%)	Q
	Refinery Expense (cents/UEDC)	M
Improve Hardware Performance	Refinery Reliability Index (%)	M
	Refinery Yield Index (%)	M
Improve EHS Performance	Refinery Safety Incidents	Q
Reducing Laid Down Cost	LDC Vs Best Comp. Supply – Gas (cents per gallon)	Q
	LDC Vs Best Comp. Supply – Dist. (cents per gallon)	Q
Inventory Management	Inventory Level (MMBbl)	M
	Product Availability Index (%)	M
Quality	Quality index	Q
LEARNING & GROWTH		
Organization Involvement	Climate Survey Index	M
Core Competencies and Skills	Strategic Competency Availability %	A
Access to Strategic Information	Strategic Systems Availability	A

Case 4.3

Exhibit CS4.3.7 USM&R brochure of strategic themes

USM&R Strategic Themes ...

will guide us to our vision and are defined above each graph.

USM&R Strategic Measures ...

that will keep us focused on achieving USM&R's strategic themes are explained in the graphs and the bulleted text accompanying them.

Win/Win Relationship

Improve Dealer/Wholesale Marketer profitability through customer-driven products and services and by developing their business competencies.

- Total profit earned at Mobil outlets and split between our dealers/ whole-sale marketers and Mobil.

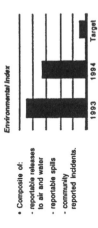

Dealer/Mobil Gross Profit

Good Neighbor

Protect the health and safety of our people, the communities in which we work, and the environment we all share.

- Composite of:
  - reportable releases to air and water
  - reportable spills
  - community reported incidents.

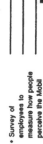
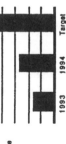

Environmental Index

Financially Strong

Reward our shareholders by providing a superior long-term return which exceeds that of our peers.

- Income divided by capital employed including all allocations.

ROCE

Delight the Customer

Understand our consumers' needs better than anyone and offer them products and services which exceed their expectations.

- The Mystery Shopper program rates how well each of our stations is delivering the "best buying experience."

Mystery Shopper

Safe & Reliable

Maintain a leadership position in safety while keeping our refineries fully utilized.

USM&R Days Away From Work

Manufacturing Reliability Index

On Spec On Time

Provide quality products supported by quality business processes that are on time and done right the first time.

- Composite of incidents of:
  - product off spec
  - order shipped late
  - business process errors
  - customer complaints
  - cost of rework.

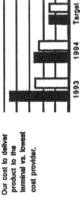

Quality Index

Competitive Supplier

Provide product to our terminals at a cost equal to or better than the competitive market maker.

- Our cost to deliver product to the terminal vs. lowest cost provider.

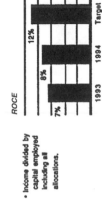

Laid-down cost

Motivated & Prepared

Develop and value teamwork and the ability to think Mobil, act locally.

- Survey of employees to measure how people perceive the Mobil workplace environment.

Climate Survey

Exhibit CS4.3.8 USM&R metric owners

Metric owner	SERVCO	Measures
Ed Mitchel	Business Performance & Analysis	Financial
Jeff Webster	Marketing Development	Share of Segment
Marty DiMezza	Gasoline Marketing	Mystery Shopper Friendly Serve
Borden Walker	Convenience Stores	Alternate Profit Center Revenues and Gross Margin
Tony Johnson	Manufacturing Services	EHS Performance Lower Cost Hardware Performance
Ted Shore	Supply Programs	Laid Down Cost Inventory Level
Carol Ellis	Supply Programs	Lubes Quality
Chuck Coe	Supply Programs	Fuels Quality
Bill Klarman	Human Resources	Organizational Involvement Core Competencies
Pierre President	Information Systems	Strategic Information Access

Exhibit CS4.3.9 Corporate performance share (CPS)

Salary groups 13–19: Metrics and awards % (of reference salary)

EPS growth relative ranking					
1 / 2	2%	4%	8%	10%	Seven major competitors
3 / 4	2%	3%	6%	8%	Amoco Arco BP Chevron Exxon Shell Texaco
5 / 6	1%	2%	3%	4%	
7 / 8	1%*	1%	2%	2%	
	8 7	6 5	4 3	2 1	

ROCE relative ranking

*Discretionary

Exhibit CS4.3.10 Metrics

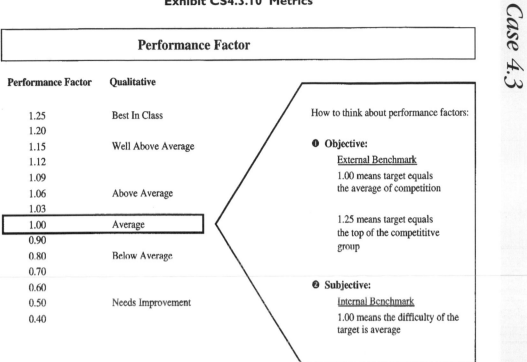

Harvard Business School	0-197-026
	September 5, 1996

Case 4.4

Robert S. Kaplan prepared this case as the basis for class discussion rather than to illustrate either effective or ineffective handling of administrative situation. Mr. Ed Lewis of Mobil's Business and Performance Analysis group provided invaluable assistance.

Mobil USM&R (B): New England Sales and Distribution

The New England Sales & Distribution (NES&D) was one of seven regional sales and distribution business units within Mobil's US Marketing and Refining (USM&R) operations (see Mobil USM&R (A) case for company description and background). NES&D responsibilities included:

- the 1,400 branded Mobil stations in the six New England states—Maine, New Hampshire, Vermont, Massachusetts, Connecticut, and Rhode Island

- Terminal operations in New England that received and stored gasoline and home heating oil brought in by ship from Canada, Europe, and Venezuela

- Three pipelines that transported petroleum products to regional terminals, such as at Boston's Logan Airport

- A fleet of delivery trucks, including the unionized truck drivers, that transported gasoline and heating oil to dealers and wholesalers

Tony Turchi, general manager of NES&D, had held a series of assignments in finance and planning before becoming marketing manager for New England region. He was appointed general manager in the 1994 reorganization (see USM&R) (A) case). Turchi recalled the rules under the old regime:

It was pretty simple. Our goals were to increase sales volume and reduce controllable costs. We had lots of measures, and they all related to these two goals.

In early 1994, shortly after NES&D had been established as an independent business unit, its senior management team conducted a strategic planning exercise. The exercise started with a traditional SWOT analysis (Strengths, Weaknesses, Opportunities, Threats). The team then developed a high-priority list of strategic opportunities that blended national strategies, such as fast-friendly-serve and On-The-Run convenience stores, with regional strategies, for example dealer development, felt to be critically important for the future success of New England. Turchi recognized, however, that the strategic planning exercise had two gaps.

We now had a vision, we had clearly identified strategic opportunities, and we had developed some strategic initiatives. But we didn't have a good measurement tool. We were still just measuring, in fact over-measuring, volume and costs. Second, a climate survey of our employees revealed a real hunger for a reward and recognition system. It didn't have to be big money, but our people wanted to understand where we were taking the business and to be more actively involved with that direction.

Turchi had not participated in the development of the USM&R Balanced Scorecard (described in USM&R (A) case).

In retrospect, I wish that could have been involved in the process to understand better why some of the measures were placed on the scorecard. On the positive side, however, it was great to be able to see the thought process spelled out: how McCool and the ELT (Executive Leadership Team) were going to be measuring the performance of the overall business, how a regional sales and distribution business unit fit into the overall picture, what measures were important to the ELT, and what strategies they were recommending. Even more important, the USM&R scorecard gave us a great template to follow. It took a lot of the guess work out of what the ELT meant by a Balanced Scorecard.

A NES&D team worked in the second half of 1994 to develop a New England Balanced Scorecard. The project team talked to employees in the terminals, the truck drivers, and to NES&D's channel partners, the independent gasoline dealers and wholesalers. The team also coordinated with USM&R servcos[1] to get their input for the New England strategic objectives and measures.

By the end of 1994, the effort produced NES&D's first Balanced Scorecard (see Exhibit CS4.4.1). The group also linked the strategic issues and opportunities it had identified from its SWOT exercise to the Balanced Scorecard measures (see list in Exhibit CS4.4.2). Turchi felt it was important for people to see how the Balanced Scorecard could measure progress along the strategic issues they had identified.

Our first opportunity (see Exhibit CS4.4.2) was to integrate the best client experience with our franchise offerings, like fast-friendly-serve, destination C-store, and our car care

---

[1]Servcos are the service organizations within USM&R, such as gasoline marketing, communications, finance and information systems that provided support services to the operating business units. The operating business units, such as NES&D, had very little internal staff support.

and maintenance services. If we succeeded in this opportunity with our dealer-partners, we would hit our targets for F5 (profitable growth), C1 (delight the customer). I1 (marketing and product development) and I2 (focus sales force). Number 3 on our hit parade was one of our most important objectives—to provide awesome new training and development for our people and partners. This was one of our most critical leading indicators. It would show up in two of our learning and growth measures (L1, L2) but it would also impact C1, the mystery shopper rating.

Turchi worked with his leadership team, consisting of the top managers in NES&D plus three representatives from key USM&R servcos, to set targets and weights for linking the scorecard measures to the bonus plan for the entire New England team (see Exhibit CS4.4.3). The linkage included the performance factor (degree of difficulty) for each measure to reflect how close the target was to best-in-class capabilities. Turchi recalled:

I had to estimate many of these factors midway through the year, because the process wasn't set up at the beginning of 1995. But I tried to be honest; I gave a 0.96 on several measures. Overall I came out with a performance factor of 1.03. I worked this through with Brian Baker, who's my ELT coach, and defended the weights at a meeting with a whole bunch of servco representatives.

Turchi felt, however, that the scorecard was too complicated to communicate to his 300 employees in the field.

In 1995, we were doing Balanced Scorecard 101. We had to learn to walk before we could run. We needed to make it simple and understandable to all our people. We also wanted to create some fun and excitement.

In late-January, the weekend after the Super Bowl, the NES&D leadership team organized a major meeting in Waterville Valley, New Hampshire. They decorated a meeting hall like a football field, gave everyone football sweatshirts, showed video-tapes of the great teams like the Green Bay Packers and Pittsburgh Steelers, and had an announcer from NFL films describe how the great football teams had all the elements—offense, defense, coaches, the support groups—working together. The leadership team then announced the New England region's Super Bowl for 1995. The team selected five critical measures from NES&D's Balanced Scorecard:

- Gasoline Volume
- Return on Capital Employed
- Customer Complaints
- Mystery Shopper Rating
- Our Commitment to Dealers

These five measures would serve as the scorecard for the New England Super Bowl. The Super Bowl metaphor became clearer when the team stretched the targets on these five measures beyond the levels of communicated to USM&R's Fairfax headquarters. For example, the official ROCE target for the New England region required a net income after tax of $22 million.[2] The

---

[2]Numbers have been disguised for confidentiality.

Super Bowl target, however, was set at $27 million in net income, an additional $5 million stretch. The team set similar stretch goals for sales volume, mystery shopper ratings, customer complaints, and dealer commitment. For the NES&D organization to win the Super Bowl, it would have to hit the stretch targets on all five measures. If it hit all five, everyone would get a cash bonus of $250 and a great weekend next winter at a resort hotel in Vermont. If it failed in any one, no reward.

The leadership team then rolled the Super Bowl program out to all the people in the field. Dan Quinn, NES&D field logistics manager, described the process:

> We talked to the drivers, the union people and took them through the strategy, the Super Bowl concept, and asked for their support to help us achieve our goals, how they could impact the measures. The truck drivers didn't believe us. They said, 'the marketing guys get all the good rewards and go out and have a good time; they never include the terminal guys.' We had to convince them that we were serious. They were going to get the same reward as the marketing people.

> Then they started to ask us about the threats and the weaknesses, and told us how they had tried to make improvements over the years. They wanted to know how they could continue to help. How do you explain ROCE to a truck driver? We talked about the components of ROCE they could impact, like how their safe driving could affect expenses and productivity. If they could deliver when there's snow on the ground, while other drivers had an accident with their trucks lying on the side of the road, that would mean a lot to our customers in terms of product availability and satisfaction. We explained the mystery shopper program and how we would rate stations that were doing great and how we would deal with stations that could be a problem.

> The only exception will be the union people, since the new contract, covering about 100 people, doesn't allow this type of variable pay compensation plan. So we will probably have another Super Bowl program for them, using a sub-set of the measures that everyone else is being paid on.

Turchi concluded:

> You can see the difference in our people. Pre-BSC, the scorecard for an area manager was pretty simple: sales, sales, and sales. For the manager of a terminal, it was cost, cost, cost, and perhaps a little safety. Now we are trying to have the people in both positions be mini-general managers, to have them think broadly about our entire business. People in these positions are the ones that are absolutely critical for the success of our organization.

Case 4.4

Exhibits

Exhibit CS4.4.1 The New England Scorecard

Objectives	Measures
FINANCIAL	
F1 Return on Capital Employed	ROCE
F2 Cash Flow	Cash Flow
F3 Profitability	Net Margin
	Net Profit After Tax
F4 Lowest Cost	Full Cost Per Gallon
F5 Meet Profitable Growth Targets	Gasoline Volume Growth Rate
	Existing Oil, Gas & Lubricant Growth Rate
	Premium Product Growth Rate
CUSTOMER	
C1 Delight the Customer	Mystery Shopper Rating
	Customer Complaints
	Customer Compliments
C2 Dealer/Wholesaler Marketer Profitability	Dealer/Distributor Gross Margins
INTERNAL	
I1 Marketing & Product Development	Sal-Op Gross Profit/Store
I2 Sales Force Focus	# of Oil, Gas and Lubricant Units Meeting Upgrade Standards
	# of 'Our Commitment' Stations
I3 Manage the Business	Gasoline Runouts
	Distillate Runouts
	Station Runouts
I4 Improve HS&E	Environmental Incidents
	Days Away From Work Due to Injuries
	Accidents
I5 Quality	% of Stations Scoring 100% in Quality Assessment Program
LEARNING & GROWTH	
L1 Organization Involvement	Climate Surveys
L2 Core Competencies and Skills	Progress on Development Plans – Focus on Leadership
L3 Access to Strategic Information	Availability of P&L, BSC, Cash Flow, Field Marketing Tools

Exhibit CS4.4.2 Linking strategic opportunities and initiatives to Balanced Scorecard Measures

Strategic opportunity	Balanced Scorecard measures
1. Integrate the Best Buying Opportunities with our Franchise Offerings	F5, C1, I1, I2
■ Target our C-stores, Service Bays, Gasoline Offering/Service to Road Warriors, Generation F3, True Blues	
2. Address the Deadly Gap (against key competitor)	
■ Aggressive Marketing Tactics	F3, F5
■ Managing Distributor Consolidation of 'Have Nots' and 'Haves'	F5
■ Partnering in Product Supply and Logistics – Improve Asset Utilization	F4
3. Provide Comprehensive Training and Development for our People	C1, I2, L1, L2
4. Improve the Climate: Open Communications, Feedback, Rewards and Recognition	L1
5. Improve Health, Safety and Environmental Performance	I4
6. Premium Products	F5
7. Legislation/Regulation – Potential Threat to Volume and Profit	F3, F5
8. Improve Cost Structure	F4, L2

Case 4.4

Case 4.4

Exhibit CS4.4.3 New England S & D – 1995 compensation linkage

	1995 Weights	'Net' Weight	1995 Perf. Factor	1995 Goal*
Financial (30%)				
Net Profit After Tax	15%	5%	1.00	22
Full Cost Per Gallon	30%	9%	1.00	6.8
Gasoline Volume Growth	15%	5%	1.06	103%
Existing OG&L Growth Rate	30%	9%	1.06	101.5%
Premium Product	10%	3%	0.96	100%
Customer (15%)				
Mystery Shopper Rating	60%	9%	1.10	82
Customer Complaints	40%	6%	1.00	824
Internal (35%)				
Sal-Op Gross Profit/Store	20%	7%	0.96	12,500
# of O.C. Stations	20%	7%	1.00	527
Environmental Incidents	20%	7%	1.06	19
DAFW	20%	7%	1.06	16
Accidents	20%	7%	1.06	20
Learning & Growth (20%)				
Climate Survey	50%	10%	1.03	4
Developmental Plans	50%	10%	1.00	2
Total Index		100%		
Performance Factor			1.03	

*Numbers disguised to maintain confidentiality

Managing People, Managing Change

Chapter

5

Introduction

Achieving significant performance improvements is not easy. As discussed in Chapter 1, many change initiatives simply have no impact; others get off to a promising start but quickly lose momentum when the novelty (or urgency) disappears; and others still, provide a one-off performance gain. Few change initiatives end up generating on-going improvements over time. The question then, is how can companies approach and manage the change process in a way that will trigger sustainable improvement and positive pace of improvement? This is the subject of the present chapter.

Various levels of analysis

The process side of change initiatives can be examined at various levels of analysis: at the corporate restructuring level, at the project level, and from an integrated point of view.

Corporate transformation

Most models of corporate transformation emphasize a three-step sequential process of restructuring, revitalization and renewal. The first step, *restructuring*, aims at rapidly bringing the profitability of the firm back to acceptable levels and laying down the foundations for future success. This is typically achieved by pruning the firm's activities and overhauling its structure, systems and processes, which then make it possible for it to proceed with significant downsizing. One of the main themes of this restructuring phase is the instilling of discipline and return to 'flawless execution'.

The subsequent *revitalization* stage is concerned with improving the firm's growth and profitability. The dominant focus switches from cost-cutting to revenue generation, and the new rallying calls include empowerment, identifying new business opportunities (both internally and externally), leveraging competencies in new ways, and developing and committing to a new vision.

In the third stage, *renewal*, the firm strives to institutionalize some of the lessons from the previous two stages. It tries to establish routines whereby unproductive assets are signalled early and new competencies are developed and shared more effectively. The aim is to strike a balance between discipline and innovation and to ensure that the changes engaged provide a platform for sustained future change. The firm may also endeavour to gain a deeper understanding of the nature of the transition it has undergone and to question what blockages prevented it from changing previously. Such reflection—on the underlying constraints and blindspots to change—is the first step towards becoming a 'learning organization'.

While acknowledging the apparent logic of restructuring as a necessary prelude to revitalization, Chakravarthy points out that this process, which he calls 'sour-first-sweet-later', is rarely observed in practice.[1] Many companies have restructured effectively, but none seems to have reached the 'renewal' stage. More worrying still, many companies have struggled with the revitalization stage. For example, between the mid-1980s and 1993 Kodak went through five separate restructurings, eliminated 40 000 jobs and spent $10 billion to diversify its activities, 'but all that produced little and the stock faltered'.[2] Phillips has also undergone several consecutive cycles of restructuring, each heralded as the last. Even in 'the exemplar case of GE, [. . .] revitalization was stalled for nearly eight years after the impressive early successes with restructuring'.[3]

Some of the cases presented earlier in this book would also belong in this category. For example, TSB's restructuring (1989–91) led to the elimination of over 4000 positions and had an impressive impact on the bank's financial performance: cost/income ratio fell from over 70% to less than 60% and operating profit before bad debt and claims more than doubled. The bank also took a number of actions to 'revitalize' its operations, including: continued process redesign; a strong Total Quality Programme; refurbishing of many branches to make more quality space available for selling; transferring hundreds of employees from back-office positions into 'sales' positions to capitalize on this additional space; and providing all of its salesforce with expert systems allowing them to convert more opportunities into sales. In spite of these efforts, the retail bank's income line remained essentially flat between 1992 and 1995. Its profit increased slightly over that period, but largely due to the continued decrease in the number of employees.[4]

Why is revitalization so difficult? Some commentators suggest that a firm actually sows the seeds of its difficulties during the restructuring phase. In particular, success requires conditions that are often destroyed, or at least are not fostered, during the restructuring phase. Revitalization is about growth; it requires new ideas, new products, new approaches, continuous adaptation to constantly evolving conditions, capitalizing quickly on market opportunities, most of which require strong involvement by people at all levels of the organization. Revitalization thus requires employees who are both *willing* and *able* to contribute ideas, projects and energy, which, in turn, requires them to be empowered by, and to trust, their management.

Successful restructuring, on the other hand, tends to require decisive actions and to involve sacrifices for many employees. Building on the American joke that 'turkeys do not vote for Thanksgiving', it is generally believed that radical change that involves sacrifices for employees requires forceful leadership rather than extensive consultation and empowerment, particularly if the restructuring phase has to show results quickly. Typically, restructuring also involves layoffs, which tends to increase fear and distrust, and may encourage managers concerned for their jobs to hold on to their authority and decision-making power, rather than delegating it to their staff and encouraging them to take risks.

---

[1] See Chakravarthy, B. 'The Process of Transformation: In Search of Nirvana,' *European Management Journal,* December 1996, pp. 529–539.
[2] L. Grant, 'Can Fisher focus Kodak,' *Fortune*, 13 January 1997, p.77.
[3] Chakravarthy, B. *op. cit.*, p.530.
[4] See Case 1.2, 'Transforming TSB Group (B): The First Wave (1989–91)', and Case 2.3, 'Transforming TSB Group (D): The Second Wave (1992–95): Focusing on Quality and Processes'.

Restructuring may also leave the firm with insufficient spare capacity to invest into growth opportunities. First, creativity and innovation require a minimum amount of slack on the human resources front. If people are devoting all their time to taking care of today, they have no time to anticipate, prepare or try to change 'tomorrow'. In many companies, several waves of 'trimming' have removed spare capacity as quickly as, sometimes even before, it was created! Such actions can plunge the company into what is sometimes called a 'death spiral', where the concern for matching people to output stifles the initiatives that might fuel the next generation of products or services.[5]

Second is the issue of monetary resources. Innovation may require investment in hardware, software, employee training or celebrations, and these outlays may not be feasible during severe restructuring periods.

Companies that fail to revitalize—that is, to foster in their employees the desire to seek challenges, cooperate, learn new skills, take risks and accept responsibilities— are condemned to keep repeating the restructuring process. Such organizations may become addicted to 'Big Bangs', with employees never getting to feel like architects of change—only its victims. This phenomenon is well illustrated by the introductory comments of participants in a workshop on organizational renewal:

'I'm from Sears and we've had five restructurings in five years.'

'I'm from Nynex and we've had *nine* in *four* years.'

'I'm from Citibank, and I've had seven bosses in four years.'[6]

Under these circumstances, it is difficult to blame employees for displaying a certain amount of cynicism (or change fatigue), and for failing to commit fully to the latest set of changes.

At the project level

The early chapters of this book dealt at some length with the *content* of process-based change initiatives. As illustrated in the cases presented in these chapters, such initiatives typically require significant cross-functional cooperation, substantial resources and sustained effort on the part of a group of individuals. Even when such projects do not trigger clear, active resistance from parts of the organization, their sheer complexity— combined with the fact that people tend to be very busy and to have precious little

---

[5] Nohria and Gulati recently provided evidence supporting this proposition. They reasoned that innovation would be inhibited by both too little and too much slack. (Too much slack breeds complacency rather than healthy discipline, which should not be good for innovation, while too little slack prevents experimentation whose success is uncertain). Data collected from 264 functional departments of two multinational corporations supported the predicted curvilinear relationship. See N. Norhria, and R. Gulati, 'Is Slack Good or Bad for Innovation', *Academy of Managment Journal*, **39**, (5) October 1996, pp. 1245–1264.

[6] Cited in '*Sustaining Change: Creating the Resilient Organization*', Marks, M. L. and Shaw, R. B. in *Discontinuous Change: Leading Organizational Transformation*, eds D.A. Nadler, R.B. Shaw and A.E. Walton, Jossey-Bass: San Francisco, 1995, p. 97.

slack time and resources to devote to 'planning and organizing for tomorrow'—make it quite easy for the project to lose momentum and eventually to die before it gets completed. The challenge is even greater when, as is typically the case, the project perturbs the established power and influence structure and hence triggers some 'resistance to change'.

The third instalment of the TSB case series (Case 5.1 Transforming TSB Group (C): Managing Change Teams) describes in detail the way TSB approached its five projects, particularly its very successful 'Branch Network Redesign' Project. This project was managed by a dedicated team of full-time change agents, based on CEO Peter Ellwood's strong belief that 'the only way to bring about large-scale radical change is to divorce the action of change from the day-to-day process of management. Responsibility for delivering change should be given to dedicated, focused project teams using rigorous project management methodology'. Quantum Corporation, a hard disk manufacturer, also uses cross functional teams, but manages them differently from TSB (as described in Case 6.2 Quantum Corporation: Business and Product Teams).

More generally, top management must pay significant attention to a series of issues, including the following:

■ Will the members of the project team work full-time on the project, or will they still work part-time in their 'regular job'? (In the latter case, what proportion of their time and energy are they expected to contribute to the project? How do we know they will be able to divert these resources from their 'regular job'?)

■ How will responsibilities be split between the 'project team' and the 'regular hierarchy' (e.g. the functions)? How will the interface between the team and the 'line' be managed?

■ Who should be appointed as project leader? Which characteristics are we looking for?

■ Who will agree on a 'contract' with the team and follow-up on its progress? How will this follow-up be performed? Who is the team (particularly the team leader) reporting to?

■ Do the team members have adequate project management skills?

■ Should we have rules about prototyping of new systems and processes?

■ At what point should the team be disbanded? Is there a plan to support the reintegration into line management of the team members that worked full time on the project?

■ How do we insure that 'line management' will follow through on the project after the project team is disbanded?

■ How do we coordinate the various concurrent projects? Are there opportunities for cross-fertilization?

Managing 'integrated change'

The last issue listed above highlights one of the major dimensions that companies must pay attention to: change projects have a greater likelihood of succeeding when they are part of a coordinated and coherent change effort. This is the 'integrated level of analysis' we mentioned at the beginning of this chapter; broader than each individual project, and more concrete and detailed than the broad 'restructuring/revitalization/renewal' sequence.

While some areas of management, such as the study of leadership, tend to generate lists of traits or necessary conditions, writers on 'change management' tend to propose models and frameworks. These frameworks generally imply that change management is a stepwise process and that the organization can make the transition from ill-adjusted to healthy by simply following the instructions. However, the profusion and variety of available models should alert readers to the complexity of the endeavour. A flavour of the variety of approaches can be gleaned from the views of three well-known specialists. They are presented by their authors and summarized here as signposts and potential guidelines:

Todd Jick's ten commandments

Jick emphasizes that successful change requires a twin focus on the organization's *readiness* and *capability* for change. Readiness for change is a function of whether there is adequate motivation for, and perceived benefit from the changes. Capability to change depends on whether there are enough supports, enablers and skills to enact the changes. Jick's 'Ten Commandments' are summarized in Table 4.1.

Price Waterhouse's 'better change' programme

Building on its consulting experience with several large corporations, Price Waterhouse's Change Integration® team proposed 15 guiding principles for 'better change' and 11 pitfalls for broad-based change projects. They are listed in Table 4.2.

John Kotter's eight-step model

Kotter proposed an eight-step model for transformation (i.e., large-scale, radical change) efforts (see Table 4.3). First published in a 1995 article, this model was expanded into a book a year later.[7] Kotter's eight steps are rooted in eight common errors he observed in many companies' approach to corporate transformation efforts:

1. Allowing too much complacency

2. Failing to create a sufficiently powerful guiding coalition

---

[7] 'Leading Change: Why Transformation Efforts Fail' *in Harvard Business Review*, March–April 1995, pp. 59–67, and *Leading Change*, Harvard Business School Press: Boston, 1996).

Table 5.1 Todd Jick's ten commandements

Increasing the organization's readiness to change	Understanding the organization's internal and external environments is a crucial step to form the basis for a tailored vision, guiding coalition, and implementation plan.
1. Analyse the situation and its need for change	Assess how much change is needed, extent of support base and resistance to be expected.
2. Create a shared vision & common direction	To crystallize multi-faceted change effort and marshall hearts and minds.
3. Separate from the past	Similar to familiar concept of 'unfreezing'; break from old ways. Key issue is how quickly/drastically and how?
4. Create a sense of urgency	Key challenge! Creating a sense of urgency before real crisis strikes, without triggering a repetitive effect that would weaken the credibility of management's calls for change. Enlarging employees' mind set through communication and visioning exercises helps.
Building the organization's ability to change	
5. Support a strong leader role	Combining strong leadership (from insider or outsider?) with widespread employee involvement. Produce quick but sustainable successes.
6. Line up political sponsorship	Acceptance of change is also a function of who supports it.
7. Craft an implementation plan	There are benefits to careful designing and communication of an implementation plan. Yet there are also benefits to 'getting started' to build momentum and gather experience.
8. Develop enabling structures	Change various systems and structures (e.g, hiring, training, career paths, measurement and reward, organizational structures, etc.). Twin goal: To make systems and structure congruent with change, but also to signal commitment to change.
9. Communicate, involve people and be honest	Communicate and communicate again, through multiple means, but in ways consistent with the 'new' goals and values (communication can be a two-way street)
Making change stick	
10. Reinforce and institutionalize change	Re-energize effort to overcome normal fatigue. Institutionalize changes *and* make change a constant part of the organization's life.

Adapted from 'Implementing Change', Jick, T. D. in *Managing Change: Cases and Concepts*, Irwin, 1993; and 'Managing Change', in *The Portable MBA in Management*, ed., A. R. Cohen, Wiley, 1993.

3. Underestimating the power of vision

4. Undercommunicating the vision by orders of magnitude

5. Allowing obstacles to block the new vision

6. Failing to create short-term wins

7. Declaring victory too soon

8. Neglecting to anchor changes firmly in the corporate culture.

Table 5.2 Price Waterhouse's 'better change' programme*

15 guiding principles	11 pitfalls
1. Confront reality	1. Failure to deliver early, tangible results
2. Focus on strategic contexts	2. Talking about breakthroughs, drowning in detail
3. Summon a strong mandate	
4. Set scope intelligently	3. Everything is a high priority
5. Build a powerful case for change	4. Old performance measures block change
6. Let the customer drive change	5. Failure to 'connect the dots'
7. Know your stakeholders	6. The voice of the customer is absent
8. Communicate continuously	7. The voice of the employee is not heard, either
9. Reshape your measures	8. Senior management wants to help, but doesn't know how
10. Use all levers of change	
11. Think big	9. 'What's in it for me' is unclear
12. Leverage diversity	10. Too much conventional wisdom
13. Build skills	11. Same old horses, same old glue
14. Plan	
15. Integrate your initiatives	

*From Price Waterhouse Change Integration® Team, *'Better Change: Best Practices for Transforming Your Organization*, Irwin, 1995

Table 5.3 Kotter's eight steps to managing radical change*

1. **Establishing a sense of urgency**	Examining market and competitive realities; identifying and discussing crises, potential crises, or major opportunities
2. **Forming a powerful guiding coalition**	Assembling a group with enough power to lead the change effort; encouraging the group to work together as a team
3. **Developing a vision and strategy**	Creating a vision to help direct the change effort; developing strategies to achieve that vision
4. **Communicating the vision**	Using every vehicle possible to communicate the new vision and strategies; teaching new behaviours by the example of the guiding coalition
5. **Empowering employees to act on the vision**	Getting rid of obstacles to change; changing systems or structures that seriously undermine the vision; encouraging risk-taking and nontraditional ideas, activities and actions
6. **Generating short-term wins**	Planning for visible performance improvements; creating those improvements; recognizing and rewarding employees involved in the improvements
7. **Consolidating gains and producing more change**	Maintain the sense of urgency; reinvigorating the process with new projects, themes and change agents
8. **Institutionalizing new approaches**	Articulating the connections between the new behaviours and corporate success; hiring, promoting and developing employees who can implement the vision

Adapted from Kotter, J. P. '*Leading Change*', Harvard Business School Press: Boston, 1996.

Our view is that managing change is part art, part science, and arguably more art than science! Yet, for all their variety and difference of emphasis, 'change management frameworks' do have a certain number of common points.

A simple, generic model

The following four-step model remains the simplest and most compact approach to change. It applies equally well at an organizational and individual level. It also has much commonsense appeal. For people, let alone a large group, to engage in a change process, they must:

1. Have enough (shared) dissatisfaction with the status quo

2. Have a (shared) vision they are trying to enact

3. Have some guidance on the strategies and tactics that will help them reach the vision

4. Believe that the benefits of the journey and/or of reaching the destination exceed the costs.

All this seems quite reasonable! Imagine you are expected to go from point A to point B, and getting to point B is not trivial. Before engaging in a potentially costly journey, you would want to have an understanding of (a) why you need to move (why A is no longer good enough), (b) where you are going and (c) how to get there. Also, if I want you to perform the journey without too much prodding on my part, you probably also need (d) the belief that going from point A to point B, this way, 'is worth it'.

Even when it is well managed, change takes time

It takes time for change strategists and change agents to identify and communicate the need for change, to craft a compelling vision and appropriate strategies for change. Employees also need to be given time to make the same mental journey. A bank teller who has spent 20 years following procedures and 'staying out of trouble' is not instantaneously transformed by the new manager who urges staff to 'Be entrepreneurial'! It is not enough to tell people they are empowered in order to reap the benefits. Nor is the CEO's pledge that 'This time we really mean it!', enough to erase magically the years of management indecisiveness and constantly shifting priorities. For those on the receiving end, it takes time to internalize the desirability and feasibility of change.

The change must be multifaceted

Well-managed process-based change efforts, the focus of this book, can produce large performance improvements. In too many cases, however, the improvements have proved unsustainable. Even in cases where they were sustainable, they often remained a 'Big-Bang', a one-off improvement without 'tomorrow'. In order to 'stick', change programmes need to be multifaceted and to address the *causes* of the underlying problems.

In Chapter 1, we introduced a framework for change efforts featuring a joint focus on process, technology, people, and structure and systems. The cases presented in that Chapter illustrated how companies worked on several, often all, of these 'boxes' concurrently.

Chapter 1 also summarized an alternative way to structure change efforts, the well known 7Ss model, first introduced by Richard Pascale and Tony Athos in *The Art of Japanese Management.*[8] The 7Ss stand for: (1) hiring the right Staff, (2) training them in the right Skills, (3) managing them in an appropriate Style, (4) selecting the values to Share with them, (5) installing the right Systems, (6) improving the Structure, and (7) mobilizing a coherent Strategy.

Change efforts often run out of steam because the company worked on symptoms of problems, rather than on their root causes. For example, many companies invest much time and money into trying to make their employees 'more entrepreneurial' through training, communication and exhortation campaigns. Such programmes can indeed be valuable, but they do not eliminate the need for the firm to wonder *why they are facing this problem in the first place.*

When a company complained to us that its entire staff of 30 000 were singularly lacking in entrepreneurial drive, we asked whether the former had put in place a special sifting mechanism that allowed it to identify, attract and retain nonentrepreneurial people so effectively! Indeed, one would expect that, in a large population, a few entrepreneurial employees would manage to sneak in! Beyond the rhetorical question, the more serious explanation is that the company did hire a number of entrepreneurs but, over time, some of these employees left and the others lost their entrepreneurial drive. Exhorting them to recover that drive may be sufficient for a few employees, but, in time, employees' goodwill will once again bump up against the forces that created the problem in the first place. These forces may include the anxiety of supervisors who do not really want workers to be entrepreneurial, rules and procedures that do not allow them to take initiative, and/or simply a lack of resources (time or money). Such forces tend not to go away by themselves.

Case 5.4 on British Airways summarizes the impressive array of change initiatives that the company implemented over a 12 year period. Frameworks such as the 7Ss model or our own 'four box' model can help categorize the multiple facets of such change efforts.

No Gain, No Pain ('what about us'?)

People trying to improve their physical fitness level have probably heard at one point the famous encouragement 'no pain, no gain'. Casual observation of exercise clubs suggests that many individuals are buying this encouragement! Too many change efforts have been presented to employees along that format. 'Here is a lot of pain, it is necessary because competitors are beating us/the economy is becoming more global/head office says so'.

This type of admonishment may work for an hour of physical exercise, it does not work as well for significant corporate restructurings that require people to absorb large amounts of stress and to make significant sacrifices. Imagine yourself walking into a class and meeting an instructor who says to you, tersely: 'This class is going to be boring, it is going to involve a lot of work, many of you will flunk, but it is necessary

---

[8] Pascale, R. and Athos, A. *The Art of Japanese Management,* Simon & Shuster, 1981.

because you will soon compete on the Global Executive Market with people who come from countries where they face similar tough educational environments.' You might accept it for a one-hour class, you probably would not for a one-semester course!

Much has been said and written about the need to redesign processes to eliminate non-value adding work, where 'value adding' is defined with respect to customers. We have also discussed in Chapter 4 how process improvements often generate, in the short run, more spare capacity than real profit increases. For example, redesigning the claim administration process may decrease average processing time from eight to four hours, which is very good news, but it creates no immediate financial benefit. We still have the same income number, the same number of people working for us and the same number of office floors occupied.

To translate the process improvement into financial value, it is necessary to eliminate the spare capacity, which can only occur through *volume increases* (the capacity is then used in value creating activities), and/or *retrenchment* (which takes the spare capacity out of the system). Going back to the average processing time going from eight to four hours, we must now increase the number of policies sold (which will, probably, also increase the number of claims), increase the price of our policies (which may be facilitated by our now improved customer service), and/or reduce the number of people processing claims.

Retrenchment strategies (accompanied by downsizing) tend to be a less uncertain and quicker way to translate spare capacity into improved financial performance than trying to grow profitably into that capacity. Unsurprisingly, process redesign efforts have very often resulted into significant downsizing, which may create value and gain for shareholders, but tend to create mostly pain for employees. Figure 5.1 represents some of the key dimensions involved is this pain–gain dynamic.

Most 'process reengineering' efforts start off in the upper left section of Figure 5.1, with a focus on creating value for *customers* and improving *efficiency* by redesigning processes. If well managed, this is likely to work once, maybe twice, but how long would you want to work for a company that is going through one wave of downsizing after another?

Figure 5.1 In search of balance

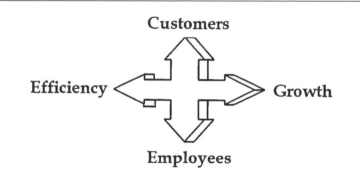

Empirical observation suggests that people are much more likely to accept the pain of restructuring if the change effort also makes it clear to them how this pain is connected to future gains. People must have some hope that the sacrifices they are making will bear fruits *for them* in some foreseeable future.[9]

Executives trying to sell pain to employees need to look hard for some gain, somewhere. Good questions to ask about 'new corporate visions' and 'strategies' are: What is the top management selling? Is there something in the 'new vision' that makes it worth while for people to make sacrifices? What is in it for them? If you cannot find any gain in the offering, then you know right away it is going to be a tough sell!

On the issue of resistance to change

Most managers have come to expect some resistance to change when they try to implement it. For example, one account of the Taco Bell journey under John Martin's leadership concluded:

> *Another lesson to come out of John Martin's experience: Expect resistance and be prepared to deal with it. People with a vested interest in the way things are will be upset when you change them. If some people are upset it's a good sign that you're doing something significant.*[10]

This perception seems to be fairly widely shared, at least in audiences of executives with whom we interact. We have developed an exercise where we ask managers their views on why people resist change. Their responses are strikingly similar across groups. The first set of reasons mentioned tends to be fairly 'negative'; people resist because:

- there is a natural propensity to resist change; it's 'natural', it's 'human', though some people do seem to display it more than others;

- people resist because they are afraid; afraid of losing power, influence, prestige, money, career prospects, or their job;

- 'they' (the resisters) don't see the big picture, they are too narrow-minded;

- 'they' do not want to abandon their 'comfort zone'; and/or

- 'they' suffer from the well-known 'not-invented here' syndrome.

Once the flow of answers is exhausted, we ask managers whether they, themselves, have ever resisted an initiative within their firm. When some of them agree, we ask why. A typical answer goes something like: 'Because I thought about what they wanted to do and I disagreed with the change; I didn't think this was the right thing to

---

[9] These success must then be publicized when they materialize, even if they represent 'easy victories'.
[10] Hammer, M. and Champy, J. *Reengineering the Corporation: A Manifesto for Business Revolution*, Harper Business, 1993.

do or the right way to do it'. This 'I thought about it and, after careful consideration, disagreed with the change' is typically conspicuously absent from the same managers' explanations for why 'others' resist!

This little exercise has proved fascinating, with executive audiences unfailingly asserting that 'others' tend to resist for largely negative reasons while 'we', the participants, resist for largely positive, or at least legitimate, reasons.

While management writers tend to portray the resistance of employees to change as natural, and something to be expected, some authors look at it from a very different angle. For example, a theory of motivation known as Self-determination Theory starts from the belief that human beings are initially programmed for growth and development. This rather 'positive' initial programming is also reflected in the three fundamental human needs posited by the theory (aside from physiological needs):

- *A need for autonomy/self-determination,* which involves initiating, monitoring and correcting one's own actions.

- *A need for competence,* where competence involves understanding how to attain particular outcomes and effectively performing the required actions,

- *A need for relatedness,* which involves developing secure and satisfying bonds with others in one's social group.[11]

Self-determination theory is often regarded as 'impractical' and 'too theoretical' for the world of business. It is true that it originated in the field of education and that most of its evidence involves children as subjects. Still, other authors have advanced similar views over the years. For example, Deming and Ishikawa, two of the most prominent advocates of the Total Quality Management movement, identify three main sources of human motivation at work:[12]

1. Intrinsic motivation: the 'joy of climbing a mountain just because it is there' (Ishikawa, 1985, p.27) and, more generally, growing, learning, and developing one's self (Deming, 1986, pp. 72–86).

2. Task motivation: the good feeling that comes from accomplishing things and seeing them actually work (Ishikawa, 1985, p.28; Deming, 1986, p.72).

3. Social motivation: the energy that comes from cooperating with others

---

[11] For more information on self-determination theory, see, for example, Deci, E. L., and Ryan, R. M. *Intrinsic Motivation and Self-Determination in Human Behavior,* Plenum Press, 1985, or Deci, E. L., Vallerand, R. J., Pelletier, L. G., and Ryan, R. M. 'Motivation and Education: The Self-Determination Perspective,' *Educational Psychologist,* 1991, pp. 325–346.
[12] Deming, W. E. *Out of the Crisis,* Cambridge, MA: MIT Center for Advanced Engineering Study (1986); Ishikawa, K. *What is Total Quality Control? The Japanese Way,* Prentice-Hall, 1995, cited in Hackman, I. R. and Wageman, R. 'Total Quality Management: Empirical, Conceptual and Practical Issues', *Administrative Science Quarterly,* **40**, 1995, pp. 325–326.

on a shared task and the incentive provided by recognition from others (Ishikawa, 1985, p.28; Deming, 1986, p.107).

Similar positive basic beliefs about human beings' attitude towards work and challenge, are also expressed by several chief executives well known for their strong sense of drive and excellence. Consider the comments of Robert McDermott, CEO of the United Services Automobile Association (USAA)—and a retired US Air Force brigadier general:

> *In the military, we found that perhaps the most important factor in esprit de corps is being needed by the other guys in your unit. [. . .] No one's happy in a slack, easy job. Working harder makes people happier. Or at least working better makes people happier.*[13]

Or again, take the example of Larry Bossidy, CEO and chairman of Allied-Signal whose stock nearly quintupled over the first five years of his tenure:

> *We've had some success at Allied-Signal because our people are thirsty for progress. They were disappointed and discouraged by the stagnation, and they recognized what was going on around them. They saw corporate cadavers all over the landscape, and they didn't want Allied-Signal to be among them. So they responded well when they were challenged and where they were made to feel that what they were doing had meaning.*[14]

Someone starting with such beliefs would not be shocked when encountering resistance to change, but rather than attributing it to a 'well-known human propensity to resist change, however promising', they would try to understand why the 'resisters' do not see the change in as positive a light as they, the change agents, do. They would then try to discover the *causes* of the resistance, rather than its *symptoms*—and to *address* the resistance rather than *overcome* it.

Too often, managers expecting strong resistance engage in a self-fulfilling prophecy. Expecting resistance they behave in such a way (guarded, secretive, untrusting) that they actually encourage the hostility they predicted would be there. For example, anticipating trouble, top management does not let employees in on a forthcoming change, but tries instead to spring it on them, with predictable consequences.

Or again, top management may expect employees to resist setting challenging targets, so it pushes for even more demanding targets with the expectation that they will make a deal and settle around the initial expectation. We can never know whether employees would have resisted top management's initial goal, but we should not be surprised that they resisted top management's opening bid!

---

[13] 'Service Comes First: An interview with USAA's Robert F. McDermott', *Harvard Business Review*, September–October 1991, pp. 116–127.
[14] 'The CEO as Coach: an interview with Allied Signal's Lawrence A. Bossidy' (conducted by N.M. Tichy and R. Charan), *Harvard Business Review*, March–April 1995, pp. 69–78.

Being tough but fair

A large pharmaceutical company decided to implement Windows 95 throughout the corporation. It considered how to proceed and, after careful consideration, reluctantly chose a unilateral approach; one morning, people showed up at the office and switched on their computers to find the Windows 95 main screen.

When challenged on the brusque nature of the decision-making and communicating process, executives explained that they had little choice, as it was clear to them that an open discussion of this issue would lead to endless delays and no agreement. This assumption can be debated, but it is probably realistic! Had the IT group tried to hold open-ended discussions on this issue of computer platform (Should we have a standard one and, if so, which one, starting when?), they would probably still be discussing. Between a 'top-down' approach and a 'bottom-up' approach, the top-down one was clearly preferable in this case.

What is not clear, however, is why the choice has to be limited to 'top-down' vs. 'bottom-up', as there would seem to be many shades in between, even in cases when bosses have to set rules and/or communicate unpleasant news. Evidence suggests that it is possible to set limits in ways that preserve reasonable choice for those on the receiving end, and without detriment to their intrinsic motivation. This approach is called *Informational Limit Setting* and involves four dimensions:

1. Being clear in stating the policy/limit	State the rule
2. Providing the reasons for it	Explain the rule
3. Acknowledging possible conflicting feelings or dissatisfactions (on the part of the receivers)	Acknowledge the conflict
4. Allowing as much flexibility as possible in *carrying out* the policy/limit	Provide flexibility where possible

There is nothing conceptually complex about Informational Limit Setting. The difficulty is in its implementation, which requires patience and respect for the other party, but also good judgment as to which part of the change should be part of the (nondiscussible) rule and which parts can be discussed. For example, with respect to implementing Windows 95, the decision about standardizing platforms could be the rule, which might then be communicated as follows: (1) We shall standardize to Windows 95, and we shall do so by such and such a date. (2) Here is why (. . .). (3) We realize that this will create problems for some of you. We also realize that we are not opening this standardization up for debate, and here is why (. . .). (4) What we would really like to discuss, however, is how we can make this process as easy as possible for you, and in particular, what we can do to help.

More generally, Informational Limit Setting is an attempt to achieve simultaneously two seemingly contradictory goals: maintaining both *order* in the process (efficiency) and *respect* for the 'change recipients' (involvement). The importance of achieving this balance is underlined by numerous recent studies highlighting a key distinction:

1. Individuals are capable of making a clear difference between procedural and distributive justice, where 'distributive justice' refers to the perceived fairness of the outcome (which is partly associated with how favourable the outcome is for the subjects), and 'procedural justice' is the fairness of the *process* that led to the outcome.

2. The perceived *fairness of the process* has a *major impact* on employees' reactions to the outcomes of that process. In particular, subordinates seem to be much more likely to accept unpleasant outcomes when the process that led to them is perceived as fair.

Evidence on the power of procedural fairness comes from many fields. In the performance appraisal context, for example, a field experiment showed that, compared to a control group, employees involved in a performance appraisal system emphasizing fair procedures displayed more favourable reactions towards the appraisal system and the managers who administered it, as well as a greater intention to remain with the organization, *even though they received significantly lower performance evaluations than the control group.*[15]

In the strategic decision-making process, Kim and Mauborgne found that procedural justice between head office and subsidiary unit top managers was associated with better acceptance and executions of decisions. The chance to influence the decision process and the right of appeal inspired subsidiary managers to go beyond the call of duty and to engage in innovative actions, spontaneous cooperation, and creative behaviour on behalf of the organization when implementing decisions.[16]

Finally, Korine found that the performance of product development teams (assessed by top management) had strong positive associations with the teams' assessment of perceived fairness in decision-making of top management and the teams themselves.[17]

The key finding is that people can see beyond their own short-term self-interest. A fair process signals respect for the dignity of employees—they are regarded as ends

---

[15] Taylor, S.M., Renard, M.K., Harrison, J.K, and Carroll, S.J. 'Due Process in Performance Appraisal: A Quasi-Experiment in Procedural Justice', *Administrative Science Quarterly*, September 1995, pp. 495–523.
[16] Kim, W.C. and Mauborgne, R.A. 'Fair Process: Managing in the Knowledge Economy', *Harvard Business Review*, July–August 1997, pp. 65–76.
[17] Korine, H. 'A Procedural Justice Perspective on Managing Innovation Teams', unpublished doctoral dissertation, INSEAD, 1997.

in themselves, not just means. From the converging findings of various authors, the key constituents of 'fair process' include:

- adequate consideration of others' viewpoint: soliciting input prior to evaluation and using it;

- suppressing personal biases;

- applying decision criteria in a consistent manner across individuals and time periods;

- providing timely feedback for decisions;

- adequately explaining the basis for decisions, which means that this basis must be (perceived to be) reasoned adequately and communicated sincerely.

The leader's role

The leader's role is, simply, essential. It is difficult to find any example of a significant change involving more than a handful of people that occurred without *someone* taking the lead and prodding the group into action. There are, however, many ways to exercize this leadership.

At one extreme is the charismatic leader. The notion of personal charisma is generally associated with an outgoing personality that can generate enthusiasm in others, which typically involves the leader displaying a high degree of self-confidence and enthusiasm him or herself. Studies of charismatic leaders also highlight these leaders' tendency to advocate an appealing and somewhat unconventional vision; take a high personal risk to support their vision; inspire a strong sense of community and team spirit (it's 'us' vs. 'them'); and communicate high expectations of, and confidence in, their followers.

While some degree of personal charisma may help a leader generate enthusiasm and commitment, leaders can also have a profound impact on change projects through their ability to display consistency of focus and a structured approach toward problems. Too many leaders, especially charismatic ones, are much better at starting up new initiatives than seeing them through to a successful conclusion. Framing the vision is an important dimension, but carrying the project through is at least as important!

This consistency of focus must be expressed in words, but also in deeds. Within British Airways, for example, Sir Colin Marshall was known to have a quasi-obsession with customer service, impressing his subordinates with a relentless attention to detail and remarkable mastery of customer satisfaction and plane timeliness data. He also paid a lot of attention to the airline's staff, which he demonstrated, for example, by taking part in a very high proportion of the deliveries of BA's first major training programme.

As an ex-BA manager put it to us: 'Everything Colin Marshall did, he found a way to relate it to customers or employees. Walking through an airport with him, for example, was an experience! He would stop every 10 seconds to shake a hand, talk to an employee or a customer.' In spite of not being as charismatic as, for example, Richard Branson (the chairman of the Virgin Group), Sir Colin Marshall communicated very clearly to everyone that the twin pillars of his strategy were 'caring for customers and caring for employees'.

Leaders can also have a major impact by surrounding themselves with people that bring complementary skills to their own. At British Airways, for example, Marshall brought in an American consultant named Mike Levin, whose brusque manners and sharp efficiency allowed the former to stay above the fray and remain untainted by some of the harsh measures that needed to be taken. At TSB, Finian O'Boyle (the branch Network Redesign' Project director) was less experienced, less visionary and probably less charismatic than the project director, Des Glover. He was, on the other hand, much more structured and systematic. Together, they constituted a very strong, complementary leadership team.

In his book *Managing on the Edge*, Richard Pascale deliberately selects Ford's low key president, Don Petersen, to exemplify transformational leadership. Petersen is the first to admit he has no master plan, but that he has orchestrated widespread pressure for change on several institutional fronts.

> *Petersen distrusts the concept of a strong, overpowering leader . . . The executive's role is that of prodder, facilitator, and catalyst. The approach taps the collective genius of the organization in order to identify and solve problems. As such, it is not as vulnerable to the imperfections of a strong leader.*[18]

Managing change, managing paradoxes

This chapter has highlighted a number of dimensions that can contribute to a more successful implementation of change. Our concluding thought is that, beyond useful frameworks and concepts, the job of a change agent involves many paradoxes.

Change agents must be self-confident and display a strong personal drive, but they must also have enough humility to listen to the concerns and objections raised by change recipients. They must have a vision, a strategy and guiding principles, but they must also be pragmatic, pay attention to unfolding events, build on initiatives that turn out to be successful even if they were not part of the original plan. They must identify 'early adopters' and 'project champions' to create a powerful coalition and generate early successes, but they must avoid leaving too many people 'behind' the pace of the change effort. They must be open and fair, but they also have to be shrewd and politically astute. They must have patience and persistence, but they must also, sometimes, act quickly and decisively.

The cases that follow illustrate various aspects discussed in this chapter. Case 5.1, 'Transforming TSB Group (C): Managing Change Teams' pertains to the

---

[18] Pascale, R.T. *Managing on the Edge: How the Smartest Companies Use Conflict to Stay Ahead*, Touchstone, 1991, pp. 66–67.

management of change projects (particularly process redesign projects) as such. Case 5.2 'Resistance to Change at Air France: Bernard Attali's Experience' describes the overall change effort at Air France over the years 1987-1993, ending with the resignation of the company's Chairman. Case 5.3, 'Pulling Air France Out of Its Dive: Air France under Christian Blanc' illustrates how leaders can sometimes get people to accept fairly drastic measures. Case 5.4, 'The Making of the 'World's Favourite Airline', reviews the transformation of British Airways over a 13-year period (1980–93) and highlights both the process and content of one of the most successful change programmes on record.

Case 5.1

This case was written by Francesca Gee, Research Associate, under the supervision of Soumitra Dutta and Jean-François Manzoni, Professors at INSEAD. It is intended to be used as a basis for class discussion rather than to illustrate either effective or ineffective handling of an administative situation.*

Transforming TSB Group (C): Managing change teams

In 1989, Peter Ellwood joined the TSB Group as chief executive of a newly formed retail banking division—TSB's core business. He found it in urgent need of radical change. The cost/income ratio had risen from 65.5% to over 72% in three years, TSB's products were not competitive enough, and the bank was losing market share. To deal with the bank's problems, he developed a four-step strategic approach:

> We created a vision of where we wanted to be, then we turned that into a mission statement to marshall hearts and minds within the company. We decided we would become the UK's leading financial retailer through understanding and meeting customers' needs and by being more professional and innovative than our competitors—not the biggest, but the best. We then evolved a strategy for achieving our mission with a double focus on driving costs down and increasing quality, sustainable income. Finally we turned that strategy into a series of specific actions.

The strategy was to be implemented through five projects. An important decision was whether to entrust the projects to line managers or dedicated teams. In Peter Ellwood's eyes, the decision was clear:

> I believe that the only way to bring about large-scale radical change is to divorce the action of change from the day-to-day process of management. Responsibility for delivering change should be given to dedicated, focused project teams using rigorous project management methodology.

Peter Ellwood took office in May 1989 and the five teams were formed six weeks later. Their initial brief was to submit an interim report to TSB's board by the end of September, followed by a blueprint for change a month later. They were then made responsible for implementing the changes. This was a daunting task.

THE FIVE PROJECTS

The five projects, described in Exhibit CS5.1.1, addressed a wide range of issues at TSB. In its 1989 form, the bank was the result of the amalgamation of dozens of nonprofit-making Trustee Savings Banks over nearly two decades. In 1986, the TSB Group, which included four banks (TSB England & Wales, TSB Scotland, TSB Northern Ireland and TSB Channel Islands) was floated on the London Stock Exchange. It acquired £1.5 billion in capital which it quickly spent on a series of expensive acquisitions. By 1989, when Sir Nicholas Goodison, former chairman of the London Stock Exchange, became the TSB Group's chairman, it was clear that the diversification strategy had failed.

The most complex of the five projects, Network Redesign, was aimed at transforming the organization of work in TSB's network of branches. The thrust of the project was to move back-office processing work from branches to dedicated customer service centres (CSCs). The result would be more efficient processing of transactions while branches would have more space and manpower to devote to service and sales. The project also involved restructuring the branch network itself by creating clusters of branches and decreasing managers' span of control, as shown in Exhibit CS5.1.2.

The Head Office Review's objective was to rationalize support functions which were spread out among the head offices of the four former banks and TSB England & Wales's six regions. This would involve building a flatter and more effective structure for the central head office as well as considering its transfer to a cheaper site. Three smaller projects were to conduct detailed reviews of TSB's strategy for corporate customers to develop customer and product profitability analyses, and, finally, to promote best demonstrated practice within branches across the UK. This case concentrates on Network Redesign, the most complex of the five projects.

The five teams each had between seven and 60 members (see Exhibit CS5.1.3). While the Head Office Review task force was made up of both TSB staff and external consultants, Network Redesign employed only TSB personnel. Launching five major projects simultaneously under tight deadlines would have been a challenge even if other hurdles had not complicated the teams' tasks. The first redundancies, announced in late 1989, created a climate of uncertainty which affected morale; staff assigned to projects were uneasy about their long-term futures. A freeze on recruitment made it harder to make up for TSB's insufficient skills in technology, project management and finance. Network Redesign faced an additional challenge in that its project office in London had to supervise dozens of teams deployed across the country.

ADOPTING A PROJECT MANAGEMENT METHODOLOGY

When the projects were launched TSB was not familiar with managing cross-functional projects. It had no culture of project management, lacked a reliable methodology and each team used its own method in the first few months. Confusion prevailed, especially in the case of Network Redesign because of its complexity. As project manager Finian O'Boyle explained:

> The project structure was an amorphous mass. We met once a week, with more than 20 people sitting around a table—everybody who had some contribution to make to the project. That continued until March 1990. Obviously, some progress was made, but how much was unclear. When the steering committee asked a question, our

answer was vague. The steering committee could see that there was a lot of activity going on, but couldn't feel confident that we were actually making progress. We were losing credibility. At that point, I met consultants who described to me a very structured, but relatively simple project management methodology.

The methodology (which is shown in Exhibit CS5.1.4) established clear monitoring and accountability rules:

■ The project director was accountable for delivering output to the chairman of the steering committee, to whom he submitted detailed terms of reference: a detailed action plan including timing, critical path analysis, and the costs and benefits of implementation.

■ Teams reported to their steering committee at least once a month.

■ The steering committee's chairman reported to the executive committee.

■ The project director and key team members had to attend a project management course.

■ Performance was monitored against timetable and budget.

The methodology, which was adapted to meet TSB's needs, forced each team to identify and involve all parties with a stake in the project. Mervyn Pedelty, chairman of the head office review steering committee, said:

Projects were no longer operating in vacuums. When there were logjams or resource problems, or when targets were unrealistic, it forced the issues on to the table. A project couldn't go wrong without clear early warning signs.

The methodology was adopted throughout the company, Mervyn Pedelty said:

It proved so successful that we decided to standardize it for use on all projects. We had proven that projects were the best agents of change and it was sensible to have a consistent approach to project management. It meant that one developed a common language system, a common set of priorities, that everybody understood. It also enabled us to tackle more change than other organizations would have the appetite for.

AN EXAMPLE OF PILOTING: PROCEDURAL AND OPERATIONAL CLUSTERING

A major objective of Network Redesign was to move part of the branch administrative workload to CSCs, a process known as clustering because each CSC would serve a number of branch 'clusters'. Finian O'Boyle explained:

To set up CSCs we needed to establish technology links between branches and CSCs. We knew that this would take longer than the patience levels of senior executives allowed. We also knew that there were going to be large-scale redundancies. To

319

provide some rationale for the redundancies and to deliver part of the movement of work to the CSCs, we divided clustering into two parts.

The first part, so-called 'procedural clustering', involved moving some staff to CSCs where they would process transactions on behalf of branches using largely unchanged procedures; it was implemented between May and November 1990. 'Operational clustering', which involved establishing a new communications network to give CSCs direct access to branch database records, was rolled out across the bank from August 1991.

An important rule imposed by the methodology was that changes should be tested through prototypes and pilots before they were implemented across the bank. 'Prototypes and pilots are a risk-controlled way of implementing change', said Finian O'Boyle. 'You take on different bits of the risk at each stage of the implementation.'

Procedural clustering, which basically consisted of distributing procedures and personnel between branches and CSCs, was expected to remove 12.5% of administrative work from each branch. Before launching a full-scale pilot, TSB tried out different procedural options in a series of prototypes such as the one described by Finian O'Boyle:

> *We set up a unit above a branch which we called the CSC. We moved work from a handful of branches into this unit, then we tried out the procedures to establish what sort of work a CSC would carry out, how this affected work in the branches, what pieces of paper had to move between the branches and the CSC. This prototype was aimed at getting the procedures right.*

Once prototyping was completed to the satisfaction of line management, the pilot could begin, he said. This required a detailed audit, assessing achievements against objectives, identifying critical success factors and completion criteria, and estimating cost savings and benefits:

> *In a pilot you had a specific set of tasks and activities documented beforehand which led to the creation of a new unit. We tested the tasks and activities which we expected to use in a full roll-out, adopting the procedures and service levels which we would expect from a unit in a full roll-out. The pilot also had to achieve specific target cost savings and benefits.*[1]

The procedural clustering pilot went live on 2 April 1990. Four months later, it had established the technical viability of the project (including technology and procedures) and all its objectives had been completed. Procedural clustering could then be rolled out, which involved installing supporting technology, briefing and training staff, and moving them from branches into CSCs. The roll-out consisted of 81 subprojects, one for each CSC. In each case, the CSC manager was the project leader with a member of the Network Redesign Project team acting as controller. During roll-out, the team measured set-up costs, operational costs and the workload transferred to CSCs.

While procedural clustering was being rolled out, the prototyping and piloting of operational

---

[1] Pilots, which took place at small branches, could fail to give clear indications of problems that would arise during roll-out at larger branches, where work methods were different.

clustering, had begun. Its goal was to further automate clerical administration, transferring an additional 11.5% of branch workload (measured before Network Redesign) to CSCs (Exhibit CS5.1.5 shows the amount of clerical work actually transferred). The technology and procedures were first prototyped, then piloted. Since operational clustering required new technology, it was implemented gradually as new systems were developed. Each stage was prototyped individually which involved installing hardware and software at three CSCs. The pilot proved that computerization of procedures worked; set-up costs and the additional workload transferred from branch to CSC were then measured during roll-out.

Management changes were synchronized with the changes in structure and the roll-out of new technology. For example, the new area directors and senior branch managers were all appointed in the summer and autumn of 1989, as areas and clusters replaced the old structure of district offices. Finian O'Boyle explained:

> This was done before we started setting up the CSCs. That way, we had people supporting us out there. New area directors who had benefited from the reorganization were more supportive of the changes. If we hadn't changed the organizational structure, there would have been resistance by people who knew they were going to lose their jobs . . . We appointed the more senior people at the very top first, then it took them a month or so to appoint the next level down. It was a phased, hierarchical process that enabled each new appointee to select his or her managers.

PROJECT TEAMS AND LINE MANAGEMENT

The involvement of line managers throughout the Network Redesign Project was seen as crucial. The project team provided the concept, but it would be line managers' responsibility to bring it to life by implementing changes. Network Redesign Project director Des Glover:

> We did get excellent support from the line in terms of making the savings happen. When we set up the project team, I made it very clear that we had to manage this through the line. We could not say to a branch manager, 'You will reduce this staff.' We had to do it through the line, so he would only deal with his line management. We would deal with the senior people, discuss it up front, then they were responsible for implementing it. The idea was that at the end of the project, the project team would disappear and the line would own the result. They would say, 'We don't know what this project team did, we did all the work.' That made it much more difficult to manage because we could not actually tell anybody to do anything, we had to persuade them. It required more skills.

Managers were given full responsibility from the beginning so they felt they had a stake in the change process. Finian O'Boyle explained that newly appointed CSC managers had to set up their own centre:

> It gave people, even at a relatively low level, a taste of, 'I have responsibility for making this change work'. This is one factor that made subsequent change easier. At all levels in the organization, people have become used to implementing change, turning things around.

The strong culture typical of a project team sometimes caused difficulties with line management, said Finian O'Boyle:

> Projects do not always mesh well with the organizational structure of the rest of the business. That was a serious issue and there was the potential for conflict between the project, which was dealing with one-off change in non-status teams, and the rest of the bank, which is very hierarchical and where the most senior managers like to have total control over their function.

In line with TSB's project methodology, the teams' internal structure had to be well defined. The most elaborate structure was that of Network Redesign, the largest task force. Sixty people worked on the project, including 12 project leaders, project controllers and project workers who reported to subproject teams. Team members came from several functional areas: technology, finance, organization and methods, personnel, premises, and marketing. External consultants provided support on modelling branch profitability and various changes linked with clustering.

The success of the project largely depended on the presence of senior executives on its steering committee. The Network Redesign steering committee, which Peter Ellwood himself chaired initially to stress its importance, included the three regional directors as well as the directors of personnel, marketing and technology.[2]

The 10 domain controllers were senior executives who were affected by the project to which they contributed resources, ideas and functional input. Their responsibility was to make sure that the right resources were supplied and that any issues raised by the project team were quickly resolved. They oversaw 'internal suppliers' and 'maintainers' from several functions, such as retail business services, regional business systems, personnel, premises and technology who delivered solutions to subprojects.

The project director was responsible for the overall organization and management of the project and reported directly to the chairman of the steering committee. The project manager worked in conjunction with the project director (as in a chairman/managing director relationship) and was responsible for managing the tasks and activities in the main project, and assisting with the planning and control of the sub-projects. He or she coordinated project reporting and quality assurance and was responsible for enforcing project management methodology and for the coordination of supporting systems.

The project director and manager headed the Network Redesign management team which grouped project leaders, each of whom managed a subproject from start to finish. The management team met monthly to approve staff appointments, recommend changes in scope and direction and approve project plans as well as any major changes to these plans.

The project support office, which was headed by Peter Ellwood's personal assistant, handled coordination of the five strategic projects, made quality checks on the reporting, prepared monthly reporting packs for the management committee and monitored progress and staff numbers.

---

[2] In 1990, once the project was well under way, Charles Love, one the bank's regional directors, replaced Ellwood as chairman and the number of members was cut to five: the directors for HR and finance, Ellwood's personal assistant, Des Glover and Finian O'Boyle.

Projects also depended heavily on line functions for support. Technology was particularly critical and some 200 technology staff initially reported to the project team.

SPECIFIC TARGETS AND MONITORING

The five task forces' initial brief in July 1989 was to submit to TSB's board an interim report by the end of September. This was to be followed a month later by a detailed blueprint setting out what each project sought to do and why, its estimated costs and benefits, its timing and a description of the consequences of each project.

In Network Redesign's case, implementation targets were as follows:

- The clustering pilot had to be fully operational; specific clusters were to be identified in all regions; key personnel to run branches and CSCs had to be identified by appropriate line managers.

- A detailed plan for refitting branches and CSC premises, and a detailed blueprint for roll-out and time schedule were to be drawn up;

- Clear plans were needed for closing down branches and area offices in each region, to include plans for dealing with the implications for staff.

- Detailed profit-and-loss and cash-flow implications would be spelled out, with a critical path analysis and a clear timetable.

Exhibit CS5.1.6 shows Network Redesign's implementation timetable, while Exhibit CS5.1.7 shows how the project was segmented into over 20 clearly defined subprojects.

The Head Office Review team's blueprint included a final cost structure and staff implications, as well as a plan setting out how and when TSB could achieve those objectives. The team was to identify key personnel in the revised structure and present detailed plans on moving head offices (including timing, premises and other aspects) and on how to run centralized regional administrative functions (including implications in terms of staff, estimated profit and loss and cash flow, critical path analysis and a detailed projected timetable). Throughout the project, the emphasis was on setting and meeting targets.

Peter Ellwood stressed the importance of close monitoring:

> We demand exacting project management standards with firmly fixed objectives and progress being tracked through senior steering groups who review the deliverables regularly with the line managers who have to use them. This degree of organization means that we know where we are with the implementation of a project at any given time.

The Network Redesign Project team was required to make a detailed report to the project support office monthly. Expenditure was budgeted and controlled using standard retail bank procedures and documentation. Des Glover's principle was that the cost and value streams should be synchronized so that whenever the delivery of benefits was postponed, so should spending:

Case 5.1

323

We carefully tracked the project's projected costs and benefits. We reported to Peter Ellwood on a monthly basis and to the board quarterly. That was managed pretty vigorously. Every month we had a big argument as to how many staff we had and how many had disappeared! These meetings were more about where we stood on the finances than on how far along we were in the process. The focus was basically, 'You've spent £20 million up to now, we should have had this much benefit, have we or haven't we?

TEAM MANAGEMENT

Restructuring the head office and branch network had produced a large pool of 'unallocated' staff from which some project team members were drawn. 'Most team members were taken from the "unallocated list", they were people who didn't have a job in the new environment', explained a manager. The extent of the changes caused nervousness, and Peter Ellwood noted: 'We had a problem at one stage in that people wouldn't go into projects because they thought that once the project was over they'd be out of a job'. He tried to counter that feeling by stressing the importance of project members:

> *I remember going to a meeting, it might have been on Network Redesign, and saying, 'I rate you guys very highly. The reason you're on this project is not because you're worse than other people, it's because you're better. You are people who can manage change of this magnitude, and we are totally committed to ensuring you're put back into the organization'.*

Finian O'Boyle:

> *A key issue within TSB is the standing of project teams and how to get them to dissolve back into line management. It is about making the best use of people so you can have people who work on a project, then move into a line job. We don't have a process to facilitate that.*

The Head Office Review project was deliberately scheduled to last no more than 12 months. Its team was wound down progressively, as shown in Exhibit 5.1.3, and disbanded by the autumn of 1990. 'We accomplished the bulk of the work in six to nine months', said Mervyn Pedelty. 'The problem with many change projects in organizations is that they take on a life of their own and go on for years.'

The Network Redesign team was disbanded at the end of 1992. By then, the regional and area offices had been set up, as well as all CSCs. Procedural clustering had been fully implemented and operational clustering had been introduced to all major CSCs. Peter Ellwood and the steering committee then decided that the project team should be disbanded and outstanding tasks handed over to line management. Finian O'Boyle recalled:

> *The final subproject was to hand over all remaining items to the line. We had to ensure that the documentation was up to date and that line management knew that they had the responsibility to continue the implementation. We wanted line management to have ownership of the successful implementation.*

Once the outstanding tasks had been handed over, however, it was difficult to ensure that they were carried out in the original spirit of the project. Des Glover remarked:

> *Experience shows that between two major initiatives, nothing actually happens. An initiative that no longer has a champion dies. If someone is not actively chasing it, people find more important things to do!*

Team leaders made efforts to ensure that team members went back to their jobs, possibly with a promotion, but this wasn't always possible, Mervyn Pedelty said:

> *We consciously decided that if project members were made redundant, it would be a disaster. We had many discussions and arguments and the vast majority of the Head Office Review team members found jobs. I was highly concerned that they did. It was up to the project director to use his influence. People who had done a good job on a project were often in great demand. They became project 'professionals' who could move from project to project.*

While some Network Redesign team workers found line management positions, many left, either immediately or within the next couple of years. The benefits of working on a project often boiled down to 'a line in your resumé', said Des Glover:

> *Working on a project can make you unemployable in the organization. The line functions feel threatened if you have to work for them once the project is over since, for two years, you were calling the shots. Whatever function you go into, you know more about the way the business functions than they do.*

A project manager explained that in the early 1990s some project leaders had probably been disappointed at not being offered more senior line positions once their project were completed:

> *I think there's been a shift in attitudes recently and people now acknowledge that there are good career opportunities in going from one project to another. For example, I have been working on projects for the last 10 years, and almost all the staff in business improvement work on projects. Also, the HR department is developing a skills database to help select team members.*

A SUCCESSFUL APPROACH TO CHANGE MANAGEMENT?

Under Peter Ellwood, TSB clearly opted for a project-based approach to change management. Both Network Redesign and the Head Office Review were successes. TSB cut the number of regional offices from six to three, avoiding duplication and creating one bank with a single focus. Operations were centralized in 'product factories'; for example, seven regional mortgage processing centres were merged into one. Some 15% of branch back-office work was moved into CSCs, improving efficiency and giving branches more space and time to serve customers. Some 300 branches were refurbished and a new telephone banking centre was prototyped. The headcount was reduced by over 5500 in three years and 1100 highly paid branch managers were made redundant. However, some questions remained.

Finian O'Boyle:

> The reason we were successful with projects was that when Peter Ellwood arrived, there was a big shake-up. Everybody realized that for the company to survive, they had to play ball. Some were able to work in this matrix structure. The project team needed line commitment to make change management work in a hierarchical structure. Subsequently TSB slipped back into a hierarchical way of doing things with big projects run within the line, from one of the functions. That is a testimony to the fact that TSB finds it difficult to have projects standing outside the line. That is not necessarily a criticism, but if you read books on organizational change they often talk of groups of people coming together quickly, achieving an objective and then disbanding back to their previous jobs. That does not currently happen within TSB.

While project management had succeeded in implementing major change initiatives soon after Peter Ellwood's arrival, it was unclear whether this would continue to be the best way to handle change. TSB knew that it had to translate accelerated change management into continuous improvement. In Peter Ellwood's words,

> We realize that change management has become a core competence of the organization. Change is not something that is tackled every five years, it requires a permanent cultural shift, so that the business can achieve continuous improvement.

Exhibits

Exhibit CS5.1.1 Five project teams presented in July 1989

Network Redesign

To identify and assess ways of improving the performance of the branch network, in particular by creating more selling space. Also responsible for the reconfiguration of branches and the provision of support from the planned CSCs. Will pay close attention to administration arrangements above branch level.

Project director: Initially Harry Read (outgoing Director of Technology), then Des Glover.

Steering committee chairman: Peter Ellwood.

Head Office Review

To review in detail head office costs across the whole of Retail banking and will examine whether certain tasks undertaken currently in regions should be centralized, e.g. mortgage processing.

Project director: Brian Cooper, outgoing finance director.

Steering committee chairman: Mervyn Pedelty, Finance Director.

Best Demonstrated Practice

Will identify the factors that promote superior branch performance and will establish the management action that is needed to ensure that best practices are adopted in all branches.

Project director: Joe Turner.

Steering committee chairman: Tony Wood.

Commercial Market Development

Responsible for formulating a detailed development plan for TSB's commercial business.

Project director: John Boreham.

Steering committee chairman: Eric Wilson.

Product and Customer

Will ensure TSB has products and services to sell that are the most appropriate for specific customer groups.

Project director and steering committee chairman: Frank Abramson, Director of Marketing.

Exhibit CS5.1.2 TSB's Retail Banking network management structure

Early 1989

TSB England & Wales	TSB Scotland	TSB Northern Ireland	TSB Channel Islands
6 regional directors			
54 district managers			
1600 branch managers			

Late 1989

TSB Bank	TSB Channel Islands
3 regional directors	
22 area directors	
159 senior branch managers	
1600 branch managers	

Note: The new structure was implemented in several stages throughout 1989. TSB Channel Islands was a private banking subsidiary. TSB Northern Ireland, which had a 56-branch network, was sold in May 1991.

Exhibit CS5.1.3 Size of task forces

	Oct 89	Oct 90	Oct 91	Oct 92	Oct 93
Project Support Office	1	4	0	0	0
Network Redesign	62	62	58	58	0
Head Office Review	27	7*	7*	0	0
Best Demonstrated Practice	8	10*	10*	10*	10*
Commercial	16	16	0	0	0
Managing for Profit	7	4	0	0	0
Total	121	110	85	78	10

* In line positions.

Exhibit CS5.1.4 Project methodology

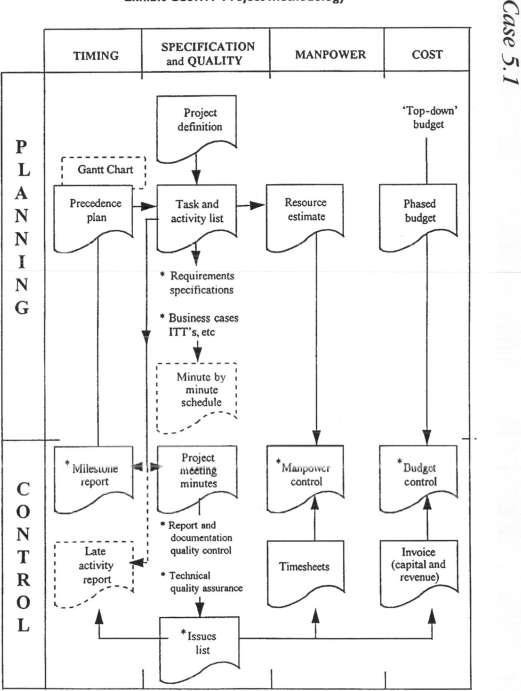

Note: ---optional; *key control document.

Exhibit CS5.1.5 Clerical effort transferred through procedural and operational clustering (by grade)

	Clerical grade 4	Clerical grade 3	Clerical grade 2	Secretaries
Procedural	16.2%	14.4%	12.0%	6.8%
Operational	9.6%	17.8%	15.8%	29.9%
Total	25.8%	32.2%	27.8%	36.7%

Exhibit CS5.1.6 Network Redesign implementation timetable (spring 1990)

Project start	8/89
Report produced	10/89
Board sign-off	11/89
Appoint area directors	completed 11/89
Appoint senior branch managers	completed 11/89
Establish brance management positions	ongoing
Transition to three regions and 21 areas	completed 5/90
Introduce sales clusters	4/90
Pilot CSC	4/90
Introduce procedural clustering	5-11/90
Move to full operational clustering	7/90-7/92
All CSCs live	9/91
Complete roll-out of back office	1992
New design branch programme	mid/late 1992
Project run down	2/92
Introduce additional CSEs	3/4 year programme

Exhibit CS5.1.7 Network Redesign project segmentation (January 1991)

Sub-Projects

Procedural clustering subprojects

Procedural clustering pilot. Handled the movement of people and tasks to CSCs, made it possible to downgrade branch managers. Included the creation of CSC processes and resulting changes in branch processes. Involved redesigning processes, documenting them, measuring them and forecasting their effect.

CSC premises roll-out. Involved setting up 77 CSCs in six months after identifying and fitting locations.

Procedural clustering roll-out. Involved 77 sub-projects, one for each CSC.

Branch Manager Re-grading. Removed a number of branch manager positions and regraded others in line with reduced responsibilities.

CSC organization and staffing review

Communications. Kept staff up-to-date on developments in roll-out and motivate them to adopt changes; ensured cooperation of senior managers; to inform external audiences of progress in improving quality and availability of services.

Network automation subprojects

Operational clustering pilot. Established telecom links between branches and CSCs to eliminate telephoning and faxing.

Operational clustering roll-out. Transferred 11.5% of branch workload to CSCs.

Super Service pilot. Expert system introduced as part of branch automation to help staff identify customers' product needs and structuring sales interviews.

Customer profiles project. Involved introducing a new standard procedure to integrate the use of customer information into sales interviews

Super Service roll-out. Implementated in all TSB Bank branches in parallel with measurement of Super Service

Telephone strategy. Involved setting up a prototype Teleservice unit to handle CSC calls and future development of telesales.

Operational clustering optimization

Regional and district restructuring subprojects

Changes in region and organizational structure (one project for each of Northern, Western and Southern regions). Involved rationalizing some regional head office functions and turning district offices into area offices. Staff cut from 1441 to 612.

Refurbishment of branches and CSE deployment subprojects

Branch refurbishment roll-out. 800 branches to be refurbished to 'new design' with more interview space to be used by sales staff.

Customer service executive deployment. 1500 sales posts to be created by 10/94

Branch refurbishment review

Other subprojects

Viability monitoring. Continuous auditing and monitoring of financial viability of each subproject against objectives

Branch refurbishment viability monitoring. To review the methodology used to define the financial viability of individual branch refurbishment proposals.

Technology coordination. To monitor Technology Division's delivery of key systems such as procedural clustering, Super Service, operational clustering.

Disbanding project team. To hand remaining items to line management.

Case 5.2

This case was written by Jean-Louis Barsoux, Research Fellow, and Jean-François Manzoni, Assistant Professor, both at INSEAD. It is intended to be used as a basis for class discussion rather than to illustrate either effective or ineffective handling of an administrative situation. It is based on publicly available sources and is intended to be used in conjunction with the corresponding INSEAD-CEDEP case on British Airways (1980–93).*

Resistance to Change at Air France: Bernard Attali's Experience

INTRODUCTION

In September 1988, Bernard Attali is named as the new boss of Air France. Hailed as a strong appointment and 'a breath of fresh air', his arrival coincides with a national economic boom, a new leftwing government and the reelection of François Mitterrand as president. France is buzzing.

Fast forward five years. It is 23 October 1993: tyres are burning on the runways, both Paris airports have been blocked for nearly a week and the strikers are rejecting Bernard Attali's plan to return Air France to solvency. The unions, overwhelmed by the groundswell of seething anger, are no longer in control of their members. Asked what they think of Air France's deficit, the strikers respond: 'It's not our problem. It's not because of our salaries that the airline has lost its way. For 10 years now, we've been the ones to pay. Now it's over. It's their mess. It's their crisis.'[1]

The following day, on the evening news, transport minister Bernard Bosson, fearing that the conflict may spread beyond Air France, announces the withdrawal of the plan, citing the lack of dialogue between top management and unions as a key factor. This triggers the immediate resignation of Bernard Attali.

This was the sad conclusion to a five-year reign that had started very promisingly. Attali remained dignified and refused to comment on the events, but many observers felt he had been seriously let down by the government.

---

*Copyright © 1997 INSEAD-CEDEP, Fontainebleau, France.
[1] Le Monde, 26 October 1993, p. 20.

As Bernard Attali left his office, on the 15th floor of the Montparnasse building in central Paris, many people were wondering what had gone wrong.

THE AIRLINE SECTOR IN FRANCE

Air France was officially born in 1933, following the merger of four French transporters and the acquisition of a fifth. During the war, the airline was placed at the service of the government, and at the end of the war the company was nationalized and given exclusive rights on popular international routes. From 1948 to 1961, under the iron leadership of Max Hymans, the company developed the 'Air France spirit' born of a strong attachment to the airline and its prestige as a promoter of French expansion.

In the late 1950s, a new company, Air Inter, started operating a few domestic flights with leased aircraft. The prevailing attitude at Air France was that as national carrier, it was busy enough fulfilling its international mission to bother with a regional service. In 1961, Air Inter was officially inaugurated, and by the late 1960s had become a vibrant success.

Two other airlines, UAT and TAI, which had been granted limited traffic rights on very long-haul routes in the 1950s, merged in 1962, forming UTA. A subsidiary of the Chargeurs Group, the parent company's president used his political muscle to renegotiate the routes attributed to the company. In 1963, the government took away some of Air France's exclusive rights—the western and southern parts of Africa, Australia and the Pacific and awarded them instead to UTA.

The result was three French companies whose networks essentially did not overlap. This avoided harmful competition between the companies, while at the same time providing sufficient expansion opportunities to all three.

In 1971, the heads of Air France, Air Inter and UTA had joined forces, with the backing of the government, to take on the union of the flight personnel. After 26 days of strike and a lock-out, the government lost its nerve and backed down. The agreement which ensued guaranteed high levels of remuneration and favourable work conditions for French pilots.

In 1974, the authorities 'invited' Air France to take over the newly built Roissy airport, north of Paris, thus forcing the company to split up its operations. Meanwhile, Air Inter maintained its base at Orly, south of Paris, together with most of the long-haul foreign carriers, making these the natural choice for transatlantic passengers travelling to and from the French provinces. The state also intervened in the purchasing policies of the airline. In the interests of French industry, Air France was obliged to boost its fleet with Airbus A300s in the mid-1970s, though the size of the aircraft was disproportionate to the carrier's needs.

In 1986, with the African market collapsing and competition in the Asian market intensifying, UTA sought access to more lucrative markets. The French authorities accorded UTA the right to fly to San Francisco, but not New York or other European destinations requested, thus marginalizing UTA.

In 1987, through stock market purchases, UTA acquired a stake of 35% in Air Inter. Air France matched this stake. The remaining third was held by various shareholders, including SNCF with

12.3% and CDC Participations and Crédit Lyonnais, each with 4%. In 1988, UTA declared it would be prepared to relinquish its stake in Air Inter against development opportunities for UTA. Meanwhile, Air Inter, too, was feeling unnecessarily constricted and was lobbying for European routes.

At this time, Air France was a fragile edifice. Uniquely, for a national carrier in Europe, it was denied access both to whole chunks of the world, as well as to its own domestic market. These represented obvious structural weaknesses for competing in a European market on the verge of deregulation.[2] In contrast to the highly centralized situation in Germany, Britain, Spain or Italy, where a single carrier covered both domestic and international flights, Air France was a 'tree deprived of roots'.[3]

By the time Attali arrived, the idea that something would have to be done about reorganizing the French airline industry had been floating around for some time. The situation had to change, but how: towards competition or complimentarity? Commentators speculated that there were two possiblities: either one huge airline incorporating all three companies; or else one large entity (Air France plus Air Inter) and one smaller one (UTA). At the time, the head of UTA's parent company, Jérôme Seydoux, (who acquired the Chargeurs Group in 1979) regarded the former proposition as 'counterproductive, given that the players in French air transport are a long way from recognizing the accumulated lag in productivity'.[4] He claimed that grouping the three companies before they had 'cleaned up their acts' would mean relinquishing power to the state and to the unions.

THE AIRLINE THAT ATTALI INHERITED: THE FRIEDMANN LEGACY

Bernard Attali's predecessor at the head of Air France was Jacques Friedmann. A close adviser of the incoming rightwing Prime Minister Jacques Chirac, Friedmann was appointed head of Air France in February 1987, thus becoming Air France's third chairman in four years.[5]

Within four months of taking over, Friedmann launched an ambitious listening exercise dubbed 'Projet Air France'. At first, the unions suspected that this initiative was intended to identify potential productivity gains or workforce reductions. But the approach was new: the personnel was widely consulted. Categories of staff who knew very little about each other (notably ground and flight staff) were assembled in small groups to discuss work processes. The initiative was well received, as one manager recalled: 'He came down among us. These weren't symbolic tours of the workplace with lots of handshaking. They were genuine work meetings, and it worked.'[6]

---

[2] Unlike US deregulation, liberalization of air transport in Europe was a gradual process, delivered in several packages: in 1988, 1990 and 1993. These measures gradually abolished many of the restrictions on discounts, capacity, market access and new entry. They also introduced a ruling whereby an airline from one EC country could carry traffic between two other EC countries. Nevertheless, the opening of European skies has not had a huge impact on competition because: (a) airport slots remain in short supply; (b) many state carriers have been kept afloat by subsidies and (c) most long-haul routes to nonEuropean destinations are governed by bilateral agreements that determine who flies where, when, how often and at what price.
[3] Attali, B. *Les Guerres du Ciel*, Fayard, 1994, p. 25.
[4] *L'Expansion*, 9 September 1998, p. 117.
[5] Friedmann's immediate predecessor was Marceau Long (1984–86) who was offered the vice-presidency of the *Conseil d'Etat*; the president before him was Pierre Giraudet (1976–84) a former head of the Paris subway system, who had reached retirement age.
[6] *L'Expansion*, October 1993, p. 85.

Another of Friedmann's preoccupations throughout his time in office was to pursue the efforts of his predecessor Marceau Long to reduce the company's debt. Recruitment was carefully regulated, and between 1985 and 1988 Air France's workforce grew at about one-third of the rate of British Airways (BA), Lufthansa, or KLM during the corresponding period. Indeed, in the period 1985–88, of the big four European airlines (along with BA, KLM and Lufthansa) Air France was the only company to reduce its number of maintenance employees—thanks to a generous early retirement scheme in 1987 which prompted significant departures among supervisors and adjusters.

Investment in aircraft was also tightly restricted, although spending did increase under Friedmann (to Ffr 3.4 billion for 1986–88 compared with Ffr 2.3 billion in the period 1983–85). By 1989, the average age of Air France's fleet was over 10 years old compared to less than 8 years old for its main European rivals.

These on-going efforts paid off. Between 1983 and 1988, the debt/capital ratio was reduced from 4.4 : 1. Financially the company was starting to look in good health. In 1988, it had recorded unbroken profits for five years (see Exhibit CS5.2.1). On the other hand, this 'policy' did result in a loss of European market share,[7] increased fuel costs (because the older planes were less fuel efficient) and, some argued, a deterioration of the company's image.[8]

In May 1988, with the uncertain outcome of the forthcoming presidential election, Friedmann told the personnel that the *Projet Air France* would be put on hold. He wanted the approval of the transport ministry before pushing further with the exercise. As it turned out, the socialist president, François Mitterrand was reelected, which quickly led to the replacement of the rightwing government. At this stage, Friedmann suspected that his days were numbered. After only 19 months at the head of Air France, he was replaced by Bernard Attali.

Friedmann's last significant action before handing over the reins was to secure a commitment from the board for an intensive fleet investment programme worth Ffr 42 billion, to cover the purchase of 48 planes and a dozen 737-500s (100 seaters) to be delivered from 1991.

With Attali set to take office, Air France's position looked healthy: it was ranked third among European carriers by turnover behind BA and Lufthansa (see Exhibit CS5.2.2 for rankings). The company's fleet stood at 110 planes against BA's 170 (not counting the fleet of the recently acquired British Caledonian). According to the standard industry measure (ATKs[9] per employee), Air France's productivity stood just behind KLM, on a par with Lufthansa and ahead of BA.[10]

There was a broad consensus in the press regarding the strategic challenges facing Attali. For example, *Le Figaro-Economie*,[11] claimed Attali would confront three major challenges very quickly: increased competition on US routes (see Exhibit CS5.2.3 on evolving shares of US traffic);

---

[7] Air France's share of European traffic (passengers/kilometres transported) dropped steadily between 1979 and 1987, going from 13.4% to 11.2% (*L'Expansion*, 9 September 1988, p. 116).
[8] Perri, P. *Sauver Air France*, Paris: L'Harmattan, 1994, p. 93.
[9] Available tonne kilometres (ATKs): The number of tonnes of capacity available for the carriage of revenue load (passenger and cargo) multiplied by the distance flown.
[10] *Libération*, 6 October 1988, p. 9.
[11] *Le Figaro-Economie*, 29 September 1988, p. 1.

Case 5.2

adapting to the new climate of European deregulation set for 1993; and the likely reorganization of the three French airlines, Air France, UTA and Air Inter.

ATTALI'S CREDENTIALS: THE PRESS'S VIEW

The immediate press reactions to Attali's appointment were fairly favourable. Of course, as the twin brother of President Mitterrand's special adviser, there was little doubt about the political nature of his nomination, but this was not new at Air France. Like Friedmann before him, Attali was a product of the French system: a top flight graduate of the élite Ecole Nationale d'Administration (ENA), with many years experience in the public sector (see Exhibit CS5.2.4 for description of l'ENA and Attali's CV).

In other respects, however, Attali's profile clearly represented a break with the past—and the press acknowledged this. For a start, at 44 he was Air France's youngest ever chairman. Traditionally, the top job at Air France had been considered something of a cushy job—'almost an honorary post'—somewhere loyal servants might land 'at the end of their careers', as one journalist put it.[12]

A lot was expected of Attali. For the financial daily, *La Tribune*, Attali was the ideal candidate: 'He is young, has big company experience with GAN and an international outlook thanks to a spell with a large British insurance group [Commercial Union].'[13] The leftwing daily *Libération* highlighted some of his past coups—notably the purchase of the CIC banks from Suez[14]—which he might leverage in his new job. *Le Monde* considered that: 'The government has bet on someone with high-level experience in both negotiation and international cooperation, indispensable skills for confronting the deregulation of European air transport.'[15] And the *Financial Times* confirmed his standing within the French business community: 'Although Mr Bernard Attali . . . is closely associated with the Socialists, he enjoys widespread esteem as a former chairman of the GAN state insurance group.'[16]

For all his credentials, Attali faced a tricky challenge, coming into an industry he knew nothing about, with the industry on the verge of momentous changes. The situation was captured by *La Tribune*: 'Attali will therefore have to make his mark as a real boss in the eyes of Air France's employees, but beyond that he will have to take key decisions which will determine the company's course for the next decade . . . Very soon he will have to make his mind up about the national carrier, before defending his position with the French authorities.[17]

TAKING MEASURE OF THE CHALLENGE

On offering Attali the job, socialist Prime Minister Michel Rocard had warned: 'You know, Bernard, it won't be easy.' Attali was aware that the state-owned airline represented a difficult social context, with its 14 unions and its history of industrial unrest (see Exhibit CS5.2.5 for a

---

[12] *L'ObsEconomie*, 28 February 1991, pp. 39–41.
[13] *La Tribune*, 28 September 1988, p. 1.
[14] This was a delicate negotiation in that it was the first time an insurance company was acquiring a bank, and the CIC did not want to fall under the control of the GAN Group.
[15] *Le Monde*, 29 September 1988, p. 38.
[16] *Financial Times*, 28 September 1988, p. IV.
[17] *La Tribune*, 28 September 1988, p. 11.

brief overview). But he was also conscious that this was a tremendous opportunity, a chance to make his mark. Attali later said: 'When, at the age of 44, you are asked to manage a group of 37 000 people, you don't have the right to hesitate.'[18]

'The company is in need of a genuine manager', Attali declared to journalists on arriving.[19] Internally, he delivered a similar message. On the day of his appointment, he assembled his board of directors and showed firm intentions: 'I will take very seriously my role as leader, and I would like to be kept informed by you of all notable events in your various sectors of activity. You will be judged on two simple criteria: your performance as managers and your loyalty towards me.' He also made it clear to them: 'As from today, there are two realms which I consider as my responsibility: relations with the authorities and all external communication.'[20]

On emerging from the meeting he addressed a letter to the entire staff: 'Be prepared: I intend to be very ambitious for Air France. Each time we have the means, we will charge forward together. And I will put everything into it.' For the press, as much as for the personnel, these were unprecedented words from a head of Air France. As *l'Expansion* remarked: 'Charge forward? There's a new and exciting idea, for all those who suffer from having been held back for too long.'[21]

In the first week, talking to senior managers one-on-one, Attali quickly took measure of the need to revitalize the company: 'I talked to them about customers, they talked to me about the company; I asked questions about the product, their answers were all about production.'[22] This misplaced focus was also reflected in the company magazine—a glossy review produced three times a year, in which more space was devoted to the French pioneers of aviation than to prevailing company issues.

Attali's early encounters left him feeling uneasy. He was struck by the apparent sense of complacency and smugness which characterized internal discussions. But before he could set to work on these issues, he was blindsided . . .

Barely 12 days after taking the job, Attali had a first strike on his hands from the maintenance employees. It turned out to be Air France's longest ever strike, lasting 100 days (from October 1988 through to January 1989) and costing an estimated Ffr 1 billion in cancelled flights and bookings.

Essentially, it was the result of earlier measures imposed on the maintenance division where numbers had been tightly regulated since the mid-1980s. The sudden upsurge in traffic in 1988 caught everybody flat-footed. In January 1988, *l'Expansion*'s headline had been: '1988: Why it will be a Dismal Year'. Nine months later the magazine was trumpeting: 'Growth in 1988: Courtesy of the Crash!' For Air France, which had anticipated modest growth, 1988 proved a year of continual upward revision, with ever-increasing demands on the fleet and a steadily worsening punctuality record. The mechanics were simply caught in the middle.

---

[18] *Le Moniteur*, 28 July 1989, p. 1.
[19] *L'Usine Nouvelle*, 21 February 1991, p. 23.
[20] *L'Expansion*, 6 January 1989, p. 48.
[21] *L'Expansion*, 6 January 1989, p. 48.
[22] Attali, B. *op. cit.*, p. 11.

Case 5.2

At one point, irritated by the contents of a union pamphlet, Attali went down to see the mechanics in their workshops, and talked with them for two hours. Such behaviour on the part of the chairman was unheard of at Air France. Eventually, Attali's efforts, which included a Christmas night spent discretely negotiating a return to work with one particular union, paid off. The negotiations led to the recruitment of 1500 new maintenance employees in the following months. Attali's interpretation of the strike was that it was 'for a love of a job well done' and that it was actually 'a growth strike'.[23]

The eventual resolution of the strike and his understanding attitude towards the strikers earned Attali plaudits from the press, the verdict being that 'he showed genuine comprehension towards the unions'.[24] Asked by one journalist if he would be able to reform the company without incurring the expense of a large-scale strike, he answered: 'Industrial unrest is always the sign of a collective failure. I will do everything in my power to avoid such jolts and to maintain and improve the dialogue within the company.'[25]

ATTALI'S EARLY PROMISE

Shortly after his appointment, *l'Expansion* observed that, 'There is a real breath of fresh air at the head of Air France'.[26] Here was a boss who was prepared to come down to the lower floors, without warning, in shirt sleeves, to ask for information or to enquire directly about the work of a particular department without protocol or reprisals. Whenever he took a plane, he always remembered to go to see the pilots.

He would even make impromptu visits to the cafeteria, commenting tongue-in-cheek: 'Why should I be the only person out of 37 000 employees not allowed to go and take a coffee with the others?'[27] Without being familiar, he was approachable. He had no team of advisers or 'inner circle', just secretaries and an assistant.

He soon distinguished himself as a glutton for work, with three secretaries working full-time just for him. He thought nothing of coming in to work on a Sunday, and he would sometimes call on his senior managers to do likewise.

Over the first six months, he studied the business conscientiously and quickly grasped the ins-and-outs of each issue he handled. He wanted to see everything, understand everything, and be involved in everything. He regularly went out into the field and dined fortnightly with pilots at one of the company's Meridien hotels. As one pilot commented: 'If you start talking to him about the price of jet fuel, say, he can give you the full cost breakdown on the spot.'[28]

Strategically, too, there were signs of a change in direction. Asked whether he intended to pursue the massive self-appraisal exercise launched, but finally dropped, by his predecessor, Attali answered:

---

[23] Attali, B. *op. cit.*, p. 13.
[24] *Le Nouvel Economiste*, 17 May 1991, p. 29.
[25] *L'Expansion*, 6 January 1989, p. 50.
[26] *L'Expansion*, 6 January 1989, p. 48.
[27] *Vogue Hommes*, July/August 1991, p. 113.
[28] *Le Point*, 10 December 1990, p. 96.

Case 5.2

> *The collective reflection exercise generated both a momentum and some frustrations, some expectations and some criticisms. I think it would be difficult to pick up this process, suspended six months ago, as though nothing had happened. In the weeks to come, I will marshall my thoughts in order to rebound in my own manner. Air France will take-off again.*[29]

Within the first few months of 1989, Attali set about making a number of symbolic changes. He established a consultative training committee, and a communications department; he introduced a weekly company newsletter 'Air France Infos' and a modest profit-sharing scheme. He gave the go-ahead for Air France's first-ever TV advertising campaign. Then, in May 1989, he assembled 500 Air France managers to talk to them about the future of the airline.

As *L'Express* commented that month: 'With 1993 and the continuing process of deregulation firmly in sight, Attali is taking an offensive stance rather than opting for caution. The company's strategy has four main lines: quality of service, growth, competitiveness and social cohesion.'[30]

This ambitious strategy was confirmed two months later by Attali himself:

> *Today, air traffic is buoyant. We need to respond to this demand and seize the chance to conquer new market shares. The time has come to step up a gear and make sure that each of our actions fits in to the strategic logic of the group. . . . In the forthcoming battle, we will need friends. There are numerous possible alliances and the partnerships should be lasting. We cannot afford to make mistakes.*[31]

The time had come to build critical mass.

FULL THROTTLE

Air France's weakness in the domestic market convinced Attali to take 35% stake in France's fourth largest carrier, the regional airline Transport Aérien Transrégional (TAT), following tough negotiations in July 1989. 'Joli coup' was *Le Monde's* verdict on the operation.

This was but the first of a succession of strategic coups aimed at consolidating Air France's domestic position and extending its global reach. In September 1989, Air France sealed a broad collaboration pact with Lufthansa. The two airlines also launched a joint training programme for pilots in the United States, as the French civil aviation authority could not meet the demands of Air France. These closer ties between Europe's second and third largest international carriers, signalled a determination to compete against BA, which was discussing a stake in United Airlines (United States) and a stake in Sabena, the Belgian carrier. Attali announced that he, too, was 'seeking to establish links with one or two strong partners in Europe, the United States and the Far East'.[32]

At the same time, Air France was pursuing a big fleet renewal programme. The company

---

[29] *L'Expansion*, 6 January 1989, p. 50.
[30] *L'Express*, 26 May 1989, p. 94.
[31] *Le Moniteur*, 28 July 1989, p. 1.
[32] *Financial Times*, 15 September 1989, p. 31.

ordered 28 aircraft in 1989 (due for delivery in 1991), and at the end of that year its fleet stood at 112 aicraft. The purchase of five Boeing 747-400 freighters (also due for delivery in 1991), was actually the largest single order for cargo aircraft received by Boeing. The Ffr 11.8 billion of investment in new aircraft, contributed to a rise in net debt payments from Ffr 216.5 million in 1988 to Ffr 804.9 million in 1989. But to meet its immediate needs, Air France was having to lease aircraft from other carriers (often with their livery on the side).[33]

Then, at the start of 1990, and without warning, Air France acquired its main French rival, UTA, for Ffr 7 billion. These negotiations lasted six weeks and were carried out in the utmost secrecy. Neither the head of UTA nor that of Air Inter were in on the talks between Attali and Jérôme Seydoux, chairman of UTA's parent company, Chargeurs Group. Inevitably, this placed a very heavy burden on the two individuals concerned, but Attali was in his element. It reminded him of his brief spell as financial director of the Club Méditerranée, and the deals he had struck with heavyweight financiers.

In the process, Air France also gained indirect control of Air Inter which had a monopoly privilege on domestic air transport and operated out of Orly airport, south of Paris. This significant move lifted Air France's share of the home market from 78% to 97%, thus putting it on a par with its main European rivals: Lufthansa with 99.8%, KLM with 97.2% and BA with 89% in their respective countries.

As a condition for EC approval of the merger, Air France was forced to shed its stake in TAT, and to relinquish some 50 international routes to independent French airlines. Nevertheless, the size of the new-look Air France Group was impressive with its network of 246 destinations, its annual sales of around Ffr 55 billion, and its fleet of 196 aircraft (52 from Air Inter), with orders for 240 more (130 firm and 110 options).

Attali was proud of his achievement. With one move he had hoisted Air France from ninth to third in the world league, in terms of turnover, overtaking BA in the process (see Exhibit CS5.2.2). Air France was now Europe's largest airline. Here was visible proof that Attali was different from his predecessors: 'After a succession of "figureheads", the airline now has, in Attali, a chairman with forceful ambition.'[34]

The acquisition resulted in a number of rapid economies, notably in terms of joint purchases of fuel, insurance contracts, and the common recruitment of flight staff. In other respects, Attali wished to keep the three companies separate, against the advice of René Lapautre, the former head of UTA, who recommended immediate integration on the British Airways–British Caledonian model.[35] Attali reasoned that:

> We are not in the same situation as BA and BCal who were competing on numerous routes. Our three companies are very much complementary. Each one has its own

---

[33] Lehrer, M. Comparative institutional advantage in corporate governance and managerial hierarchies: The case of European airlines, Unpublished doctoral dissertation, INSEAD, 1997, p. 167.

[34] L'ObsEconomie, 28 February 1991, pp. 39–41.

[35] When British Airways acquired the financially troubled British Caledonian in 1987, it immediately suppressed the airline and made 2500 redundancies. The remaining BCal personnel were put through an existing BA training programme. Within weeks the route structures had been integrated and the BCal livery switched to BA's colours.

Case 5.2

history and identity. Merging the three at once, just for the sake of a neat organizational chart, would be very risky. I have chosen a middle path.[36]

Attali was concerned that a forced merger might engender social unrest. His concern was reinforced when Air France decided to transfer some of its Nice personnel to Air Inter's colours, while maintaining their seniority rights, salary and status. The decision was part of the first rationalization programme, but after five days of strike, the management backed down.

Soon after this setback, Jean-Cyril Spinetta, a former adviser to the transport minister, was named as the head of Air Inter to replace Pierre Eelsen. Eelsen had wanted to resign over the handling of the takeover, on which he was not consulted. He was persuaded by the transport minister to postpone his decision to avoid destabilizing the newly constituted group. The appointment of Spinetta, who did not have a 'history' with Air Inter, further strengthened the position of Attali.

As *Le Point* saw it, this left Attali holding a strong hand: 'It remains for this ambitious man to prove that he is a leader of stature. The situation of his group, which is in mid-upheaval, is ripe for it.'[37]

PLAN FOR ADJUSTING TO THE ECONOMIC CONTEXT (JULY 1990)

In May 1990, Attali was starting to sense some troubling signs. After three years of rising profits, American carriers were announcing lower figures for the year ending March 1990. There were similar signs in Europe, partly as a result of a 20% rise in fuel prices. Looking at the figures more closely, Attali realized that charter traffic within Europe had collapsed, and that, although passenger loads on international lines remained steady, yields were dipping. Attali's diagnosis was that this heralded the start of an overcapacity crisis.

Attali's response, in June 1990, was to initiate a 'Plan for Adjusting to the Economic Context' (Plan d'Adaptation à la Conjoncture). Described by some as a mini-recovery plan, one French financial daily actually headlined: 'Attali's Caution'.[38]

Launched in July, this initiative mainly involved a reduction in the number of unprofitable flights as well as spending controls and cutbacks. Some of the necessary route closures were obvious, like the direct flight between Lille–New York with its average of 27 passengers! But the airline reliance on a route-by-route accounting system meant that it was far from clear which routes were genuinely profitable in terms of their overall contribution. The reason for this is that airlines often run costly short-haul flights which feed their high margin long-haul flights.

That summer, Attali made it clear in the company magazine that there was no question of diverging from the strategic course he had set:

I understand that one might feel a gap sometimes between the strategy defined for the long term and the management of day-to-day difficulties. It's a bit inevitable. I

---

[36] *ObsEconomie*, 28 February 1991, p. 40.
[37] *Le Point*, 10 December 1990, p. 96.
[38] *La Tribune de L'Expansion*, 21 June 1990; cited in Attali, B. *op. cit.*, p. 102.

Case 5.2

have set the course for reaching the goal. It is now up to the different hierarchical levels, by delegation, to translate them into daily practice.[39]

At the same time, he held up three reasons for the sag in load factors and yields—Hurricane Hugo in the Carribean, visas and taxes in Algeria, and intense competition on North Atlantic routes.

GULF CRISIS (AUGUST 1990–MARCH 1991)

On 2 August 1990, Iraqi troops invaded Kuwait. Very quickly, the sharp rise in the price of aviation fuel and air insurance premiums, together with a sharp drop in passenger traffic driven by the fear of terrorist attacks, provoked a deep crisis in the airline industry. The plan, which Air France had introduced in July, was reinforced in September 1990, with a set of austerity measures. These included the suspension of numerous investments, a freeze on recruitment, the nonreplacement of natural departures, the nonrenewal of fixed-term contracts, and reductions in overtime. In all, 1500 ground staff jobs were to be shed through natural wastage by the end of 1991.

Unlike numerous rival airlines, Attali refused to entertain the idea of layoffs. This prompted one journalist to comment: 'He denies himself cost savings where most capitalists would have sought them.'[40] Instead, Attali reinforced a number of the measures launched in the *Plan d'Adaptation à la Conjoncture*, closing nine routes between Paris and Europe, and 50 routes outward bound from the provincial airports.

Yet getting the unions to accept the necessary changes proved time-consuming. Attali considered that he devoted some 50% of his time to the unions. In conversation with one Air France veteran, Attali asked: 'How many times have we convinced each individual union representative, face-to-face, later to find that the only thing they could all agree on was to shoot down the top management.'[41]

For their part, the unions were still not sure what to make of Attali. They readily acknowledged his willingness to avoid conflict, but were not sure whether 'this reflects the consummate art of manipulation or whether it stems from a genuine social concern . . . They get the impression that they are listened to, but there is no concrete outcome at the other end.'[42]

On top of all this, Attali had to deal with a succession of transport ministers (five in five years), and was constantly at pains to make them understand the tremendous social complexity of the company. He told them what a melting pot of categories the company represented, and how explosive the mixture could prove. He talked of the 'paralysing mistrust' which reigned in union circles, and particularly of the 'psychosis of the pilots'.

Several European airlines posted losses for 1990. Air France, too, recorded its first losses since 1982, in spite of substantial sales of assets, including 14 aircraft for a total of Ffr 2.4 billion (see

---

[39] *France Aviation*, August 1990, p. 4.
[40] *Le Point*, 10 December 1990, p. 97.
[41] Attali, B. *op. cit.*, p. 75.
[42] *L'ObsEconomie*, 28 February 1991, pp. 39–41.

Exhibit CS5.2.1 for evolution of financial situation of Air France). During the Gulf Crisis numerous airlines imposed stringency measures. Even, British Airways, one of the few profitable companies in 1990, announced 4600 voluntary redundancies at the start of 1991 (see Exhibit CS5.2.1 for workforce comparisons).

In January 1991, Air France announced that it was cutting 2000 flights from its schedules, around 6% of its total, as well as postponing the delivery of seven Airbus A340s for a year, and cancelling options on four Airbus A310s.

A month later, a new set of stringency measures was introduced following the declaration of war in the Gulf. Anxious to cut costs without shedding jobs, Attali opted for the solution of reducing hours across the board by 6% and net pay correspondingly, freezing all salaries at their 1990 levels, and introducing measures to encourage the early retirement of 200 managers aged over 53, causing one union representative to comment: 'For the first time in its history, Air France has broken the guarantee of employment.'[43]

There were also fears among the union representatives that the Gulf War would serve as a pretext to introduce harsher measures: 'I can see the current circumstances being used to pile on everything at once', claimed one union leader.[44] Nevertheless, at a time when most of the European and American airlines were making redundancies, some union representatives privately conceded in interviews that an enforced reduction in working hours was actually the lesser of two evils.[45]

At the start of March 1991, the conflict ended and there was immediate pressure from the unions to revoke the measures agreed three weeks before. The management accepted, reinstating the full hours, but maintaining the salary freeze.

The intensity of this pressure showed Attali that the company was still a long way from accepting the reality of the situation. To bring it back to its senses, Attali decided to call upon the services of Arthur Andersen, a consultancy company with considerable experience of helping large, private companies adapt to the market. As Attali saw it: 'It was necessary to begin something like a collective psychotherapy . . . to hold up a mirror to the company, to get us to look at the reflection, and to draw the necessary conclusions.'[46]

THE ARTHUR ANDERSEN STUDY (MARCH–AUGUST 1991)

The Andersen audit exercise lasted six months and was undertaken in close association with the unions and works council. It involved meetings with over 1000 executives and staff at all levels. It provided an opportunity to review the whole organization, the totality of procedures and to allow everyone to participate.

Although the unions cooperated, there was a latent feeling that much of this ground had been covered in the abortive Friedmann initiative. There was also a suspicion that the consultants

---

[43] *Le Nouvel Economiste*, 22 February 1991, p. 29.
[44] *Le Nouvel Economiste*, 22 February 1991, p. 29.
[45] *Le Nouvel Economiste*, 22 February 1991, p. 29.
[46] Attali, B. *op. cit.*, p. 106.

involved could not quite seize the specificities of a French state-owned group, and were focusing exclusively on the productivity issue. Nor did it help that the Andersen study was very costly (Ffr 38 million) and that the cost was later leaked.

Meanwhile, the annual report for 1990 trumpeted the fact that Air France was ranked third in the world for the number of international passengers transported; fourth biggest exporter in France, and top exporter of services in France.

In the midst of the audit exercise, in May 1991, sensing a certain amount of doubt and anxiety, top management made an effort to communicate more openly with the unions. A hundred or so union representatives attended a huge video-presentation with graphs and tables analysing the results of the company in detail. This exercise was intended to raise the sense of involvement of the unions and was later reinforced by a new corporate advertising campaign destined to reach both employees and customers: 'Air France has chosen to turn the spotlight on its personnel, its most valuable resource, whose professionalism and commitment to passenger service are the key to our efficiency', ran the message.

In July 1991, a change in the financial structure of the airline with the state-controlled BNP bank taking a 9% stake in the airline (amounting to Ffr 1.25 billion). Along with a state injection of Ffr 2 billion of fresh capital for development purposes, this significantly improved Air France's balance sheet, and allowed it to pursue its three-year fleet renewal programme (66 planes for the group at an estimated cost of Ffr 14.5 billion in 1991, Ffr 12.1 billion in 1992, and Ffr 12.5 billion in 1993).

In August, an internal bulletin was issued announcing first results of the Andersen audit:

> The preliminary observations of the Arthur Andersen report suggest that, while Air France is essentially a well-run airline, there is room for improvement both in terms of structure and processes . . . The company's operating style is characterized by a striving for technical excellence. It is possible to push further still in terms of adapting to customer needs and in terms of focusing on profit.

In the financial press, reports on the Arthur Andersen findings were rather more scathing. *Le Figaro-Economie* announced: 'Air France needs to reform itself', going on to criticize 'the inadequacy of the strategic direction which is not properly understood within the company'.[47] *Les Echos* (13 August 1991) talked of 'A dismal diagnosis of Air France's situation' and highlighted the coexistence of centralized and decentralized organizing principles: 'We know who can say no, but it is less clear who can say yes'.[48] And *Le Quotidien* headlined 'Air France: Hunting down the dinosaurs' noting that 'the structure of the airline is at once too complex, too fragmented and too monolithic'.[49]

As Attali himself recalled in his memoirs, the verdict of Arthur Andersen intervention was simple: 'Over the years, Air France had grown heavier, had developed new processes to handle new

[47] *Le Figaro-Economie*, 14 August 1991, p. 1.
[48] *Les Echos*, 13 August 1991, p. 6.
[49] *Le Quotidien*, 14 August 1991, p. 1.

tasks, but without reviewing the existing structures and processes'.[50] On the basis of these recommendations, a new plan dubbed 'Cap 93' (destination 93) was devised.

In August 1991, Attali was elected head of the International Air Transport Association for the coming year. Taking over from Bob Crandall of American Airlines, it was a real endorsement of Attali's new-found stature and credibility in the industry.

THE CAP 93 PLAN (SEPTEMBER 1991)

In September 1991, the new Cap 93 plan was launched. Billed by Attali as the most ambitious reorganization in the company's history, he made it clear to the employees: 'We can no longer rely on the help of the State to mop up our deficits. It's up to us to find the road to recovery.'

Besides an emphasis on quality and a strong sales and marketing initiative, including the introduction of a frequent-flyer programme, the plan required a substantial productivity drive. To finance its investment programme, the Group would need to make up a shortfall of some Ffr 1.5 billion. This would mean further reducing the workforce—about 1000 of the 1500 reductions announced in 1990 had already been implemented. The new plan involved a reduction of the ground staff by an additional 2500 before 1993, and a review of the work practices and careers of flight staff (established 20 years previously when the company basically used only four types of aircraft). Thanks to careful planning, there were no enforced redundancies. Some voluntary redundancy measures were introduced and heavy use was made of training and redeployment to fill posts vacated by natural wastage (over 800 a year).

Alongside these reductions in staff, there was a reorganization effort aimed at simplifying the structures and the decision processes, with the suppression of at least two levels of hierarchy across the board, and as many as five in some staff functions. In parallel, the reward structures for the ground personnel were reduced from seven to five salary bands. These delicate measures were implemented without any significant social unrest[51] thanks to Attali's 'negotiating talents and his discrete concessions'.[52] Attali was fully aware that: 'In an airline, more than in other businesses, you can never succeed against the personnel, only with them'.[53]

The Cap 93 measures were also intended to accelerate the integration of the group—an objective reinforced by the simultaneous decision to relocate the head office from central Paris to Roissy airport (to the north of Paris) by 1995. The Montparnasse building was sold off for Ffr 1.6 billion.

Meanwhile, Attali was still pursuing his negotiations with other companies. In September 1991, the *Financial Times* announced that Air France was about to complete a strategic alliance with Sabena, thereby thwarting a year-long attempt by BA to invest in the loss-making Belgian carrier. The cost of Ffr 670 million was barely the price of a new aircraft. Until late in the summer of 1991, BA had been regarded as the front-runner. The significance of the deal was that Brussels, along with Roissy-Paris, was the only continental airport not yet saturated. It would have afforded

[50] Attali, B. *op. cit.*, p. 106.
[51] The same changes had produced massive disruptions at the state railways just a few months before.
[52] *L'Expansion*, 3 October 1991, p. 129.
[53] *L'Expansion*, 3 October 1991, p. 129.

BA with a foothold in mainland Europe. Instead, it was Air France which could boast two significant international hubs in Europe.

Yet these significant strategic advances did little to ease the tension internally. When Attali met with Henri Krasucki, the veteran national union leader, the latter offered him a well-intentioned warning: 'I've listened to your people on the front lines and I have to tell you: take care of your middle managers, they're in bad shape.'[54]

In October 1991, *Aviation International* noted that: 'The prevailing climate within the company is not serene. Perhaps because, confronted by a wave of emergencies, the chairman has acted rather quickly, without really taking the time to explain his strategy, including to those who should normally have been the first informed.'[55]

In November 1991, as Attali emerged from a meeting, UTA strikers trapped him inside the headquarters for several hours until he promised to discuss grievances over planned job cuts. While travelling with Jack Welch on the GE chairman's private plane, Welch warned his friend Attali that he should move faster. As Welch saw it: 'For major restructuring, you think you can ease the pain by taking your time. In fact, it's the uncertainty which causes suffering. You have to move fast.' Attali responded by citing the Arabic proverb: 'Those in a hurry are already dead.'[56]

As 1991 drew to a close, nearly 50 000 jobs had been lost in the industry as a result of restructuring. The hangover resulting from the Gulf crisis was exacerbated by a recession in many world markets. Industry capacity continued to outstrip demand.

On 1 January 1992, UTA's employees were transferred to Air France as the companies merged. The salaries and terms and conditions of work for the flight staff were immediately aligned with the more favourable terms prevailing at Air France. Air Inter continued to operate under its own identity.

Externally, Attali pursued his international strategy. In March 1992, he led a group of investors to take a 40% stake in CSA, the Czech state carrier, at a cost of Ffr 90 million, providing a second potential bridgehead (with Brussels) at the heart of Europe, and he continued to look for partnerships further afield.

In July 1992, Attali's mandate was renewed by the French socialist government for another three years and he pledged a more aggressive, market-oriented stance from the company. He told journalists that the image of Air France as 'a big, sleepy, monopolistic elephant' was a thing of the past: 'I'm a chess player, and the first lesson in chess is never understimate the opposition. Many of our competitors are clearly making that mistake.'[57]

Attali forecast that things would get worse before they got better in the industry: 'We haven't seen anything yet. It's going to be a terrible earthquake and we are all going to be engaged in a frightful war for the next 10 years . . . The problem in Europe is that there are about 60 airlines

---

[54] Attali, B. *op. cit.*, p. 86.
[55] *Aviation International*, 15 October 1991, p. 18.
[56] Attali, B. *op. cit.*, p. 92.
[57] *Financial Times*, 9 July 1992, p. 28.

that matter, and that's 40 too many.'[58] He rounded-off by asserting defiantly: 'We are number one in Europe in terms of turnover and our ambition is to remain number one.'

In September 1992, Attali concluded a strategic alliance with the national carrier of Canada, the fastest growing of the G7 countries. This offered possibilities of wider openings in the North American market, with Air Canada on the verge of buying Continental Airlines and having sealed a commercial agreement with United Airlines. Two months later, Attali signed an exclusive commercial agreement with Aeromexico.

THE RETURN TO BREAK-EVEN PLAN (OCTOBER 1992)

At the start of 1992, the Cap 93 plan was destined to shed 2500 jobs over two years. But by October 1992, with worse-than-expected losses for the first half of the year, it became clear that more was needed. The company launched a new initiative labelled the Return to Break-even Plan (Programme de Retour à l'Equilibre, known as PRE) intended to restore the company to profit by 1994; these measures supplemented those already taken under the Cap 93 plan. The new measures involved 1500 additional job cuts among the ground staff, which included the company's first-ever forced redundancies—36 in all. The unions called a 24-hour strike to protest against the latest job cuts.

At the same time, Attali decided that he needed to start tackling the employment conditions of the flight personnel, and warned the prime minister of his intention to try to reduce the costs of this category of personnel by around 10%. Attali proceeded with caution, mindful of the maxim of one his predecessors: 'With the flight crews, you can count on the worst.'[59]

In October 1992, a dedicated team was named to handle the negotiations. These lasted six months: six months of posturing, stalemates and false walkouts. Eventually the pilots union agreed to modify the previously sacrosanct 1971 agreement. In particular, they renegotiated the procedures relating to forced redundancies, previously based on the LIFO principle. However, there was an important condition; that the company would not try to get mileage out of the concessions, this being the only way the union representatives could convince their members. 'No triumphant celebrations, otherwise we will all be done for.'[60] Attali stood by his promise.

A chartered accountant was commissioned by the Central Works Council of Air France to look into the likely consequences of the latest plan. He concluded that:

> Numerous elements of the PRE look more like simple accounting decisions than a veritable strategy. The productivity gains seem to result essentially from squeezing the numbers of ground personnel. This cannot be regarded as a productivity gain. If those people were doing work considered necessary for the airline, who will do it when they leave?[61]

Twenty-three percent of the job reductions targeted the sales function.

---

[58] *Financial Times*, 9 July 1992, p. 28.
[59] Attali, B. *op. cit.*, p. 182.
[60] Attali, B. *op. cit.*, p. 183.
[61] Perri, P. *op. cit.*, p. 118.

The on-going productivity drive produced some spectacular improvements. For example, in terms of freight, the number of baggage handlers required to unload a Boeing 747 was lower at Air France than in any other world airline. Other initiatives were proving less successful. For example, one element of the Cap 93 reorganization was to create a post of aircraft manager, responsible for coordinating the 11 functions involved in preparing a flight. Their effectiveness had been severely hampered by the fact that they had no one reporting to them and were not therefore considered 'real managers'.

The reorganization had also generated some very visible inefficiencies. For example, the repeated directives to maintenance to reduce its labour costs, especially its high-cost night-time work, led to greater immobilization of aircraft and therefore did not improve the bottom line.[62] Similarly, the rationalization of stocks had occasionally required spare parts to be ferried by taxi from Orly to Roissy which helped crystalize opposition in certain quarters. Attali was shocked by the reaction of one normally reasonable representative of the management union (CGC) who commented in one circular that: 'Slashing costs is driving the airline to suicide!'[63]

Meanwhile, across the Channel, BA remained highly profitable. One journalist asked Attali what he thought their secret was. Attali replied:

> There is no miracle. Certainly BA has made good progress. Firstly, the concentration of English companies began earlier than in France: BA is the result of two successive mergers. I'll add two reasons that weigh more heavily. The social charges in Great Britain represent 17% of the wage bill. In France, it represents 37%. If we had the British system, the charges of Air France would be lightened by Ffr2bn . . . Secondly, by virtue of the 1977 agreement, called Bermuda II, BA is protected on the North Atlantic by the limited number of designated American carriers.[64]

In December 1992, while acknowledging the sacrifices ahead, Attali remained upbeat about Air France in the company magazine *France Aviation* telling employees:

> I have no worries about the long-term future of our Group which has numerous strengths: its network, its size, the structure and modernity of its fleet, its recognized quality of service, the professionalism of its personnel and, of course, the hub we will develop at Roissy airport.[65]

This tone was very much a reflection of Attali's preoccupation with not overdramatizing the situation because of the lasting damage it might have on the image of the company. As he later commented: 'That's the whole paradox for a large service company. It should never do anything which harms its image, however difficult the circumstances.'[66]

Losses for 1992, amounted to Ffr 3.5bn, while productivity gains, in terms of sales per employee, over the period 1988–92 were up 12% to $168 000. This compared with corresponding gains of 22% to $173 000 for Lufthansa and 32% to $190 000 for BA.

---

[62] Lehrer, M. *op. cit.*, p. 180.
[63] Attali, B. *op. cit.*, p. 73.
[64] *L'Expansion*, 19 November 1992, p. 102.
[65] *France Aviation*, December 1992, p. 7.
[66] Attali, B. *op. cit.*, p. 173.

In spite of the launch of the PRE plan, the losses continued to accumulate. By April–May 1993, the press was starting to put pressure on Attali. *Le Nouvel Economiste* headlined, 'Can Attali save Air France?'[67] while the *Financial Times* asked how long he could 'keep flying'.[68] There were even rumours that Peugeot boss Jacques Calvet[69] was in the running to take over from Attali, especially given that the incoming centre-right government Air France had officially tabled for privatization in May 1993 (along with 21 other nationalized companies).

PRE2: THE RETURN

In July 1993, Attali concluded a marketing partnership with Continental Airlines. At the same time he started devising a new recovery programme—known internally as PRE2—involving 3000 new job reductions. Attali presented this plan to representatives of the new centre-right government headed by Edouard Balladur who criticized it as not radical enough and told Attali to come back with tougher proposals.

They eventually approved a plan to shed 4000 jobs—3000 among the ground staff and 1000 among the flight staff—on top of the 5000 implemented since January 1991 (see Exhibit CS5.2.6 for details of plan). The plan also included the closure of several routes and the disposal of noncore activities (such as the Meridien hotels chain and its duty-free shops). Twenty-five per cent of the 4000 job reductions were announced as compulsory layoffs. In his memoirs, Attali says that this was just a negotiating ploy which would have been conceded in exchange for an agreement on 4000 job cuts through attrition. As part of the deal, Air France would receive a one-off endowment from the state of Ffr 5 billion.

The announcement of the plan on 15 September 1993, coincided with the announcement of similar redundancy measures from Peugeot, Bull, Thomson, and Snecma. *Libération*'s headline the next morning was: 'The day 13 317 jobs disappeared'. The press dubbed it 'black Wednesday' for employment.

As soon as this plan was announced the unions retaliated: 'It is necessary to fight this plan, not negotiate'.[70] Meanwhile the transport minister pledged that the government would 'do its duty as a shareholder'.

ATTALI'S LAST STAND

After announcing the plan, Attali attended dozens of meetings in discrete talks with the representatives of individual unions, but made few public pronouncements.

With sporadic protests against the job cuts already under way from ground staff, the management set a meeting with the unions for 16 October. The management would propose some 'accompanying measures' relating to the plan. The strikers were hopeful that these measures would be favourable.

---

[67] *Le Nouvel Economiste*, 30 April 1993, p. 16.
[68] *Financial Times*, 19 May 1993, p. 25.
[69] At Peugeot, Calvet had already proved that he could rationalize a business and cut borrowings.
[70] *Financial Times*, 16 September 1993, p. 29.

As expected the management confirmed a one-off bonus for the lowest paid workers, then went on to announce its plans to reduce the overtime payments for ground staff working nights or on statutory holidays, and to cut work-related travel expenses. Given that the company could not legally reduce salaries, the management was looking to reduce bonuses which were out of line with market rates. The estimated savings would amount to Ffr 130 million.

The fact that this announcement came out of the blue and was contrary to expectations was explosive enough. The fact that elsewhere, the pilots had just received bonuses of Ffr 3000–5000 as a result of a prior agreement to reduce cockpit crews to two, simply exacerbated the issue. Once again, the ground staff felt they were being treated more harshly than the flight crews—and with the new measures hitting the lowest paid hardest, the latent feeling of injustice surfaced with a vengeance.

18 October: with several hundred employees on the runways, AF had to cancel its flights departing from both Paris airports. Bernard Attali indicated that the plan would be implemented *'sans faiblesse'* (without fail).

19 October: in an uncompromising letter to employees, Attali commented on the previous day's events: ' . . . certain irresponsible individuals blocked passengers and sabotaged equipment. They detained and threatened executives. All these actions . . . will not be sanctioned . . . there is no excuse for recourse to violence or sabotage of equipment.'[71]

The protests continued in spite of the intervention of the riot police and appeal from the prime minister Balladur to the strikers, promising that there would be a minimum of forced redundancies. Balladur's pledge left Attali facing an uphill struggle as he privately conceded: 'Excluding the possibility of straight layoffs, has left me completely hamstrung. People have to feel that the worst is possible in order to become reasonable.'[72]

20 October: Attali asserted on a French national radio station (RTL) that: 'Through consensus and dialogue with the unions, we are capable of reducing jobs without too much breakage.' Meanwhile, transport minister Bernard Bosson, on emerging from ministerial consultation announced that 'There is no going back on the plan devised by the top management of Air France.'[73]

The strike was costing an estimated Ffr 50 million per day plus Ffr 20 million lost by the freight operations.

21 October: disgruntled farmers joined Air France employees on runways to protest against trade talks threatening farm subsidies. There was a real fear that the strikes would escalate into industrywide protests against a reform of the employment bill, aimed at increasing flexibility in French labour law.

22 October: the conflict was widely covered in the French press, and not just in unfavourable terms. In the newspapers, radio and TV as much was said about the impact on the salaries of

---

[71] Perri, P. op. cit., p. 20.
[72] Perri, P. op. cit., p. 21.
[73] Both cited in Perri, P. op. cit., p. 24.

low-paid employees as the annoyance to users. A SOFRES survey of 800 people commissioned by the current affairs TV programme *7 sur 7* on 22–23 October showed that: 44% of interviewees supported the movement; 27% were sympathetic towards it; and only 16% were actually hostile to it.

Night of 22–23 October. The transport minister announced a meeting with the ground staff unions. Attali was told he did not need to attend, as this agenda-setting meeting would lead to formal negotiations at Air France the following day. The government representative unilaterally withdrew the two measures deemed to have triggered the conflict. The unions demanded the withdrawal of the entire plan.

24 October. On the evening news the transport minister, Bernard Bosson, announced the withdrawal of the plan, triggering the immediate resignation of Bernard Attali.

As the *Financial Times* saw it, the day of reckoning had only been delayed.[74] The *Economist* commented: 'Union leaders know that, as one of them said, they have "won a battle, but not yet the war." That implies they are still fighting the war. It also suggests that, having tasted victory, they do not mean to let up, either now or in three months' time, when the government has promised a new plan to restructure the firm.'[75]

Leaving his Montparnasse office, Attali felt betrayed by a government which had lost its nerve. He could be proud of the fact that he had presided over the company's biggest makeover, massively improving its global reach under very tough conditions, and lifting to third position in the world rankings by turnover.[76] (See Exhibit CS5.2.7 for a summary of the key events under Attali's leadership.)

Critics focused more on the missed opportunities and the lack of dialogue which had characterized his presidency—and the sorry state in which he left the company.

Later, reflecting on events, Attali said: 'I am utterly convinced that if the government had shown more composure, we would have weathered the crisis. Some minor concessions would have been needed, of course, but the plan need not have been withdrawn.'[77]

---

[74] *Financial Times*, 25 October 1993, p. 3.
[75] *The Economist*, 30 October 1993, p. 33.
[76] According to the December 1993 issue of *Fortune* magazine. Cited in Attali, B. *op. cit.*, p. 128.
[77] Attali, B. *op. cit.*, p. 219.

Case 5.2

Exhibits

Exhibit CS5.2.1 Evolution of Air France vs BA (1980-93)

	Turnover		Net Profit (Loss)		Workforce	
	AF (Ffr bn)	BA (£bn)	AF FFr m	BA (£m)	AF (000's)	BA (000's)
1980	15.7	1.92	10	11	33.6	56.1
1981	19.2	2.06	(380)	(145)	33.2	53.6
1982	22.0	2.24	(790)	(545)	34.3	47.8
1983	24.3	2.50	90	89	34.3	40.0
1984	27.6	2.51	530	216	34.6	36.1
1985	30.3	2.94	730	174	35.0	36.9
1986	27.8	3.15	670	181	34.9	38.9
1987	29.0	3.26	720	152	34.8	39.5
1988	31.3	3.76	1,170	151	35.6	44.0[a]
1989	34.9	4.26	850	175	37.0	50.2
1990	35.1	4.84	(720)	246	38.4	52.1
1991	34.4	4.94	(690)	95	38.0	54.4
1992	12.4[b]	5.22	(3200)	395	44.0[c]	50.4
1993	38.7	5.66	(8500)	178	41.4	49.0

[a] Including 1828 employees from British Caledonian and consolidation of airline figures into group total (adjustment of plus 1260).
[b] This exercise integrates the sales made through the year by UTA and those made in the 4th quarter by Air France.
[c] Includes UTA workforce after merger.
Source: Company annual reports.

Exhibit CS5.2.2 Airline rankings in 1989 and 1990

Rankings			Turnover ($m)	Vairiance 1990–89 (%)	Results ($m)	Net Profit (in %)	Fleet size	Workforce
1990	1989							
1	1	American Airlines	11 720	11.8	−39.6	−0.3	552	87 300
2	2	United Airlines	11 038	12.7	94.5	0.9	462	74 000
3	9	Air France	10 466	3.3	−132.1	−1.3	202	64 894
4	6	Lufthansa	8 963	10.7	9.4	0.1	220	59 654
5	4	British Airways	8 813	2.0	169.6	1.9	229	52 809
6	3	Delta Airlines	8 582	6.1	302.8	3.5	433	61 675
7	5	Japan Airlines	7 762	5.5	65.1	1.2	99	21 156
8	7	Northwest	7 257	10.9	−10.4	0.1	399	45 200

Source: L'Expansion, 3 October 1991, p.124.

Exhibit CS5.2.3 European airline's share of transatlantic traffic (vs. US carriers)

	1980 %	1984 %	1988 %	1992 %
United Kingdom	45	39	44	54
Germany	54	52	47	43
France	52	46	38	32
Netherlands	93	93	89	79

Source: Lehrer, M. Comparative institutional advantage in corporate governance and managerial hierarchies: The case of European airlines, Unpublished doctoral dissertation, INSEAD, 1997, p. 61.

Case 5.2

Case 5.2

Exhibit CS5.2.4 Bernard Attali's CV

Born 1943: Twin brother of Jacques Attali (later to become personal adviser of François Mitterrand).

1964–66: Degree from Institut d'Études Politiques de Paris

1966–68: ENA élite postgraduate civil service training school (see below for details)

1968–72: Auditor at the Cours des Comptes

1972–74: Chief of staff to the planning commissioner

1974–80: Financial adviser, DATAR, territorial planning agency

1980–81: Finance director, Club Méditerranée

1981–84: Head of DATAR

1984–86: President of GAN, nationalized insurance group

1986–88: Head of French subsidiary of Commercial Union (UK insurance group)

1988–93: President of Air France including president of the AEA (1991)

L'ENA was founded in 1945 to renew France's high administration, compromised by wartime collaboration. It is a finishing school whose tiny élite dominate French administration, politics and industry. Entrance to l'ENA is extremely competitive and is exclusively for postgraduates, though few of the students have any work experience (the favoured route is via the Paris Institut d'Etudes Politiques). The position achieved in the final examination at l'ENA determines the choice of civil service appointment, and the spirit of rivalry is maintained throughout the course, as an individual's whole career may hinge on a quarter of a mark in the final assessment.

The highest ranked students of l'ENA earn the right to choose which of the élite *grands corps* they wish to join, these being the bodies which 'administer France'. The traditional preference for the top four graduates is the Inspection des Finances, followed by the Conseil d'Etat (for the next four), the Cours des Comptes (for the next three) and so on. L'ENA graduates are then obliged to spend several years in state service.

Admission to these *grands corps* pretty much guarantees career success. It could be argued that the products of the Ivy League colleges in America or Oxbridge in Britain also dominate various sectors of the establishment. The difference is that they turn out more graduates in a year than l'ENA has done in the 50 years of its existence—around 5000 in all. Between 1985 and 1993 the proportion of *énarques* at the head of France's top 200 companies doubled and now stands at 23%.[78] Rotations between public sector firms and government are common—regulators and regulated are recruited from the same pool.

---

[78] Bauer, M. and Bertin-Mourot, B. *L'Accès au Sommet des Grandes Entreprises Françaises: 1985–1994*, CNRS, 1995, p. 33.

Exhibit CS5.2.5 Overview of the union situation at Air France

There are 14 labour unions within Air France. Among the ground staff, the main ones are Force Ouvrière (about 40% of votes), the CGT (25%), the CFDT (20%), the CGC (7%), and the SNAMSAC (6%). The flight staff each have their own unions—one for pilots and one for stewards and hostesses.[79]

The French labour unions are distinctive from an Anglo-Saxon viewpoint in that they are not divided along trade lines so much as along political and ideological lines (and heavily influenced by Marxism and from the point of view of the class struggle). Loosely speaking, the CGT is affiliated with the communist party, CFDT with the socialists, FO is centre-left, and CFC is a managerial union which is typically centre right.

The biggest union is the communist dominated Confédération Générale du Travail (CGT) which remains staunchly opposed to 'collaboration with capitalism' on the German model.

Employers and union representatives alike are therefore accustomed to conflict. Even when there is consensus, the unions are not very keen to publicize it. A case in point: Attali was asked by union leaders to not to broadcast the agreement reached on flexible working with the pilots.[80] Industrial relations in France are often marked by stalemates and violent protests, with the government periodically called in as referee.

Exhibit CS5.2.6 Attali's plan (back to break-even Plan—Phase 2)

In exchange for a one-off state endowment of Ffr 5 billion, Attali's Ffr 5.1 billion cost-cutting package involved:

- Making 4000 jobs cuts (3000 ground staff and 1000 navigation staff) with a minimum of involuntary redundancies (the 800 straight layoffs announced at the outset being a negotiating ploy);
- maintaining the wage freeze (established two years earlier under Cap 93 Plan);
- reducing overtime pay and bonuses;
- suspending 15 of the airline's 203 destinations;
- selling off part of the group's 57% stake in the Meridien hotel chain;
- disposing of its Saresco duty-free shops;
- bringing in other shareholders into its Servair catering subsidiary.

---

[79] Attali, B. *op. cit.*, p. 74.
[80] Attali, B. *op. cit.*, p. 183.

Exhibit CS5.2.7 Key events and actions under Attali's leadership (abridged from Attali, *op. cit.*, pp. 229–54) (*continues*)

Date	Event
Oct. 1988 to Jan. 1989	Strike by the maintenance workers over working conditions
Nov. 1988	Introduction of a profit-sharing scheme (accord d'interessement) to last three years
Jan. 1989	Creation of a communications department
Mar. 1989	Board approves orders for three more Airbuses, the early delivery of a Boeing 747-400 and options on six Boeing 737-500s (in addition to the 12 already on order)
July 1989	Air France takes a 35% stake in the French regional airline TAT
Sept. 1989	Board approves order for 10 new Boeing 747-400s (five firm orders and five options)
Sept. 1989	Air France signs a broad collaboration pact with Lufthansa.
Jan. 1990	Air France takes control of UTA and indirectly of Air Inter by acquiring 70.95% of UTA's capital
Feb. 1990	Creation of strategic council grouping presidents and managing directors of the three companies
July 1990	Board approves further options on six Boeing 747-400s
July 1990	Launch of a programme dubbed 'Adaptation to the economic environment' in response to a deterioration in outlook for air transport
Aug. 1990	Iraqi troops invade Kuwait
Sept. 1990	Reinforcement of the Adaptation Plan following the start of the Gulf Crisis. Closure of nine routes between Paris and Europe, and 50 routes outward bound from the provinces
Oct. 1990	European Commission approves union of Air France, UTA and Air Inter on condition that Air France relinquishes its stake in French regional carrier TAT
Nov. 1990	Strike movement at Nice airport against the transfer of Air Inter personnel to Air France
21 Nov. 1990	Jean-Cyril Spinetta is appointed president of Air Inter
23 Nov. 1990	Bernard Attali is named president of the AEA for 1991
Jan. 1991	Air France postpones by one year the delivery of the Airbus A340s and relinquishes its options on four Airbuses

Exhibit CS5.2.7 *(continued)*

Date	Event
16 Jan. 1991	First military strike on Iraq
Feb. 1991	New set of stringency measures following the outbreak of war in the Gulf
Mar. 1991	Air France takes delivery of its first Boeing 747-400
Apr. 1991	Air France starts selling off its stake in TAT
May 1991	Air France's first corporate advertising campaign
July 1991	State injects Ffr 2bn of fresh capital for development purposes (approved by Brussels in Nov. 1991)
Aug. 1991	Boeing 767 makes its first appearance in Air France's fleet. Air France also receives its first of its twelve Boeing 737-500s (with options on six more)
28 Aug. 1991	Bernard Attali replaces Robert L. Crandall (American Airlines) as president of IATA for 1992
Sept. 1991	Air France decides to transfer its head office from central Paris to Roissy Charles-de-Gaulle airport by 1995. The Montparnasse building is sold off for Ffr 1.6bn
Sept. 1991	Launch of Cap 93, strategic plan destined to reorganize Air France's structure, to drive new productivity improvements and to accelerate the integration of the Group
Dec. 1991	Change of visual identity (new livery for planes, agencies, airports and publications)
1 Jan. 1992	Transfer of 4300 UTA employees to Air France
Mar. 1992	Air France leads a group of investors to take a 38.2% stake in CSA, the Czechoslovakian state carrier
Apr. 1992	Air France takes a 37.5% stake in Belgian airline Sabena
July 1992	European Commission approves BNP taking a stake in capital of Air France (Ffr 1.25bn shares)
23 Sept. 1992	Attali presents project for full merger between UTA and Air France to the board
Sept. 1992	Strategic alliance between Air France Group and Air Canada
Sept 1992	Launch of a new campaign with the slogan 'Ask us for the world'
1 Oct. 1992	Launch of the 'Return to Break-even' Plan (PRE1) to add to the measures implemented for Cap 93

Exhibit CS5.2.7 *(continued)*

Date	Event
Nov. 1992	Exclusive partnership agreement between Air France Group and Aeromexico
29 Dec. 1992	Shareholders general assembly approves merger between the two airlines
Dec. 1992	Synergies generated in the first three of the Group's existence account for annual savings of Ffr 992 million against a purchase price of Ffr 7bn
1 Apr. 1993	Six agencies in France become joint-agencies for Air France and Air Inter: Aix-en-Provence, Cannes, Clermont-Ferrand, Mulhouse, Nîmes and Paris-La Défense
May 1993	Government lists Air France among the 21 nationalized companies it wants to privatize
July 1993	Agreement between Air France Group and Continental Airlines
Sept. 1993	Announcement of the second phase of the Return to Break-even Plan (PRE2) with the firm intention of restoring financial health by 1995
Oct. 1993	Air France and Air Inter facilitate the exchange of points between their frequent flyer schemes
16 Oct. 1993	Announcement of PRE2 measures
17 Oct. 1993	First strikes at Charles-de-Gaulle (Roissy) airport
18–21 Oct. 1993	Conflict spreads, with strikers invading the runways and barring access to terminals
22 Oct. 1993	Unions meet with representatives of the transport ministry who withdraw measures relating to working hours and bonuses
23 Oct. 1993	Unions decide to pursue the strike
24 Oct. 1993	Transport minister announces the cancellation of the PRE2 on the evening news. A few minutes later Bernard Attali announces his resignation

This case was written by Jean-Louis Barsoux, Research Fellow, and Jean-François Manzoni, Assistant Professor, both at INSEAD. It is intended to be used as a basis for class discussion rather than to illustrate either effective or ineffective handling of an administrative situation. It is based on publicly available sources.*

Pulling Out of Its Dive: Air France under Christian Blanc

INTRODUCTION

In late October 1993, with Air France employees occupying the runways, annual losses heading for Ffr 8 billion, and the unions exercising a stranglehold over the company, the government withdrew the restructuring plan which had triggered the massive protests. Unimpressed by the government's climbdown, Bernard Attali resigned as CEO.

This left the government with the tricky task of finding a new CEO to take over the state-owned airline. Several external candidates were approached but none seemed keen to take up the challenge. One CEO told the transport minister in no uncertain terms: 'This company is now completely ungovernable. There is nothing left to negotiate.'[1]

Two days later, Christian Blanc was named CEO. At the time of his appointment, no one knew how to tackle the issue. Attali had initially tried a progressive approach to change, and had eventually switched to a more forceful one, and both had proved ineffective. Moreover, Blanc's options were further limited by recent events. First, the press was speculating that the opposition of the unions would be intensified now that they had 'tasted blood'. Second, the government was making it clear that Blanc would have to stem the massive losses without the threat of either salary reductions or forced layoffs. Third, competitors were moving fast under the impetus of air transport deregulation in Europe. *Le Nouvel Economiste*'s headline said it all: 'Air France, Mission Impossible'.[2]

Over the next five months, Blanc put together a proposal which was arguably even tougher than that of his predecessor. The three-year plan called for 5000 job reductions, a pay freeze, and

[1] *Le Nouvel Economiste*, 29 October 1993, p. 23.
[2] *Le Nouvel Economiste*, 29 October 1993, p. 20.

measures to raise productivity by 30%. Faced with opposition from the majority of the airline's 14 unions, Blanc appealed directly to the company's 40 000 employees.

The gamble paid off. More than 81% of the employees who voted gave their approval to Blanc's plan—and the unions were obliged to give their retroactive agreement to the package. The man nicknamed '*le Sorcier Blanc*' (a play on words, meaning 'the white sorcerer') seemed to have done it again.

BLANC BEFORE AIR FRANCE

While short on 'flying hours', Christian Blanc had a number of useful cards up his sleeve when he came into the Air France job. As former head of RATP, the Paris public transport service, he had experience in managing a public sector group and of implementing reforms in the face of manifold union opposition (12 unions at RATP, 14 at Air France).

During his spell at RATP (1989–92), Blanc implemented a substantial reorganization, decentralizing the group's operations, reducing the number of management layers from seven to three, and rationalizing pay scales, a particularly sensitive issue for the unions. Having set RATP back on the rails, he eventually stumbled when trying to guarantee a minimum level of service during industrial disputes and to reduce pay for striking employees. In November 1992, after failing to win government backing on these issues, he resigned 'with panache'.[3] Reflecting on Blanc's approach during the 1992 strike, one RATP executive commented: 'He is determined to introduce management reforms and he will not compromize.'[4]

Though agitated, Blanc's experience at RATP paled in comparison with his previous assignment in the South Pacific island of New Caledonia. In 1988, the socialist prime minister, Michel Rocard, called in Blanc to defuse a conflict sparked by a militant independence movement in the French colony. It was here he first met Christian Kozar, who later became one of Blanc's closest associates at RATP and then at Air France. Building on key relationships he had struck up on an assignment four years earlier, Blanc was able to reconcile the Caledonian communities in just one month, with the help of a referendum.

Prior to that mission, Blanc had twice been a regional prefect, establishing his reputation as a leftwing reformer, and had worked in numerous civil service jobs, though not always high profile ones. Indeed, he had turned down opportunities to become the minister for overseas territories and to become head of France's secret service. As Blanc once reflected: 'I've only ever taken up posts that really interested me.'[5] The significance of his Air France appointment by a centre-right government was not lost on commentators who put it down to his reputation as a tough negotiator. On the other hand, there was a certain wariness of Blanc and his 'method' from union quarters. Reacting to the news of Blanc's appointment, the general secretary of the CGT union warned: 'Mr Blanc should not try to do at Air France what he sought to do at the RATP'.[6]

[3] *Le Monde*, 5 July 1996, p. 11.
[4] *Financial Times*, 27 October 1993, p. 3.
[5] *Le Monde*, 5 July 1996, p. 11.
[6] *Financial Times*, 27 October 1993, p. 3.

AIR FRANCE BEFORE BLANC

Bernard Attali, the twin brother of President Mitterrand's closest adviser, was appointed to head Air France by the incoming socialist government in 1988. At 44, he became the youngest-ever head of Air France, and the company's fourth chairman in six years.

When Attali took over, there were several challenges awaiting him. He had to make up for the overcautious fleet investment policies of his predecessors, especially with traffic growth unexpectedly doubling compared with previous years. The impending deregulation of European air transport also meant that he needed to build critical mass so that Air France would be big enough to withstand the expected industry shake out.

In 1989, the company ordered 28 aircraft due for delivery in 1991. It also started forging alliances with partners across the world, notably a broad collaboration pact with Lufthansa. In 1990, Air France took over two smaller airlines operating in France, the long-haul UTA and the domestic carrier Air Inter (operating out of Orly). This move lifted Air France's share of the home market from 78% to 97%, putting it on a par with its main European rivals: Lufthansa with 99.8%, KLM with 97.2% and British Airways (BA) with 89% in their respective countries. The newly constituted group became the Europe's largest—and the West's third biggest—airline by turnover.[7] (See Exhibit CS5.3.1 for evolution of Air France's financial health).

Just a few months later, in August 1990, Iraqi troops invaded Kuwait. Very quickly, the sharp rise in the price of aviation fuel and air insurance premiums, together with the drop in passenger traffic driven by the fear of terrorist attacks, provoked a deep crisis in the airline industry. Air France introduced stringency measures in September and reinforced them in February 1991 with the outbreak of war. Contrary to many of its rivals, however, the airline did not resort to redundancies.

With losses for 1990 amounting to Ffr 720 million, Attali called on Arthur Andersen to conduct a review of the organization. The consultancy's recommendations led to a plan, Cap 93, launched in September 1991. The plan focused on a strong productivity drive, and proposed reducing the ground staff by 3000 before 1993 (through attrition and voluntary redundancies). By July 1992, the airline had posted losses of Ffr 690 million for 1991, as well as recording worse than expected first half results.

In October 1992, the company launched a new initiative, the 'Return to Break-even Plan', intended to restore the company to break-even by 1994. Supplementing the measures announced under the Cap 93 plan, the new measures involved 1500 additional job cuts among the ground staff, including the company's first-ever forced redundancies—36 in all, which was more symbolic than anything.

Meanwhile, Attali continued to pursue international alliances, securing stakes in the Belgian carrier Sabena and the Czech airline CSA, which further burdened the airline's debt. Losses for 1992 amounted to Ffr 3.5 billion.

---

[7] UTA was eventually merged with Air France two years later, but Air Inter remained a separate entity.

In July 1993, Attali proposed 3000 new departures—in order to secure a one-off endowment from the state of Ffr 5 billion—and was told by ministers from the new centre-right government to come back with stiffer measures. They eventually approved a tougher plan concerning 4000 jobs, including an 'opening bid' of 800 straight redundancies (see ExhibitCS5.3.2 for details). Shortly after this plan was announced, in September 1993, the unions decided to fight.

Very quickly, the conflict degenerated with strikers bringing chaos to both Paris airports, blocking runways and forcing flights to be cancelled. Attali remained adamant that protests would not affect the implementation of the plan, but after five days of intense conflict the recovery plan was withdrawn and Attali resigned. As the *Financial Times* saw it, the day of reckoning had only been delayed.

BLANC TAKES THE CONTROLS

Early, on his very first morning in the job, Blanc went to Roissy and Orly to meet with the strikers in the freight and the maintenance areas. He reassured them that Attali's plan would be scrapped completely, and that there would be no legal reprisals against the strikers. In a statement to employees, Blanc said that he had been given *carte blanche* by the government to negotiate a new recovery plan for the airline. Nothing would be decided on the plan before discussions with the employees. He also made it clear that if he did not feel he had the means to save the airline he would not hang around: 'I may be here for eight days or for four years . . . I didn't ask to come.'[8]

That same afternoon, Blanc called Robert Génovès, the head of Force Ouvrière, the dominant union representing 35% of the airline's ground staff at Roissy. Blanc told him that he could not accomplish his mission without help. Realizing that Air France may not have another chance, Génovès decided to support the new CEO. In regular contact, the two men quickly switched to the informal *tu* (you) form of address with each other.[9]

Blanc made similar overtures to the dominant pilots union, the SNPL. Blanc's pledge in both cases was that there would be no straight layoffs, but that the unions would propose which sacrifices they were prepared to make in line with the productivity targets. In the words of one union representative, they would 'choose which feathers to lose'. These were to prove crucial allies in the subsequent discussions with unions.

Sensing that he might be vulnerable to the manipulation of insiders with vested interests, Blanc knew he needed to get a broader picture of the airline. To hear from as broad a cross-section of viewpoints as possible, Blanc invited managers to write to him at home to propose changes and analyse the conflict.

Blanc quickly called on a number of outsiders to help him. He brought in several trusted executives—including four ex-RATP managers, an HR manager from Thomson, a former aide to prime minister Balladur as company secretary, and a state auditor as finance director—to strengthen his reform team. From the outset, he warned them that, like him, they should not travel first class.

---

[8] *Le Nouvel Economiste*, 15 April 1994, p. 14.
[9] *Le Nouvel Economiste*, 16 February 1996, p. 62.

Blanc also solicited a variety of experts to prepare reports. First, he encouraged Gilles Bordes-Pagès, an executive director and former pilot to prepare a report on the management of the Group, and particularly the competitiveness of its fleet. Secondly, he commissioned Cofremca, a specialist consultancy, to undertake a survey of consumer air travel expectations. Thirdly, he hired François Dupuy, a colleague of Professor Michel Crozier, a pioneer in the study of bureaucratic organizations, to assist the 40 000 employees in carrying out their own audit of the problems of the airline.

AIR FRANCE ON THE COUCH

In December 1993, Dupuy and his small team started interviewing throughout the airline, but with particular emphasis on the most conflict-ridden areas, notably maintenance, freight and flight personnel. Initially they met with cynicism, partly driven by employees' previous experience with consultants, but Dupuy's team distinguished themselves from their predecessors. They did not talk about productivity. They listened. They tried to understand the reality of the situation. Altogether, they conducted 156 unstructured interviews.

In parallel with this interview process, a questionnaire was sent out to the homes of all 40 000 employees (in December 1993). Drafted by Dupuy on a BA flight, its purpose was to take the pulse of the personnel and assess the concessions they were prepared to make. It included two or three open-ended questions where respondents had more freedom to express themselves. Insiders warned him not to expect more than about 2000 replies.

Late on the night of 3 January, Dupuy presented the preliminary findings of the interviews to the top management. He started by revealing that those they had spoken to were neither manipulators nor extremists; they were simply disoriented and disillusioned, as captured in the following quotes.

> For years now, we have been living without an optimistic message. Top management's policy has been to instil a sense of permanent insecurity, without any real message of hope.

> The more efforts we make, the worse things get. We have never worked as much, and they talk of overmanning. I don't understand where the deficit comes from.[10]

Dupuy explained: 'When employees invade the runways, they are doing the same things as workers who break their machines. When people reach that stage, it is because they are desperate and don't know how else to express themselves.'[11]

Dupuy's findings also showed that the organization was too hierarchical and centralized, with little coordination between activities. 'We found individual links but no chain, a process organized around the product and the technical imperatives, and not towards the client.'[12] An example he cited to demonstrate this lack of coordination between the various divisions was the last evening flight from Stockholm. The aircraft would park near the hangar for overnight maintenance,

---

[10] Both quotes from Le Journal du Débat, no. 4, 20 January 1994, pp. 2–3.
[11] L'Expansion, 7 November 1994, p. 90.
[12] L'Expansion, 7 November 1994, p. 90.

without the slightest consideration for the passengers who required an extra half an hour to reach the terminal. 'People were only concerned with their own job. There was no consideration for the final product.'[13]

Blanc quickly understood that the existing organization needed changing in order to make people more accountable and to focus better on the customer. But Dupuy suggested that top management should wait to devise their new plan for fear of undermining the on-going questionnaire exercise. There were also numerous reflection groups whose proposals should be integrated into the final project.

To accompany the listening exercise, the company created an internal newsletter, *Le Journal du Débat*. Starting in late December 1993, it came out about every ten days to communicate the emerging findings and to stimulate internal reflection—notably proposing the case of Renault as a model transformation. In its second issue, the newsletter commented on the shortage of space in the questionnaires for 'further comments' which had prompted several respondents to attach additional pages. This posed an unexpected problem. The closed questions would be subjected to statistical analysis, but there were not enough consultants to categorize the spontaneous comments. The issue was resolved by asking volunteers to perform the analysis. Sixty Air France employees were assigned to this task.

Issue six of the newsletter, in early February 1994, trumpeted the fact that 14 677 questionnaires had been returned, three-quarters of these with additional comments. This represented a 35% response rate, considered high for this type of survey, particularly in view of the pessimistic predictions. Among the key findings, 51% of respondents considered 'acceptable' the idea of reducing salaries; 64% were in favour of share participation in exchange for longer hours or reduced salary; and 92% considered the organization too heavy and bureaucratic.

The subsequent examination of the open-ended responses (in issue seven of the newsletter) brought further nuance to these findings. For example, it became clear that opinions were very divided over the idea of reduced salaries: some thought it should apply primarily to the highest paid; others thought it should apply across the board; and others still would only agree to it against certain guarantees on job cuts. Moreover, it clarified the opposition to this measure, with some considering that there were better alternatives, and others feeling that it was neither fair nor motivating.

Blanc's verdict on the process: 'I wanted to listen to you. And you have responded . . . It was essential for us to diagnose the situation jointly and for us to determine the strategy to pursue. For we can only meet the challenge facing Air France together.'[14]

THE PRESSURE MOUNTS

From the outset, Blanc made no attempt to hide the gravity of the situation. 'The state of the company is exceedingly worrying', he told employees. 'We can't even cover our current expenses (salaries and fuel) with our current revenues (ticket sales).'[15] To increase cash flow, the

---

[13] *Financial Times*, 16 January 1995, p. 8.
[14] *Le Nouvel Economiste*, 15 April 1994, p. 14.
[15] *Le Nouvel Economiste*, 5 April 1996, p. 53.

airline was increasingly forced to resort to 'sale on lease back' operations, whereby planes were sold off then rehired.[16]

In February 1994, the report by Bordes-Pagès was released. Described by the press as 'explosive', it showed that the Group had got it wrong for years, both in terms of fleet and commercial policies. In particular, the extreme diversity of the fleet—the culmination of underinvestment, merging with UTA, and government intervention on purchasing choices—generated inflexibility and high operating costs. For example, for its six flights per week to Hong Kong, Air France employed four Boeing 744s, one Boeing 742 and one Airbus A340. Since pilots typically specialize in one type of aircraft, this meant that three different pilot crews were needed, some of whom remained blocked for several days in Hong Kong, as well as several maintenance teams. British Airways operated two flights a day to the same destination with a single type of aircraft, the Boeing 744.

The findings of Cofremca went in the same direction, showing that Air France had not kept pace with the changing expectations of customers. Their report established that contemporary air travellers had become more complex and disloyal in choices, with the act of consumption responding to multiple and sometimes contradictory motivations. As the newly appointed head of human resources put it: 'Customers are asking Air France to change, or else they will change airlines. Air France must attract customer loyalty by offering a distinct and impeccable service . . . Everything depends on the professionalism of individuals right through the service chain.'[17] Tristan Benhaïm, who had led the research subsequently joined Air France as marketing director.

On 9 March 1994, the government approved Blanc's project for the firm, called 'Rebuilding Air France'. The following day, the entire workforce received draft copies of the 50-page document.

On the opening page, Blanc reminded everyone that 'the company's losses, in 1993, account for half the net losses of the world's 212 major carriers'.[18] The report then showed how Air France's inability to adapt to a changing context had given rise to several handicaps in relation to the competition—notably a weak debt/capital ratio,[19] a high cost structure, and ill-adapted sales and pricing policies.

The report went on to propose a number of measures to make up this competitive lag. These fell into three broad categories: reduction of costs and increase in revenues; a new organizational structure and a change in work processes; and participation of the employees in the capital of Air France. The key financial measures involved losing 5000 jobs (through attrition and voluntary retirements not layoffs), freezing salaries and selling assets such as aircraft and hotels (see Exhibit CS5.3.2 for details). These measures applied only to Air France. Air Inter remained a separate issue.

The report rounded off with: 'If we unleash individual initiative, if we learn to recognize individual contributions, thus giving meaning to people's efforts, then the company will regenerate.'[20] It

---

[16] *L'Expansion*, 7 April 1994, p. 42.
[17] Michelland-Bidegain, M. 'Formation: elle ne peut rien sans volonté de changer', in *Business, Le Journal des Nouveaux Produits*, Document Air France, 16 February 1995, no. 2, p. 2.
[18] *Reconstruire Air France*, March 1994, company document, p. 3.
[19] At the end of 1993, Air France had debts of Ffr 40 billion against a capital base of Ffr 4 billion.
[20] *Reconstruire Air France*, March 1994, company document, p. 46.

Case 5.3

talked of the need to go beyond economic measures and to mobilize the 'collective intelligence' of employees.

A week later, on 17 March 1994, Blanc assembled the company's 2800 managers in a floral park east of Paris for a day of reflection on the proposals. Letting them know they were all in this together, he pledged: 'If the project fails, the top management team will resign. But I am sure I will stay with you until the airline's future is secured.'[21] His objective, he told them, was not to restore Air France, but to rebuild it: 'This requires a revolution. If I were to capture the essence of this revolution, I would simply say: to turn Air France into a business. A normal business. Not banal, but normal.'[22]

On 26 March, Blanc received representatives from all of Air France's 14 unions for final discussions on the outline agreement presented in the project.[23] Blanc had already made it clear to the unions that the injection of fresh capital from the state (worth Ffr 20 billion over three years) hinged on their commitment to the redundancy measures. By the 31 March deadline, only six of the 14 unions had signed the agreement. The others had refused, arguing that these measures were essentially the same as those proposed by Attali. Blanc told them that unless they reconsidered their positions he would 'go to the people'.

GETTING TO YES

To gain support for his reform plan, Blanc resorted to a referendum, which was set for the first week in April. Appealing directly to the airline's 40 000 employees was a big gamble, but it paid off. The 35 600 voters (84% turnout) voted overwhelmingly (81%) in favour of the plan. Anxious at having been outflanked, all but two of the unions subsequently gave their approval to the package, allowing Blanc to proceed with the implementation of his rescue package, and opening the way for transformations which were inconceivable only a few months before.

A ONE-WAY TICKET

In May 1994, Air France initiated talks with constructors Airbus and Boeing to renegotiate aircraft contracts placed since 1990 and thus reduce investment commitments. Top management also started discussions with the pilots' unions to raise the threshold of hours at which pilots claimed overtime payments, and with the ground staff unions to increase the working week by one hour, to 39 hours (without extra pay). The key, as Blanc saw it, was not so much to work 'harder', as to work 'smarter'.

At the start of June 1994, the new organizational structure was put into place. This involved breaking up the previous structure based on a handful of large functions, such as the personnel

---

[21] *Le Nouvel Economiste*, 15 April 1994, p. 14.
[22] *Géopolitique*, Autumn 1995, p. 82.
[23] The fourteen unions concerned were: le Syndicat National des Pilotes de Ligne (SNPL); le Syndicat National des Officiers Mécaniciens de l'Aviation Civile (SNOMAC); le Syndicat des Pilotes de l'Aviation Civile (SPAC); le Syndicat des Mécaniciens au Sol de l'Aviation Civile (SNMSAC); le Syndicat Général du Personnel Air France (SGPAF); le Syndicat des Ingénieurs, Techniciens, Cadres et Agents de Maîtrise Air France (SITCAM); la CFTC, l'Union Syndicale d'Air France (USAF); le Syndicat des Personnel Assurant un Service Air France (SPASAF); la CGC, Force Ouvrière, le Syndicat des Cadres-Force Ouvrière d'Air France (SCFOAF); le Syndicat National des Personnels Navigants Commerciaux (SNPNC); and le Syndicat National Unitaire des Navigants de l'Aviation Civile (SUNAC).

division, which covered the flight crews, or the commercial division, responsible for ticket sales and marketing. Instead Air France was reorganized into 11 separate profit centres, a mix of self-contained geographical and business units. Five profit centres covered geographical zones (such as Europe, the Americas, Asia-Pacific) and six covered activities (such as freight, maintenance and the Paris airport services).[24]

The aim was to increase accountability and efficiency, to break down the large functional bureaucracies, and to instil momentum for change. An immediate consequence of the reorganization was that one manager in three changed his or her reporting relationship. The new structure stripped out several layers of managers. As Christian Kozar, head of the profit centre for airport operations Paris and a close aide to Blanc, explained: 'Hierarchical management is finished. We have cut the management levels in our profit centres from eight to three and have created a much more responsive operation.'[25]

Outside the airline industry, the merits of decentralization were widely recognized, but within Air France the reorganization had some detractors, notably Gilles Bordes-Pagès, the author of several critical reports on the running of Air France. As he saw it: 'Either the profit centres will behave like independent businesses and spend more time exchanging services than fighting in the marketplace; or else recourse to arbitration will become the norm and general management will resume its daily operational role, thus undermining the logic of accountability.'[26]

In July, Air France received its first slice of fresh capital, approved by the EC—with subsequent slices of the Ffr 20 billion package to be paid out only if the airline delivered on its restructuring plan.

On the commercial side, Air France dropped its prices to try to win back some of the market share lost as a result of the strikes. It also created a joint frequent-flyer programme with Air Inter, which had long been promised to customers. Elsewhere, Air France actively considered which yield management software to purchase in order to optimize revenues from each category of passenger.

As part of the company's plan to focus more on its airline interests, Air France sold off its controlling stake in the Meridien Hotel chain to the British company Forte in September 1994. The following month, the group sold off its building in the Champs-Elysées for Ffr 400 million. The airline also announced its intention to relinquish its stake in the Belgian carrier, Sabena.

Meanwhile, Blanc had significantly streamlined the top management structures. The executive committee had been reduced from 48 people to 18, while the board of directors now included just eight people.

At the end of August 1994, Steven Wolf, orchestrator of the recovery programme at United Airlines was appointed as Blanc's adviser. He was the 11th external senior appointment to Air

---

[24] The five geographical profit centres include: Europe, America, Asia, Africa, and the Caribbean. The six activity based profit centres include: freight operations, Paris airports, maintenance, industrial services, IT and telecommunications, and centralized services.
[25] *Financial Times*, 16 January 1995, p. 8.
[26] *Le Nouvel Economiste*, 13 January 1995, p. 51.

Case 5.3

France since the announcement of the rationalization, and the CFDT denounced such hirings as 'shocking' at a time of job losses. Three months later, Wolf lured his compatriot, Rakesh Gangwal as head of scheduling. The two men quickly decided that some of the activities delegated to the profit centres needed recentralizing, notably scheduling, income optimization and management accounting. Blanc explained to the workforce that 'Air France will be a networked organization, with decentralized responsibilities, but with centralized integration'.[27]

By October 1994, the *Financial Times* was able to report that Air France had reduced losses for the first half of the year to Ffr 2.6 billion, from Ffr 3.8 billion for the same period in 1993. Moreover, first half sales had stabilized, after the sharp decline of 1993, and totalled Ffr 27.5 billion—this figure masked an increase in volume sales of over 10% which was offset by price reductions aimed at winning back passengers from rival airlines. Operating profits had more than doubled, rising from Ffr 657 million to Ffr 1.6 billion.

At the close of the year, the company offered employees shares in the airline, in exchange for a salary reduction. 12 000 employees (35.6%) took up the offer, raising staff ownership to about 5%.[28] This innovative scheme had the dual advantage of reducing costs while strengthening employees' commitment to recovery measures. Blanc subscribed to the maximum salary reduction under the share scheme, reducing his salary by 15%. This relegated him to 255th position in the airline's salary rankings—and he made no secret of it.

CONCLUSION

In January 1995, with just over a year in the job, Blanc was able to reflect on progress so far. With all the goals for 1994 having been met, he reckoned that Air France was about one-quarter of the way towards the objectives set for January 1997.

There were a number of encouraging signs. For example, unit costs in 1994 had dropped 9% after a steady increase of 2% per year between 1988-93. Also, productivity was up by 12%, against an average of 2.5% per year between 1988–93. And the fall-off in revenues per kilometre flown had slowed down—dropping only 1.6% (in line with industry norms) versus 6.7% in 1993.

On the other hand, there were also causes for concern. A few sporadic strikes had broken out in October 1994 in protest over the new working conditions. One union representative reflected: 'Dupuy and his team were essentially brought in to determine how management could make us swallow the pill.'[29]

There were also problems with the management tools: 'We don't have a robust cost accounting system in the profit centres, which is a real obstacle when you want to install productivity measures.'[30] Blanc also identified another potential problem: the relocation of the headquarters to Roissy, to be accompanied by a 30% reduction in personnel.

---

[27] *Le Nouvel Economiste*, 13 January 1995, p. 51.
[28] A 20% ceiling had been set by the company.
[29] *L'Expansion*, 7 November 1994, p. 92.
[30] *Le Monde*, 26 January 1995, p. 17.

Blanc was conscious that the effort ahead remained immense, but he was optimistic: 'The difference is that we have taken altitude. In our profession, taking altitude means discovering a horizon. The French people hope and expect Air France to recover its status as one of the country's leading companies. Our responsibility is to deliver on those hopes.'[31]

Asked how he would describe his approach, Blanc answered: 'It is not a soft approach, but it is a recovery without drama, founded on a respect for individuals and collective energy. It is a negotiated restructuring.'[32]

---

[31] *Le Monde*, 26 January 1995, p. 17.
[32] *Le Monde*, 26 January 1995, p. 17.

Exhibits

Exhibit CS5.3.1 Evolution of financial situation of Air France (1986–93)

Air France	Turnover Ffr bn	Net Profit FFr m	Workforce 000's
1986	27.8	670	34.9
1987	29.0	720	34.8
1988	31.3	1170	35.6
1989	34.9	850	37.0
1990	35.1	(720)	38.4
1991	34.4	(690)	38.0
1992	12.4[a]	(3200)	44.0[b]
1993	38.7	(8500)	41.4
1994	39.9	(3500)[c]	38.3

[a] This exercise integrates the sales made through the year by UTA and those made in the 4th quarter by Air France.
[b] Includes UTA workforce after merger.
[c] Exceptional 15 month exercise to align with industry norms.
Source: Air France annual reports.

Exhibit CS5.3.2 Attali's Plan vs Blanc's Plan

Attali's plan (Back to Break-even Plan—Phase 2)

In exchange for a one-off State endowment of Ffr 5 billion, Attali's Ffr 5.1 billion cost-cutting package involved:

- Making 4000 jobs cuts (3000 ground staff and 1000 navigation staff) over two years. 1000 of these jobs would be 'externalized' with the outsourcing of certain activities (such as bus transport, telecommunications) and there would be a minimum of involuntary redundancies (the 800 straight layoffs announced at the outset being a negotiating ploy).

- Maintaining the wage freeze (established two years earlier under Cap 93 plan).

- Suspending 15 of the airline's 203 destinations.

- Selling off part of the Group's 57% stake in the Meridien hotel chain.

- Disposing of its Saresco duty-free shops.

- Bringing in other shareholders into its Servair catering subsidiary.

Blanc's plan (Rebuilding Air France)

The French state's recapitalization to the tune of Ffr 20 billion was in three slices—the second and third slices to be paid out only on condition that the intermediary targets were met.

The economic measures

- Fleet reduction from 166 to 149 planes by 1999.

- Purchases reduced by Ffr 2 billion (-16%) in three years.

- Workforce reduced by 5000 in three years.

- Freeze on salaries for 1994, 1995, 1996.

- Promotions blocked in 1994.

- 30% improvement in productivity over three years.

- Disposal of noncore subsidiaries and assets such as the Meridien hotels.

The changes

- Reorganization of the airline into 11 profit centres: five by geographical zone (such as Europe, the Americas, Asia-Pacific) and six by activity (freight, maintenance and the Paris airport services).

- Constitution of a holding company incorporating both Air France and Air Inter.

- Part of the company's capital to be sold off to the employees in exchange for a reduction in salaries.

- Simplification of the product: merger of first and business classes on long haul, and possibility of converting two-class cabin into single-class (depending on times and destinations) on medium haul.

Case 5.4

This case was written by Jean-Louis Barsoux, Senior Research Fellow, and Jean-François Manzoni, Assistant Professor, both at INSEAD. It is intended to be used as a basis for class discussion rather than to illustrate either effective or ineffective handling of an administrative situation. It is based on publicly available sources (notably those listed in the references) and is intended to be used in conjunction with the corresponding INSEAD-CEDEP case on Air France, 1988–93 (Case 5.2).*

The Making of 'The World's Favourite Airline': British Airways (1980–93)

INTRODUCTION

In February 1993, shortly after stepping down as chairman of British Airways (BA), 75-year-old Lord King received an unexpected phone call. It came from the former British prime minister, Margaret Thatcher. 'I just wanted to tell you', she said, 'You were the greatest achievement of my years in office. I am so proud of you.'[1]

The fulsome tribute was not undeserved. Over 12 years, Lord King, now honorary president of the airline, had overseen a remarkable turnaround. In 1980, when Margaret Thatcher had first approached him about taking on the job, British Airways had been in awful shape. It was an unwieldy, loss-making, state-owned carrier with a dismal reputation for customer service.

In 1980, a survey by the International Airline Passengers Association showed that BA was rated by 33% of the poll as *the* airline to be avoided at all costs—ahead of Aeroflot, Nigeria Airways, and even teetotal Arab carriers.[2] In the late 1970s, the airline's punctuality record was dire (see Exhibit CS5.4.1 for key operating statistics). In the summer of 1979, only 38% of its long-haul flights left within 15 minutes of their scheduled departure time, making BA *the* most unpunctual airline in Europe out of its own home base.[3] Among disenchanted passengers, BA was alleged to stand for 'Bloody Awful'.

Once a standing joke, BA came to be regarded as an industry benchmark, under the leadership of King and chief executive Colin Marshall. They injected professionalism and a commercial sense

---

*Copyright © 1997 INSEAD, Fontainebleau, France.
[1] *Financial Times*, 1 March 1993, p.15
[2] Campbell-Smith, D. *The Struggle for Take-Off*, Hodder & Stoughton, 1986, p. 11.
[3] Corke, A. *British Airways: The Path to Profitability*, Printer, 1986, p. 120

into the airline. They revitalized its service levels, image and revenues. By 1993, BA was firmly established as Europe's most efficient and profitable carrier, as well as a worldwide reference in terms of customer service. Half a dozen Harvard cases described various facets of BA's dramatic transformation, and the company was dominating airline award competitions.

BACK FROM THE BRINK (1980–82)

British Airways was the product of a merger in 1972 between two state-run airlines: the long-haul British Overseas Airways Corporation (BOAC) and the short-haul British European Airways (BEA). Immense economies of scale should have been forthcoming, but they were neither sought nor obtained. Instead, the two airlines continued to issue separate financial reports until 1974, when their full merger was announced. Even then, an integrated management structure did not come into effect until 1977 and failed to obliterate attitudes and rivalries inherited from the predecessor airlines.

In the early 1980s, two distinct cultures continued to exist within BA—with Heathrow's Terminal One dominated by former BEA people and Terminal Three by BOAC people. Each terminal dealt with very different passengers and time perspectives, and the relative isolation of both sets of employees tended to perpetuate the contrasting cultures: the resourceful, high frequency, quick turnaround (short-haul) vs. the cosmopolitan 'representative of Great Britain overseas' (long-haul) cultures. The one common legacy of the two companies was that they both suffered from inefficiency,[4] overmanning and a heavy dominance of former Royal Air Force officers.

Back in 1979, internal estimates suggested that British Airways was at least 20–25% overmanned. The previous year, management had highlighted this problem by publishing a prominent table in the annual report, comparing productivity with that of eight unnamed foreign airlines. Expressed in available tonne/kilometres per airline employee, this showed that BA's performance was still under 60% of the foreign average. BA was competitive only by virtue of its strikingly low unit labour costs.

At that time, BA's top management was forecasting volume growth (in double figures) especially in leisure traffic—and this growth was expected to absorb the overmanning identifed in the 1978–79 review. It was envisaged that BA would have to shed no more than 2000 jobs in order to reach an acceptable level of productivity (see Exhibit CS5.4.2 for evolution of workforce numbers).

In May 1979, the conservative government was elected to power under the leadership of Margaret Thatcher, with a commitment to sell into private ownership many of Britain's major nationalized industries, including BA.[5] Besides instilling greater financial discipline into the airline, denationalization would avoid the £1 billion addition to public sector borrowings required by BA's massive fleet expansion and renewal programme, partly driven by the need to meet stricter noise regulations.

Contrary to BA's plan, the expected growth in traffic volumes failed to materialize. Instead of rising by 8–10% as forecast, traffic fell by 4–5%. The industry was about to enter a recession with

---

[4] Before BA was nicknamed 'Bloody Awful', BOAC was said to stand for 'Better On A Camel'.
[5] Gabel, H.L. *The Privatization of British Airways*, INSEAD Case, No. 2000/3033/00027, 1986.

the market turning in the spring of 1980. By 1981, the company was losing money fast and it was clear that drastic action was needed if the company were ever to be in a position to attract private shareholders (see Exhibit CS5.4.2 for BA's financial figures). In February 1981, Thatcher appointed Sir John King Chairman of BA.[6] The press referred to BA as 'technically bankrupt, with only government assurances of cash support keeping it aloft.'[7]

Though short on 'flying hours', King was a veteran of industrial restructuring in the ball bearings and power engineering industries, and a conservative party loyalist. His brief was to knock the airline into shape for privatization. From the outset, he ensured that senior managers understood this, making a point of starting meetings by asking managers when the company would be ready for privatization.[8] King commissioned a detailed report from Price Waterhouse to help determine what needed to be done to make privatization possible.

A process of redundancies, through attrition and early retirement, had been engaged prior to King's arrival by chief executive, Roy Watts, but by September 1981 it had become clear that more was needed. The board revealed the extent of the problem in a special bulletin to British Airways staff (issued by Roy Watts):

> British Airways is facing the worst crisis in its history . . . we are heading for a loss of at least £100m in the current financial year. We face the prospect that, by next April, we shall have piled up losses of close to £250m in two years . . . Even as I write to you, our money is draining away at the rate of nearly £200 a minute . . . No business can survive losses on this scale . . . That is why I have decided to take tough, unpalatable and immediate measures to stem the losses.[9]

The remedial action took the form of a Survival Plan which proposed: a cut in the workforce, from 52 000 to 43 000 over nine months; the selling of assets such as properties and surplus aircraft; and the pruning back of the route network (see Exhibit CS5.4.4 for a complete list of disposals). The company offered generous severance terms, for a limited period, to volunteers willing to leave the airline soon—so generous, in fact, that when the deadline was extended, the company found itself overwhelmed by volunteers, with close to 16 000 applications. The strength of employee demand, allied to the government's hostile stance to unions, meant that there was hardly any industrial unrest.

By the end of 1981, the recession was bottoming out. In March 1982, Price Waterhouse submitted its 500-page analysis of BA. Based on these recommendations, King proposed a Recovery Plan, to supersede the Survival Plan. The measures announced in the new plan included a rescheduling of orders for Boeing 757 aircraft, a renewed staff severance scheme to help reduce staff numbers by another 7000, to 36 000 by 1986, and a restructuring of the company into three self-contained service divisions, each with a new managing director: International Services Heathrow (long-haul), European Services Heathrow (short-haul) and Gatwick Services (domestic).

---

[6] He became Lord King in 1983.
[7] *Business*, March 1986, p. 90
[8] Campbell-Smith, D. *op cit.*, p. 27.
[9] Corke, A. *op cit.*, p. 82. See Exhibit CS5.4.3 for full text.

Case 5.4

The new organization, implemented in May 1982, had strong echoes of the former BEA–BOAC one, but had a much more forceful commercial focus, with each of the three divisions divided into subsidiary profit centres—the aim being to increase incentives and accountability.

Having had a chance to assess the top management talent, King warned the government that he would need to recruit at least a couple of top-flight executives whose salaries would fall outside usual nationalized industry rates. Between March and August 1982, King axed four of the top directors including both the financial and commercial directors. The first of his new recruits was a tough Scottish finance director, Gordon Dunlop, appointed at the start of June 1982.

Dunlop and King quickly decided that the publication of the report and accounts for the year ending March 1982 (due out in July) would have to be delayed to make sure 'that adequate provision had been made for the redundancies'.[10] When it finally emerged, in late October, the 1981–82 report showed an operating surplus of £6 million (compared to the previous year's deficit of £104) but this was overshadowed by a jumbo deficit (after taxes and extraordinary items) of £545 million. This included interest charges on borrowed capital (£111 million), the total sum set aside for severance payments (£199 million—anticipating the compensation to be paid out for 7000 redundancies still to be made), and the decision to write down the book value of certain of the airline's aircraft and other assets (£208 million) to ease future depreciation.

In the autumn of 1982, the government reaffirmed its intention to offer BA for sale to the public by the end of 1983. Anxious to give the company a new image, King switched BA's worldwide advertising account. The £17.5 million account was taken from the US agency Foote Cone & Belding (a relationship going back over three decades) and given to Saatchi & Saatchi, the British agency handling the advertising for Margaret Thatcher's conservative party, and partly credited for her election victory in 1979.

The first commercial, called 'Manhattan', was inspired by the film *Close Encounters of the Third Kind*, featured the island of Manhattan being guided in to land at Heathrow by air traffic controllers—to symbolize the fact that the airline flew more people across the Atlantic each year than the entire population of Manhattan. Although it cost several hundred thousand pounds, the highly ambitious commercial ran in 33 countries and was acknowledged to have broken new ground in airline advertising, in that the same ads were used worldwide with no more than a change of voiceover. The commercial also trumpeted British Airways as 'The World's Favourite Airline'—a quantitative claim, based on the number of international passengers carried, rather than a seal of passenger satisfaction.

This said, there were several signs that British Airways was starting to shake off its reputation for poor customer service and aircraft delays. Punctuality improved dramatically throughout 1981 and 1982 (see Exhibit CS5.4.1). The number of flights leaving within 15 minutes of their scheduled departure time rose by around 15%. Paradoxically, the downsizing improved a number of performance indices besides punctuality, such as a reduction in out-of-service aircraft, less time 'on hold' for telephone reservations, fewer lost bags and so on. There was also a 20% reduction in the number of customer complaints compared to the previous year, and by 1983, these were running at their lowest level in five years.

---

[10] *Financial Times*, 10 July 1982, p. 4.

After Gordon Dunlop, the second of King's key external recruits was Colin Marshall, appointed chief executive in February 1983.

MARSHALL'S CREDENTIALS

Though not outwardly charismatic, Marshall had an impressive international career in various service businesses. On leaving school at 18, he had joined the Orient Steam Navigation shipping line as a cadet purser (a ship's purser being responsible for accounts and for the comfort and welfare of passengers). Seven years later he moved to the United States and joined Hertz as a management trainee in the early days of the vehicle renting and leasing business. Customer service and equipment quality played an important role in the competitive car rental market. Hertz sent him to Toronto, Mexico, and New York as assistant to the president, then back to Europe to head up the UK operation where he negotiated numerous 'fly-drive' contracts with the airlines.

In 1964, Marshall was poached by rival Avis, just before it was acquired by ITT, to start its European operation. He transformed Avis's service image, giving male staff, including himself, red jackets and badges saying 'We Try Harder'. Within eight years, by going for big volume sources, Marshall helped Avis overtake Hertz in European sales and earnings. His reward, in 1976, after a series of senior appointments, was to become the only Englishman to head a US corporation.

Marshall's 15-year spell with Avis also exposed him to two influential corporate leaders. From 1965 to 1972 Avis was actively managed as a subsidiary of ITT, then headed by the notorious Harold Geneen with his fantastic appetite for management discipline and hard data. The other key figure was the head of Avis, Bud Morrow, who insisted that every executive in the company should occasionally spend time working on the front line—which, at Avis, meant behind the rental counters and underneath the car bonnets.

When the US conglomerate Norton Simon acquired Avis in a contested takeover in 1977, Marshall moved over with the company, assuming overall responsibility—as executive vice-president—for four Norton Simon companies, including Avis and the group's major food-processing business in southern California. He finally returned to the United Kingdom in 1981 to take up the post of deputy chief executive with Sears Holdings, the British retailer and shoe manufacturer. It was from this post that Sir John King recruited him.

A NEW MARSHALL IN TOWN: 1983

Marshall's philosophy, rooted in the intensively competitive car-rental business where the products offered are very similar, was that 'the initial transaction and the after-sales service are what make the difference'.[11] For someone with that customer service obsession, British Airways was about to prove quite a challenge. As he later recalled:

> When I came to British Airways, we had no one on this side of the Atlantic who had the word marketing in their title at all. There seemed to be a general view in the

---

[11] Prokesch, S. 'Competing on customer service: an interview with British Airways' Sir Colin Marshall', *Harvard Business Review*, November–December 1995, p. 111.

Case 5.4

organization that marketing was about sales and advertising and that's about it. Really the orientation was very much to operations and the whole process side of the business.[12]

Marshall quickly set about making his presence felt. He started by closing the senior officers' mess, a dining room for the top executives of the airline. Within three weeks of taking office, Marshall had already set up a core marketing team manned by four internally promoted managers, all in their thirties. As Marshall saw it at the time:

This is the group that will drive BA forward to success in our highly competitive industry. Their job will be set the policies that BA will follow in its whole marketing effort around the world, and also to see that they are implemented throughout the organization.[13]

Realizing the need for reliable data in order to improve service performance, Marshall also established a *marketplace performance unit*, a ten member team responsible for tracking hundreds of measures of performance from aircraft cleanliness and punctuality to check-in standards.

Marshall then called on an external consultant, Michael Levin, with whom he had forged a close relationship at Avis, to help him work on further organizational changes. Occupying the adjoining office to Marshall, but without an official position, Levin set about screening the upper management ranks and reflecting on the airline's structure. Operating outside the normal hierarchy, without functional allegiances, Levin's incisive and impartial advice was often acted on by Marshall.

By the time Marshall arrived, King had already decimated the staff which was down to around 40 000. But the company was still overmanned, a legacy of the BOAC–BEA merger almost a decade earlier. There were still about 4000 people scheduled to go, and Marshall decided to accelerate the pace of voluntary redundancies. The original target date of 36 000 by 1986, was advanced to March 1984.

In July 1983, Marshall launched a sweeping reorganization, splitting the operations into 11 profit centres: eight geographic 'market centres' for passenger operations, plus cargo, charter and package tours. The new organizational structure was designed to reduce layers of management, to allow more efficient communication across functions, and to ensure coherence between operations, short-term budgeting and longer term strategic management. Moreover, the newly formed profit centres were 'entrusted to a group of managers in their thirties and forties, often promoted three to four levels overnight. All profit centres reported to a newly appointed marketing director.'[14]

The reorganization was accompanied by the sudden removal of many executives deemed unlikely to adapt to the new customer focus in the company: 'In one twenty-four hour period, on

[12] *Marketing*, 2 February 1995, p. 21.
[13] Corke, A. *op cit.*, p. 104.
[14] Lehrer, M., and Darbishire, O. 'The Performance of Economic Institutions in a Dynamic Environment: Air Transport and Telecommunications in Germany and Britain', paper presented to the 4th ESF-EMOT Workshop on Economic Performance Outcomes in Europe, Berlin, 1997, p. 5.

11 July 1983, Colin Marshall terminated the employment of one hundred and sixty-one managers and executives.'[15] Over a decade later, one senior executive commented: 'To this day, the perceived overnight disappearance of battalions of managers is known as the "night of the long knives".'[16] While there was substantial turnover, BA made efforts to retain people who were technically excellent in their work, as well as those who shared the values Marshall and his team were trying to introduce.

In November 1983, Marshall introduced a profit-sharing programme for employees. Though common in the US, such programmes were novel in the UK at the time—and unprecedented in any nationalized industry. Announcing the forthcoming scheme, Marshall commented:

> An operating surplus of £200m, for example, would produce a bonus for all of one week's pay. Surpluses of £250m would ensure a two-week bonus, £300m a three week bonus, and £350 a four-week bonus . . . There are no catches. There is no ceiling. The more we bring in, and the more we save, the more we'll get. I hope the bonus at the end of the year is going to be a large one . . . It's one bill that British Airways will find it a pleasure to pay.[17]

December 1983 saw the launch of the two-day customer service course which Marshall had announced to employees in July. The training programme, Putting People First (PPF), was essentially aimed at staff in direct contact with customers and laid stress on personal development and motivation as the key to improving the airline's image and performance. The course was attended by employees in groups of up to 150, each containing a complete spread of jobs within the airline. It was designed and run by an outside consultancy, the Danish firm Time Manager International, which had carried out a similar exercise for SAS under Jan Carlzon.

Marshall explained: 'The title is significant. It deliberately refers to putting people first rather than putting customers first. We want to remind our staff that their colleagues are people, and the way employees treat each other is just as important as their treatment of customers.' He also suggested that 'reward systems may need to be adjusted so that managers see more clearly that handling people successfully is an essential part of their work.'[18] The programme ended up running continuously for two years and Marshall himself closed over 40% of the seminars with question and answer sessions, and when he was unable to attend, he made sure there was a suitably senior replacement.[19]

Marshall also spent a lot of time talking with front-line staff, and 'working' the planes whenever he flew. He would roam the aisles, introducing himself and asking the passengers how BA was doing. He had an eye for detail and a powerful memory, and executives could be sure that information he picked up (in person, from press clippings or off his computer screen) would be used to make some point about customer service at subsequent meetings. After the toilet door

---

[15] Bray, J. 'The British Airways Story', in *Mobilizing the Organisation: Bringing Strategy to Life*, G. Litwin, J. Bray and K. L. Brooke, eds., Prentice-Hall, 1996, p. 29.

[16] *Across the Board*, January 1995, p. 55.

[17] *Financial Times*, 5 November, 1983, p. 1.

[18] *Financial Times*, 25 May 1984, p. 20.

[19] Other accounts credit Marshall with attending a higher proportion of the PPF closing sessions—one even claimed 97%. The figure of 40% comes from a lengthy interview with Marshall in *Director*, June 1988, p. 48.

came off during take-off on one Concorde flight, Marshall 'went on and on about the hinges to the engineers at every meeting'.[20]

THE ROCKY ROAD TO PRIVATIZATION: 1984–87

At the start of 1984, BA's improved financial health made it look like a more realistic candidate for privatization. The UK Secretary of State for Transport confirmed that BA would be privatized with a target date of early 1985, then approached the Civil Aviation Authority to conduct a review of airline competition policy. The ensuing report, published in July 1984, recommended a reduction in the size of BA relative to other British airlines, essentially by surrendering a number of prime routes to the private carrier, British Caledonian, BA's chief domestic competitor.

Incensed, King made a very public stand against the Civil Aviation Authority's proposals, telling the government that if they endorsed the recommendations he 'would regard it as a resignation issue'.[21] The fight over routes mobilized BA's entire staff with over 26 000 BA employees organizing, signing and presenting a petition to the prime minister.

In October 1984, after a long fight, the pugnacious King managed to obtain a compromise which was highly favourable to BA. The government policy document largely rejected the recommendations of the CAA, stating that any route exchanges between the airlines should be mutually acceptable. Besides the financial savings, estimated at around £70 million a year, the battle over routes served as a tremendous rallying call for BA's employees still divided by their allegiances to BA's predecessor companies (BOAC and BEA) and still coming to terms with the recent downsizing exercises. King's vociferous protests were all the more remarkable in that he was going against a government he had always supported.

In December 1984, BA unveiled the results of the extensive corporate redesign exercise to an invited audience of 200. BA had caused something of a controversy by commissioning Landor Associates, a US firm (which had done the same thing for SAS), to undertake this exercise. The new livery—pearl grey, midnight blue and dark red—confered a consistent identity not just to the fleet, but also to all interiors, printed materials (timetables, tickets, baggage tags), ticket desks, lounges, shops, and was later complemented by Roland Klein designed uniforms. The tailfins also acquired a new motto—'To fly, to serve'—which captured the need to reconcile the dual challenges of operations and customer service.

There was a new mood in the air. In 1984, a hundred 'Customer First' teams had been created (three-quarters surviving into the 1990s). Also many new units had sprung up, with odd names such as futures audit, segments, performance improvements, user requirements, marketing research and agency affairs. These were run by rising stars, often hand-picked by Levin, who were placed in sink or swim positions.[22] Promoting young managers to high posts, switching managers across functions, and removing poor performers, became hallmarks of Marshall's style.

---

[20] Young, D. 'British Airways: Putting the Customer First', in *A Sense of Mission*, Campbell, A., Devine, M. and Young, D. The Economist Books, 1990, p. 119.

[21] Campbell-Smith, D. *op cit.*, p. 146.

[22] Lerher, M. '*Comparative Institutional Advantage in Corporate Government and Managerial Hierarchies: The Case of European Airlines*', Unpublished doctoral dissertation, INSEAD, 1997, p. 195.

Case 5.4

By 1985, BA had acquired a new HR director, Nick Georgiades, on the recommendation of Mike Levin.[23] A former professor of industrial psychology, Georgiades had acted as a consultant to BA for over 10 years. He was not without misgivings on the way the redundancy process had been handled, sensing that insufficient attention had been given to planning:

> As a result, many of those who volunteered to go were the ones who had the best chance of either starting up on their own or finding another job—almost by definition, the sort of people who should have been encouraged to stay . . . BA did not check thoroughly enough on those skill areas where the company was 'fat'.[24]

Inevitably these past failings had repercussions on his new role. The airline needed to rebalance its skill mix. To do this, he assumed full control of hiring, especially for front-line employees, and introduced a range of assessment techniques to select people who really enjoyed providing help and service to other people. But the findings from an on-going BA research programme, showed that providing improved customer service was not just a matter of recruiting better. It also depended on how BA managed the people who delivered the service.

The research programme in question focused on the problems faced by 'emotional labourers', that is people whose job is to make people feel good and emotionally satisfied. The research sites included hospices, hospitals, banks and social services. Some key findings emerged: the best workers had close group bonds, they felt well cared for by their supervisors, and they had high discretion in problem handling.[25]

Georgiades and Levin considered how to foster these conditions for BA's front-line staff, and devised a number of initiatives. First, they reconceived the rostering process, that is the way people were assigned to shifts on locations. Instead of bringing together groups of staff in an *ad hoc* way, they created 'families' of staff who came on duty together and developed a better understanding (with increased possibilities of exchanging shifts).

Second, they created the position of passenger group coordinator (PGC), who would be responsible for creating and maintaining a nurturing climate for those actually in contact with customers, while coordinating flight and terminal operations. PGCs were primarily selected for their proven interpersonal skills, not their technical expertise. Of the 400 candidates assessed for these positions, only 10 were deemed suitable.

Third, in conjunction with leading-edge academics, they conceived a new training programme for managers called *Managing People First* (MPF). Launched in April 1985, the five-day residential course represented a significant departure from previous practices. Each programme brought together managers of different ranks and functions—in groups of about 25—and used a range of small-group intensive learning methods, likened by some to group-therapy sessions. Participants received lots of individual feedback about their management behaviour and were obliged to reflect on how to communicate better with peers and subordinates. It introduced managers to the notion that they must begin to sustain the motivation of staff without exposing them to

---

[23.] Lehrer, M. *ibid.*, p. 114.

[24] Corke, A. *op cit.*, p. 112.

[25] Hampden-Turner, C. 'The Ascent of British Airways', in *Creating Corporate Culture*, Addison-Wesley, 1992, pp. 84–5.

'emotional burnout' (experienced by those dealing with the public under intense conditions). The aim of the programme, was not to develop skills, but rather to prepare people for change—and therefore, focused more on underlying beliefs as drivers of behaviour. Once the course ended, the small groups then functioned as informal 'support groups'.

British Airways bought the Chartridge Centre, a 25-acre country estate north of London, in November 1985 specifically to serve as a venue for MPF. Initially intended for 1000 managers, by the time it was completed in 1988, more that 2700 managers had attended the programme.

There were signs that MPF was engendering a new ethos. One example was the initiative of a group of managers based in New York to redress transatlantic passenger traffic devastated by terrorist threats. In late April 1986, they came up with a $7 million public relations, advertising and sales promotion campaign. Within one week they had received full approval for the programme, dubbed 'Go for it America' which was launched in mid-May. By September traffic was back to normal levels. The return to BA was around 10 to 15 times the investment in terms of redressing the bookings shortfall and gaining favourable media coverage.

One operations manager pointed out that, prior to Marshall's arrival, there was a very strict way of doing things at BA:

> *No one could step out of line. You were scared to put a foot wrong and so tended to forget what the objective was. But there are 101 ways of doing things. Managers now know they will not be punished for trying what seems to them the best way of accomplishing their task.*[26]

Alongside the new training programmes, BA introduced a new evaluation programme. Pilat, an Israeli-based organization which had designed and implemented the performance appraisal process used by the Israeli army, was brought in to develop new bases for performance appraisal and compensation.[27] The idea was to develop a system which combined a result-oriented measurement (60%) with a measurement of how people were going about their jobs (40%) and particularly, the extent of their commitment to the fundamental corporate values as judged by boss and peers. Thus, each year a manager's performance is evaluated by his or her boss according to a list of 60 'statements of behaviour' taught during MPF, identified as key behaviour characteristics necessary in a customer service business.

These changes were then reflected in the compensation system. It changed from an MBO system in which people people were judged exclusively on numbers, to a more qualitative system whereby managers received lump sum bonus payments worth up to 20% of base salary, half determined by what the manager achieved and half by the manner in which it was achieved. According to Georgiades: 'The revolutionary thing is that we are not paying only for the 'what' of their performance, but also for the 'how'.'[28]

By 1986, BA was looking pretty healthy. Between 1982 and 1986 passenger volumes had gone up by nearly 2 million to 17 million, which was reflected in the sharp increase in turnover from

---

[26] *International Management*, March 1987, p. 38.
[27] Bray, J. *op cit.*, p. 38.
[28] Corke, A. *op cit.*, p. 115.

£2.2 billion to £3.1 billion in the same period. Productivity, as measured by the volume of available tonne kilometres of passenger and cargo capacity produced per employee (ATKs), rose from 157500 in the year ended March 1982 to 220900 for the year ended in March 1986. Passenger complaints over the same period had dropped from over 3000 per week to around 400 and the number of layers of management had been compressed from nine to three. It was time to set a date for privatization.

So far the privatization had been postponed several times for various legal and technical reasons. First, the government insisted that the public sale of British Telecom needed to be properly digested by the investment community. Then BA's top management's had been sidetracked by the need to deal with the challenge over air routes by British Caledonian. Finally, there had been the anti-trust suit brought against BA and nine other carriers for conspiracy to drive Sir Freddie Laker out of business.

Laker himself had not actually sued anyone, the action having been brought by Lakers' creditors. But as pragmatists, Marshall and Dunlop, the chief financial officer, decided that any settlement with the creditors would have to include a payoff for Laker to keep the brash discounter quiet. This irritated Laker's creditors who considered that Laker was not entitled to any compensation. So Marshall and King made the rounds persuading plaintiffs and defendants alike of the need to resolve the matter. Finally, by agreeing to pay $15 million of the proposed $64 million settlement, they forced everyone in line. The out-of-court settlement reached in July 1985 cleared the way for BA's privatization.

Back in 1982, the *Financial Times*'s Lex column had commented that a privatized BA must have potential investors because 'every market sports a few masochists'. By the time it actually was privatized, in February 1987, it was 11 times oversubscribed, and its shares almost doubled in price on the first day's trading. One of the chief benefits of privatization, according to Marshall, was internal: 'Our staff had the opportunity of becoming shareholders in the business for which they work, and 95% of them took up that opportunity.'[29]

The following month, BA was placed 15th in a survey of favourite airlines league and Marshall warned competitors: 'We expect no favours—but then neither can any of our rivals. To put it bluntly, we are now purely out for the interests of BA and those associated with us.'[30]

THE BATTLE FOR BRITISH CALEDONIAN: 1987

BA's dominance over its domestic market was reinforced soon after privatization when, after a hard fight with a determined SAS, BA acquired the financially troubled British Caledonian. The merger gave BA 95% of the British market for international flights. In November 1987, the UK government approved the merger. For its £246 million investment, BA inherited a fleet of 28 aircraft (and 7000 personnel). More significantly, BA thus obtained precious landing slots which have become the key commodities at congested airports such as Heathrow (the world's busiest airport) or Gatwick (the world's second busiest). Slots are the precise times at which airlines are allowed to launch aircraft from crowded runways into crowded skies.

---

[29] *Director*, June 1998, p. 49.
[30] *Marketing*, 21 May 1987, p. 39.

The BCal managers who joined BA were put through an unaltered MPF programme whose impact 'was immediate and effective'.[31] Details of the integration programme were worked out by a diagonal task force, which had become increasingly common at BA.[32] The task force in question considered all aspects of the merger—routes, equipment, personnel, including the 2000 anticipated redundancies.

In their first consolidated accounts, for the year ending March 1988, BA recorded pretax profits of £228 million in spite of absorbing a £32 million loss from BCal. By February 1989, King was boasting that 'the former BCal operation has been successfully merged and we can look forward to the benefits of the acquisition being reflected in future results.'[33] In May, King's forecast proved accurate with BA posting a 17.5 per cent rise in group profits. BA also announced that its 49 000 employees would receive a staff profit-share bonus of just over two-and-a-half weeks' basic pay and an additional week's pay as a reward for the results and the successful integration of British Caledonian.[34]

IT AND MARKETING SLIP INTO OVERDRIVE: 1987–89

While BA's privatization and the merger with British Caledonian attracted considerable media attention, a far more significant change for BA's future had gone more or less unnoticed. In September 1986, BA underwent its third reorganization in five years—after decentralization, it was time for major recentralization.

The new structure altered the basic principle of the airline's operation and was much more closely aligned with the strategic positioning of the airline. The sales organization was centralized, control of the planning functions was transferred to the marketing department and state-of-the-art information systems were introduced with the overall aim of optimizing the network.[35] In short, the new organizational structure institutionalized the marketing orientation and, on Levin's advice, gave a central role to IT in order to support marketing.[36]

IT's prime contribution to marketing took the shape of powerful capacity and yield management systems. The capacity management system enabled BA to monitor flights and adjust available capacities on a daily basis. Having the right-sized planes at the right places at the right times helped to increase average occupancy rates (from 67% in 1986 to 71% in 1989).[37] Revenues were also boosted by a state-of-the-art reservation system developed by BA's head of IT, John Watson.

The CRS was critical in helping BA to optimize revenues. On average, every airline seat is booked two-and-a-half times, and cancelled one-and-a-half times, in the three months before the flight. BA's sophisticated computing programmes juggled ticket prices and discounts to fill the

---

[31] Bray, J. *op cit.*, p. 35.
[32] These task forces were 'composed of individuals from different functions and at different levels of responsibility to deal with various aspects of the change process—the need for MIS support, new staffing patterns, new uniforms and so on', (Goodstein and Burke, 1991, p. 12).
[33] *Financial Times*, 16 February 1989, p. 34.
[34] *Financial Times*, 24 May 1989, p. 29.
[35] Lehrer, M. and Darbishire, O. *op cit.*, p. 4.
[36] Lehrer, M. and Darbishire, O. *ibid.*, p. 6.
[37] Lehrer, M. *op cit.*, p. 117.

maximum number of seats with full-fare passengers. This translated into profits per seat that were higher than those of most other European airlines.

These IT innovations were driven by huge investments. Annually, BA was spending around 3% of its gross revenue on information technology, and the airline boasted one of the fastest and largest computer systems outside the defence industry, with 200 000 terminals attached to 15 mainframe computers and 200 mid-range computers. By 1989, 150 graduates out of the 200 graduates recruited each year were destined to work in information technology.

The centralization of sales made it easier to coordinate BA's short-haul routes in order to feed its more lucrative long-haul operations. Flight scheduling was no longer organized on a route-by-route basis to maximize local revenues, but rather to maximize revenues across BA's network. That network had received a significant boost from the acquisition of BCal which allowed BA to siphon traffic more effectively on to connecting flights.

The dismantling of the geographic market centres also resulted in centralized control over the 'product', namely cabin interiors and service levels. BA started looking more closely at the intangibles which made passengers prefer one airline to another—and started repositioning its 'product'. The airline started introducing marketing techniques drawn from the world of consumer goods and applying them to its services. The 'branding' of BA's passenger classes had a repercussion on hiring policies with BA recruiting executives with experience in confectionary products and detergents to head the new service categories. Their job, as Marshall put it, was 'to keep the shine on the product day in and day out.'[38] Club (business) Class was relaunched in January 1988, followed by First Class, and Traveller (economy) in November 1990.

Finding specific improvements for each class of service entailed considerable research and analysis of what generates satisfaction for air travellers, so much so that the phrase 'research shows . . . ' became something of a running joke internally. For example, crews were presented with a ranking of the 'emotional expectations of passengers'. Research showed that just seeing crew members created higher customer satisfaction levels. Thus, cabin crew routines were changed so that the crew came into greater contact with passengers than before. Branding teams realized that by offering starters, main course, and coffee on separate trays, the number of 'caring for customers' contacts (what SAS CEO Jan Carlson had memorably dubbed 'moments of truth') could be increased. Mike Batt, who became head of the branding operation explained: 'On a service to New York, you will get 16 'touches' from a stewardess.'[39]

However much the airline revamped the offering, it was still reliant on its staff to deliver the service. Marshall (who had become *Sir* Colin in June 1987) was fully conscious of these difficulties, explaining. 'We therefore have to 'design' our people and their service attitude, just as we design an aircraft seat, an in-flight entertainment programme, or an airport lounge to meet the needs and preferences of our customers.'[40]

New programmes such as 'To Be the Best' and 'Leading in the Service Business' took up where PPF and MPF had left off. As Marshall once observed:

---

[38] *Financial Times*, 6 February 1993, p. 6.
[39] *Business*, February 1989, p. 54.
[40] Prokesh, S. *op cit.*, p. 110–11.

I am a great believer that when you embark on these type of programmes, you are stuck with it for ever more, because if you let it stop for a period of time, you will see a very marked decline in the staff attitudes to the delivery of service. You've got to keep them 'hyped' to some extent on an on-going basis.[41]

BA also took a number of steps to develop its own in-house management potential. It set up a dedicated MBA programme in conjunction with Lancaster University in 1988, devised the 'seeds' programme to identify and develop potential change agents, and created its *Top Flight Academies* to help promising managers at every level make the transition to the next level.

Meanwhile, the IT department continued to find ways to support marketing. In 1988, the drive to generate brand loyalty led IT to establish a database of business travellers so as to respond in a more customized way to their needs and expectations. IT director, John Watson explained: 'It won't just tell BA how often a customer flies. It will detail likes and dislikes—beyond habits like smoking or non-smoking, favourite drinks and so on.'[42]

The successful interplay between marketing and IT did not happen overnight. It took some time to operationalize: there were technical systems to develop, new routines and selling practices to understand; and shifting relationships between the functions to manage.[43] Fortunately for BA, European deregulation of air transport was only just getting off the ground, so the context allowed experimentation. Once the new structure really started to deliver on its potential, BA became what Watson later described as 'a very, very powerful selling system'.[44]

In September 1988, BA finally broke Swissair's four-year grip on the World's Best Airline Award, handed out by *Business Traveller* magazine, and BA repeated its performance the following year. It seemed that what had started out as a hollow boast—'The World's Favourite Airline' had become reality. BA had grown into its slogan.

WEATHERING THE TURBULENCE: 1989–93

In 1989, two new strategic priorities emerged: cost efficiency and globalization. The need to control costs was apparent from pessimistic projections on yields (as a result of deregulation measures) and stiffening competition on European routes from low-cost British carriers. At the start of 1989, Marshall notified managers that they should attempt to cut their planned spending by 10% for the upcoming summer. Department heads were asked to outline in the company newspaper areas where they were planning to make savings. By November 1989, what Marshall called BA's 'cost crusade' had been integrated into the updated mission statement.[45]

In the spring of 1990, with the first signs of recession showing, a Margin Improvement Programme (MIP), headed by John Watson (former head of IT) was launched. Most MIP projects had time frames of 18–36 months.[46] In August 1990, this drive gained an added sense of urgency

[41] Kotter, J. 'British Airways: Sir Colin Marshall, Question and Answer Session with AMP Participants', HBS Video, 9-493-503, 1991.
[42] *Business*, February 1989, p. 53.
[43] Lehrer, M. *op cit.*, p. 95.
[44] Lehrer, M. *ibid.*, p. 124.
[45] Lehrer, M. *ibid.*, p. 127.
[46] Lehrer, M. *ibid.*, p. 129.

when Iraqi troops invaded Kuwait. Air transport had traditionally been a sensitive barometer of economic activity, turning down very quickly in a recession to recover just as quickly with the first signs of an upturn. The effect of the Gulf Crisis was to drive up the price of jet fuel and insurance premiums, while slashing worldwide air traffic under the threat of terrorism, thus turning a deepening recession into the industry's worst slump since the Second World War. The extent of the downturn was visible in January 1991 when a BA Concorde flew across the Atlantic with only two paying passengers.

The dual pressures of the Gulf War plus economic recession caught all carriers off guard. Yet BA already had a system to produce the savings in place and numerous projects already under way. For example, the number of suppliers was reduced from 10 000 to 3500 and inventories were streamlined through the introduction of new information technology.[47] Management also had a good idea of where it could afford to make cuts and, as a result, announced 4600 voluntary redundancies in February 1992—this time, far more targeted than those in the early 1980s.

Yet BA's efficiency drive was not just about cutting costs. Efforts were also made to improve revenues. Under the dual impetus of IT and marketing, BA set up a work-flow system dubbed CARESS (for Customer Analysis and Retention System). Based on the estimation that it cost five times as much to win back a lost customer than to retain one, the system was used to register and respond faster to complaints. The system helped staff track complaints and enabled them to deliver customized compensation. It helped customers to feel listened to and customer complaints staff to feel gratified rather than defensive.

The thinking behind the system was echoed in a wider training programme, *Winning for Customers*, launched in April 1992 which had customer retention as its main focus and showed employees ways of handling mishaps.

Besides endeavours to improve productivity, BA was also gearing up for globalization. In the late 1980s, Marshall had predicted that aviation would be dominated by a handful of very large airlines. Marshall's aim was to make BA an end-to-end carrier, able to both pick up and deposit passengers in virtually any part of the world with minimal baggage handling and reticketing, and without forcing people to queue up in airports.

Heavy protection of national carriers meant that BA needed to forge international partnerships and investments, so the airline set out to spin a web of alliances in its three leading markets: Europe, North America and Asia. This was very much in line with industry thinking, but few airlines had the financial resources to act on this strategy in a meaningful way.

In 1989, BA attempted to take a 15% stake in United Airlines (UAL) of the US. The bid eventually collapsed in October 1989, but BA had already got its share of the deal's financing in place with a £320 million rights issue. Thus BA was well positioned to pursue an ambitious expansion programme.

BA experienced two further setbacks when lengthy negotiations with Belgium's Sabena, then the Dutch carrier KLM, proved inconclusive. Then, in March 1992, BA acquired a 49% stake in a

---

[47] Lehrer, M. *ibid.*, p. 128.

small German carrier called Delta and renamed it Deutsche BA to strengthen its position in the German market. The following month, it secured a large minority stake in Transport Aérien Transrégional (TAT), a French regional airline, to help it challenge Air France in its home market. BA's attempts to find a US partner finally came to fruition in January 1993, when it concluded an agreement with beleaguered USAir, the sixth largest US domestic carrier in terms of revenue passenger miles flown. In March 1993, BA rounded off an intense 12-month period of spending, by buying 25% of Australia's Qantas, to provide a foothold in the fast-growing Asian market (Exhibit CS5.4.5 shows details of the purchases).

The resulting code-sharing agreements allowed passengers to travel on different airlines with minimal reticketing or baggage-handling procedures, and often with better co-ordinated flight transfers. This strategy allowed BA to double its proposed number of destinations.

On top of these industry pressures to develop global mass while trimming operating costs, BA faced a company-specific challenge: the on-going row with Richard Branson's Virgin Atlantic.

At the start of 1992, BA became embroiled in a 'dirty tricks' row with Virgin Atlantic. In his open letter to BA's non-executive directors, Branson alleged that BA staff had been involved in bugging, rubbish stealing, computer hacking, press smearing and passenger poaching.

BA's formal statement on the Virgin scandal that none of its board members 'implemented or authorized' the campaign, was considered to raise issues of corporate governance. The *Financial Times* commented: 'The board is in dire need of outside talent, not only to run the business but to correct the inward-looking and obsessive tendencies of the BA culture.'[48]

In January 1993, the libel case was settled with a public apology from BA and the award of £610 000 in damages to Branson. Later, reflecting on the debacle, the *Financial Times* noted that opinions were divided. Some thought that 'the offending actions . . . were merely those of a small group of individuals who had overstepped the bounds of proper behaviour in their eagerness to foster BA's interests', while others 'felt that, whether or not the employees had really acted in isolation, the real rogue was BA's abrasive corporate culture.'[49]

While the affair generated a great deal of unsympathetic media coverage, BA's customers showed no sign of deserting the airline. When the *Financial Times* surveyed customer reactions, it found that factors other than BA's guilt in conducting 'dirty tricks' were uppermost in travellers' minds. Even customers of Virgin stated that BA's behaviour would not stop them using BA in future.[50] This was a robust testimony to the quality of service, safety record and brand loyalty which BA had built up over the last decade.

THE SUCCESSION

In February 1993, when the 75 year-old Lord King stepped down as chairman, he was able to look back on a job well done. BA had now been voted world's best airline for five years running in the annual *Business Traveller* survey. Through a combination of robust systems, branding,

[48] *Financial Times*, 22 January 1993, p. 13.
[49] *Financial Times*, 4 March 1994, p. 13.
[50] *Financial Times*, 22 January 1993, p. 13.

Case 5.4

customer service and rigourous cost-cutting, the airline had withstood the greatest airline slump ever. BA continued to generate healthy profits at a time when virtually every other Western carrier was posting heavy losses (see Exhibit 5.4.6 for comparative results). In 1992, the world's airlines accumulated losses of over US$4 billion for the second year running.

Of course, BA's transformation had involved some painful measures, notably the waves of redundancies implemented in the early 1980s and more recently in 1991. Nevertheless, severance payments had been generous, and the number of employees was quickly climbing back towards initial levels.

According to its staff, BA was now a radically different place to work: more professional, customer-led, adapting, innovative and focused.[51] One executive whose time at British Airways coincided almost exactly with the King period, characterized the transformation under King: 'I joined a poorly run airline . . . and I left a thriving business which happened to be in the airline sector.'

In terms of the future, King knew he was leaving the airline in safe hands, with Sir Colin Marshall stepping up to succeed him. He also knew that the airline was well-positioned, both in terms of control over slots and strategic partnerships, to exploit the postderegulation marketplace.

---

[51] Young, D. *op cit.*, p. 132.

Exhibit CS5.4.1 Evolution of key operating statistics

Year	Passengers carried (millions)	Passenger load factor (%)	ATK: available tonne kilometre (m)	ATK per employee (000s)	Yield: Revenue per passenger km (pence)	Punctuality (within 15 mins) (%)
1977†	14.5	62.1	6 555	121	2.98	79
1978†	13.4	59.2	6 793	123	3.24	64
1979†	15.8	64.5	7 557	135	3.28	65
1980†	17.3	67.4	8 153	145	3.35	68
1981†	15.9	62.6	8 243	154	3.74	81
1982†	15.2	66.7	7 522	158	4.20	78
1983†	14.6	66.5	7 208	182	4.89	84
1984	14.2	64.1	6 699	199	5.57	84
1985	16.0	68.5	7 275	213	5.87	85
1986	17.0	68.0	7 956	221	5.80	82
1987	17.3	67.0	8 141	222	6.00	81
1988	*20.2	70.2	9 427	236	5.82	80
1989	22.6	69.6	11 404	243	5.96	72
1990	23.7	71.5	12 035	247	6.37	72
1991	24.2	70.1	12 929	253	6.27	73
1992	23.8	70.2	13 379	274	6.50	79
1993	25.9	70.8	14 695	315	6.13	81

* Includes 600 000 passengers carried by British Caledonian.

Glossary
Passenger load factor — RPKs expressed as a percentage of (ASKs).
Revenue passenger kilometres (RPK) — The number of revenue passengers carried multiplied by the distance flown.
Available tonne kilometres — The number of tonnes of capacity available for the (ATK) carriage of revenue load (passenger and cargo) multiplied by the distance flown.

Source: †Gabel, H. L. 'The Privatization of British Airways', INSEAD Case, 1986; and Company Annual Reports.

Case 5.4

Exhibit CS5.4.2 The evolution of BA's financial and workforce figures

Year ending March	BA workforce (000's)	Group turnover £bn	Operating result (airline only) £m	Pretax profit (Group) £m	Net profit (Group) £m
1977*	54.4	1.25	96	96	35
1978*	55.4	1.36	57	54	52
1979*	56.0	1.64	76	90	77
1980*	56.1	1.92	16	20	11
1981*	53.6	2.06	(104)	(140)	(145)
1982*	47.8	2.24	6	(114)	(545)
1983*	40.0	2.50	174	74	89
1984	36.0	2.51	274	185	216
1985	37.0	2.94	303	191	174
1986	38.9	3.15	205	195	181
1987	39.5	3.26	183	162	152
1988	42.7	3.76	241	228	151
1989	†49.7	4.26	340	268	175
1990	52.1	4.84	402	345	246
1991	54.4	4.94	166	130	95
1992	50.4	5.22	344	285	395
1993	49.0	5.66	310	185	178

† Includes employees from British Caledonian.

Source: *Gabel, H. L. 'The Privatization of British Airways', INSEAD Case, 1986; and Company Annual Reports (*Note:* Some figures were subsequently restated, typically to reflect changes in accounting principles. The numbers presented in this exhibit are those that appeared in the given year's annual report.)

Exhibit CS5.4.3 Watts' special bulletin to the BA staff (delivered 10 September 1981)

'British Airways is facing the worst crisis in its history ... unless we take swift and drastic remedial action we are heading for a loss of at least £100m in the current financial year. We face the prospect that, by next April, we shall have piled up losses of close to £250m in two years. Even as I write to you, our money is draining away at the rate of nearly £200 a minute.

No business can survive losses on this scale. Unless we take decisive action now, there is a real possibility that British Airways will go out of business for lack of money. We have to cut our costs sharply, and we have to cut them fast. We have no more choice, and no more time.

That is why I have decided to take tough, unpalatable and immediate measures to stem the losses that are threatening the jobs of every one of us. Many of those steps are drastic because nothing less will do. They have to be taken swiftly because we can no longer afford the time to debate our plans at leisure.

This is both a rescue and a recovery operation. We have to do it ourselves. Nobody else is going to do it for us.'

Source: Corke, A. British Airways: The Path to Profitability, Frances Pinter, 1986: London, p. 82.

Exhibit CS5.4.4 The 1981 Survival Plan. Disposal of activities in 1981

- Suspension of 16 international passenger routes from Heathrow, Gatwick and regional airports;

- Withdrawal of all cargo-only services and the disposal of the fleet of cargo aircraft;

- Disposal of a number of 747s and 707s and one Tristar 500, and the accelerated retirement of a substantial number of Tridents and Viscounts;

- Restriction of capital expenditure to 757s and helicopters;

- Disposal of two engineering bases and several other properties;

- Transfer of all catering to outside catering firms;

- Closure of eight BA international online stations (including Prestwick and a number in Europe);

- Closure of two engineering bases.

Source: Corke, A. British Airways: The Path to Profitability, Frances Pinter, 1986, p. 83.

Exhibit CS5.4.5 BA's Global Expansion

Date	Strategic action	Cost
March 1992	Deutsche BA, set up by BA and German banks, buys small German airline.	Undisclosed
September 1992	BA buys 49.9% of TAT, French regional carrier based at Paris Orly, from where TAT has started international operations in BA colours.	£15m
November 1992	BA pays £1m for virtually bankrupt Gatwick-based Dan-Air.	£45m (in debt and other liabilities taken on)
January 1993	BA gains stake in USAir via purchase of convertible preference shares with 24.6% voting rights. Stake maintained through support of rights issue.	£265m (£189m plus £67m)
March 1993	BA buys 25% of Qantas, of Australia.	£304m
Total cost		£629m

Source: Financial Times, 8 March 1994, p. 23.

Exhibit CS5.4.6 Comparison of BA net profit with main rivals

	1990 ($m)	1991 ($m)	1992 ($m)
British Airways	174.5	444.0	400
Air France*	−277.5	123.0	−500
Lufthansa	5.4	−272.4	−1100
AMR (American)	−76.8	−239.9	−935**
Delta	−22.3	−239.5	−565
UAL (United)	95.7	−331.9	−957
Cathay Pacific	373.0	380.0	250
JAL	97.1	−22.2	−400
Singapore	499.2	568.7	600

* Includes UTA and Air Inter
** Includes one-time accounting charges
Source: Financial Times, 18 May 1993, p. 17.

Organizational Structure and Process Orientation

Chapter

6

Introduction

In small companies, coordination tends to be facilitated by the few individuals involved, because employees are often colocated (rather than being geographically dispersed) and by the fact that employees are often 'cross-trained' to perform several jobs; small companies cannot systematically afford to hire full-time specialists. Last, but not least, employees in small companies tend to be closer to the end-customer, and there are fewer colleagues and hierarchical layers standing between them and customers.

Large corporations tend to find it more difficult to create the right conditions for encouraging and sustaining *process-oriented behaviour*. In part, this is because existing organizational structures frequently make end-customers invisible to most employees within the organization; as a result, employees' behaviour is often guided more by *internal* objectives than by *external* customer-focused ones. In addition, years of conditioning and nurturing within *functional silos* often encourage behaviours which go *against* cooperation across functions to create effective processes.

As an example, consider what happened when Friends Provident, a leading UK insurance company reorganized its back-office function. Prior to 1991, Friends Provident organized its administrative functions, much like other insurance companies, around product and functional lines such as new business, claims, premium administration and servicing. Such a structure often caused problems and inefficiencies as described by a senior Friends Provident manager:

> *The functional demarcation lines . . . were quite dysfunctional. No one had a clear responsibility for the customer and this led to a degradation in the level of customer service we could provide. For example, assume that a customer sent us a claims letter with a request for premium adjustments. This letter would first be processed by the claims group, and then passed to the premium administration group where it would wait to be processed. The delays added up, leading to a degradation of the overall level of customer service we could provide.*

As part of a larger organizational change process, Friends Provident renamed its back-office administrative function the Customer Services Division, to emphasize that its function was *service to the customer*. It then reorganized this division into a number of multifunctional process teams, with each process team dedicated to servicing all needs of one particular customer group. The creation of these multi-functional teams changed the process flow dramatically, particularly at the interface between subunits, which went from an *ad hoc* hands-off across departments to a 'focused' flow across colocated teams. Soon thereafter, interestingly, each process team formed an overall plan to determine how to best service the needs of their customers. A process team manager commented on the motivation behind these plans:

> *It all seems simple, but it was the first time we had ever done it. Why had we not done it before? I think that it was the fact that now people had a clear customer they were trying to achieve for.*

A very old debate: integration vs. differentiation

As for most fields in which research is conducted, management research features periodic 'breakthroughs'. Some researcher or team of researchers 'discover' a new idea, a concept or a tool that quickly gets publicized and popularized. The last few years have certainly produced a number of such 'discoveries': activity-based costing and management, lean production, concurrent engineering and process reengineering, just to name a few.

Strictly speaking, however, some of these ideas, or at least parts of them, had been around for quite some time. Such is certainly the case of organizational structure and the current debate between functional and process-based structures. With a bit of perspective, it is clear that this debate is the latest incarnation of what used to be called the 'integration vs. differentiation' debate. These two words, and the awareness that organizations need to manage the tension between the two dimensions, have been around for quite some time. Management researchers just keep rediscovering these ideas under different guises.

First, what do these terms mean? Over 30 years ago, in a book that has become one of the all-time 'management classics', Lawrence and Lorsch[1] defined these two terms as follows:

■ Differentiation captures *'the difference in cognitive and emotional orientation among managers (operating) in different functional departments'*,

■ Integration captures *'the quality of the state of collaboration that exists among departments that are required to achieve unity of effort by the demands of the environment'*.

Lawrence and Lorsch also positioned the dilemma very succinctly. For a number of reasons that we will come back to in the rest of this chapter, organizations as they grow tend to group employees into 'departments', which have over the years typically been created along *functional* lines (e.g. sales, production, R&D). The problem is that once we group people and ask them to specialize, they progressively develop specialized working styles, mental processes and orientations towards a variety of issues. In particular, said Lawrence and Lorsch, members of different functions tend to differ along several important lines:

■ The goals they pursue, and their orientation toward these goals. This issue was already discussed in Chapter 4 when we discussed the measurement and evaluation of employee performance,

---

[1] Lawrence, P. R. and Lorsch, J. W. 'Organization and Environment', Harvard Business School Press, 1967, p. 11, italics in original.

- Their orientation towards time: some functions must have a longer term outlook than others (e.g. R&D vs. manufacturing),

- Their interpersonal orientation: some functions or departments tend to be more task driven, while others are more 'people-oriented',

- The degree of formality of the structure within the department/ function, including, for example, the number of hierarchical levels, the availability of clear measures of performance and the formality of reporting relationships within the department.[2]

Today, we could add to this list the notion of 'back office vs. front office': is the employee part of a group that is in direct contact with 'end-users', or is he or she part of a unit that is never in contact with external customers?

The dilemma comes from the fact that, on one hand, the company wants to foster 'differentiation' of employees because this specialization creates expertise and economies of scale, and it facilitates communication *within* each subunit. On the other hand, this differentiation makes it more difficult to achieve good 'integration', i.e. it makes it harder for people to communicate and coordinate *across* subunits.

Thirty years or so ago, when managers talked about 'integration', they did not use the term 'process orientation'; they simply referred to 'coordination needed to meet the demands of the environment'. The concepts, however, are very much the same. Today, the benefits of departing from the traditional functional structure to foster a greater degree of process orientation are well documented in the literature. In addition to Friends Provident, Xerox and Quantum, all described in case studies included in this book, many other examples are available:

- By reorganizing all its supply management activities around a single process for each commodity group, Motorola's Government Electronics Group was able to reduce cycle time by 80%, reduce late deliveries by 30% and improve supplier quality performance by a factor of 10,

- By reorganizing all functional manufacturing activities around a single build-to-order process, General Electric's Salisbury plant improved productivity by 50%, reduced the manufacturing cycle from three weeks to three days, and reduced customer complaints from 2% to 0.2% of orders,

- Ford's Customer Service Division increased productivity by 20% and improved customer satisfaction performance by 20–60% when they reorganized their Customer Assistance Centre around four core processes.[3]

In spite of these promising outcomes, managers must consider two sets of questions before embarking on a journey of building a 'process-oriented organizational structure':

---

[2] Lawrence, P. R. and Lorsch, J. W. *op. cit.*, chapter 1.
[3] Boehm, R. and Phipps, C. 'Flatness Forays', *McKinsey Quarterly*, **3**, 1996, pp. 128–143.

- Do we really need explicit process-oriented structures in our organizational design? Every organization, including strict functional hierarchies, can identify working processes. Managers must consider that process-oriented behaviour can sometimes be increased without radical changes to the formal structure, for example—and as discussed in Chapter 4—by changing the way employee performance is measured.

- If we decide to modify the structure in some way, how much 'process orientation' should we incorporate into our organization design? How far should we go? In practice, process orientation is not a zero-one decision; there is a continuum of organizational structures ranging from a strict functional hierarchy to a completely process-oriented organization. A firm can choose different positions along the continuum depending upon its unique needs.

The next section describes and analyses this continuum and some of its major components.

A continuum of opportunities

Most organizations are complex entities and are difficult to represent accurately by simple structure charts or other similar abstract representations. For example, few organizations claiming to have matrix structures actually make each manager report to multiple superiors. Similarly, in many functional organizations it is common to find some managers reporting to more than one boss. In practice, most firms really have several types of structures coexisting in different parts of the organization. These structures also tend to change over time.

Having said that, the 'organizational climate' of most firms features a dominant focus, which tends to be shaped by the systems and structures in place. For the sake of simplicity and clarity in the following discussion, we shall consider such dominant themes as representative of specific structures while discussing a possible continuum of organizational structures ranging from the classical functional hierarchy to pure process-oriented organizations where no functions exist (see Figure 6.1).

Today, the modern version of balancing differentiation and integration can be expressed as the challenge of balancing the dual needs of *developing functional knowledge*, while at the same time *employing this functional knowledge for creating value for customers*. While the relative importance of these two needs can vary across organizations, both of them are critical for business success. The two needs are inter-related because the ability of a firm to create customer value is dependent to a large degree upon its ability to develop and harness functional knowledge. Unfortunately, the two needs often impose opposing requirements on organizational structures.

Functional homes, which bring together groups of like-minded individuals, have proven over the years to be very beneficial for creating and fine-tuning functional

Figure 6.1 A continuum of organizational structures

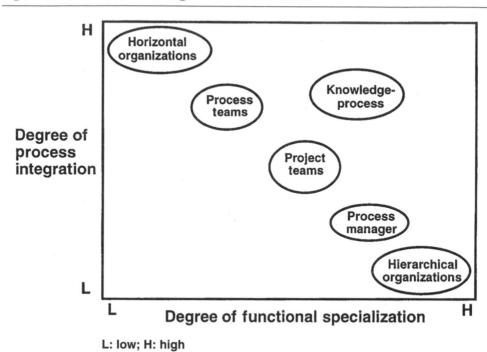

L: low; H: high

knowledge, especially in technically complex areas such as engineering and actuarial science. At the same time, functional divisions also tend to spawn communication gaps that cause fragmented processes and lead to a decrease in value delivered to customers.

What can a firm with a classical functional hierarchy do to increase its degree of process integration? One approach is to retain the functional hierarchy and create a locus of responsibility for increasing the degree of integration within specific processes. The locus of responsibility typically takes the form of a small group of managers in a staff function, who are responsible for enhancing the level of awareness of specific processes across different functional groups. This staff group typically has limited control over the execution of the process and has to rely on its ability to influence and persuade line functions to coordinate their operations for effective process performance.

Such an approach, labelled in Figure 6.1 as the *process manager* structure, has several advantages. In particular, it requires minimal change to the existing organizational structures and systems while still promoting an increased level of 'process awareness' across the different functions. On the downside, effective process integration is only possible if the different functional groups agree to co-operate. If the process manager group is not credible and lacks the requisite skills to influence possibly entrenched functional groups, the entire effort will come to a standstill.

As an example of such an approach, consider Alcatel, a leading international telecommunications manufacturer. Until the late 1980s, Alcatel was driven by a country-based organizational structure. Each country group enjoyed considerable independence in operations and was further subdivided into various functional divisions. Communication gaps were substantial, both across countries and across functional divisions. As a result, the company observed several dysfunctional effects, such as redundant versions of key products, each promoted by competing country groups and requiring its own development and support activities; processes were fragmented and customer complaints were reaching acute levels; and last, but often most embarrassing, different subunits within the Alcatel group frequently submitted competing bids for the same customer tender.

In response to these problems, Alcatel decided in the early 1990s to create a corporatewide focus on business processes by starting a corporate staff group called BPIS (Business Process and Information Systems). The mandate of BPIS managers was to map the key corporatewide business processes of Alcatel, increase awareness about these processes within Alcatel divisions and act as a support group to the line organizations in implementing process reengineering. Despite the best efforts of the BPIS managers, the unit never gained credibility within Alcatel. The fact that the unit was headed by three different senior managers in almost as many years probably did not help. By and large, the country-based groups remained like disparate kingdoms and thwarted most attempts to cooperate and harmonize processes. The results were increased losses for Alcatel during the early 1990s, until events climaxed with the departure of the CEO from the organization. The new CEO had to opt for a radical reorganization of the company in which the country-based groups were disbanded and restructured into global product-oriented divisions. As process orientation was part of the far-reaching structural changes, the BPIS group was eventually disbanded in the mid-1990s.

Teams do work . . .

Organizations have historically found cross-functional teams to be very effective for achieving specific objectives. The most common type of teams which have been in existence in most organizations over the past several decades are *project teams*. Typically, project teams are cross-functional, formed for a specific time period and for a specific business goal. They are created by assigning employees from various functions to the team and supporting them with adequate resources for achieving the assigned business objective. Team members can be assigned full time to a project team or can divide their time between the team and their functional home. In most situations, team members return to the latter when the project team is disbanded, either after a specific time period or after having achieved the stated objective.

Project teams are very popular because they provide an excellent organizational mechanism to pool cross-functional knowledge for a specific business process. Project teams can be responsible for the end-to-end execution of a corporatewide business process, or for a specific subprocess of a larger process. Companies in engineering and technology sectors, such as ABB and EDS, routinely function on the basis of project teams taking responsibility for the entire customer order fulfilment process. Other sectors, such as consultancy and law, also make extensive use of project teams.

Companies often rely on project teams to stimulate creativity. For example, IBM created a highly successful project team when it was trying to catch up with the competition and create its first personal computer. Another notable example is that of the Midland Bank in the UK, which relied on a project team to come up with the concept and implementation of its telephone banking concept (which later evolved into the highly successful First Direct Bank).

While project teams can be very effective, they have a disadvantage from a process perspective. Due to their temporary existence, knowledge about the *process* by which the team executed the project is 'lost' when the team is disbanded. Organizations have traditionally tried to overcome this limitation by documenting the process in manuals, rotating staff through successive project teams, and/or assigning key employees part time across multiple project teams dealing with the same process. For example, Intel has developed a competence in rotating key engineers across development teams as it develops multiple generations of microprocessors in parallel. Thus lead engineers in the team developing a particular generation of a microprocessor move to the team developing the next generation midway through the development cycle, and so on.

To emphasize the knowledge required for process integration and process learning, some organizations have started creating permanent team structures which are overlaid on the existing functional structure, an organizational form labelled as *process teams* in Figure 6.1. Kraft Foods, for example, realized in the early 1990s that its continued success was dependent on its ability to integrate the entire organization, from R&D to marketing, packaging, manufacturing and distribution, to create value in its key processes. Cross-functional teamwork was already prevalent within Kraft in the form of brand managers acting as process integrators for key product lines, but further emphasis on cross-functional teamwork was deemed essential to meet market demands. Boehm and Phipps describe the ensuing changes within Kraft:

> *Kraft responded to this challenge by developing a network of 'interlocking' discrete teams, overlaid on the existing functional structure, each focused on a critical element of a company's value delivery process, and linked through cross-team membership, integrated performance objectives and incentive compensation linked to business results ... Each of the individual teams is responsible for establishing performance targets, developing short- to medium-term operating plans, and prioritizing improvement initiatives for each of their respective product or customer areas. The functions are responsible for the day-to-day execution of these plans and targets, but the teams also play a coordinating role in the execution. Decision-making authority has been pushed down on to the team, and decision flows have been mapped out to clearly establish the limits of team empowerment. As additional resources are required team leaders make requests to the functional groups that house and train resources.*[4]

---

[4] *Ibid.* pp. 135–136.

Quantum, a leading manufacturer of computer hard disk drives also opted for such process teams (Case 6.2 describes the company's experience). In the late 1980s, Quantum had barely survived as it missed a technology shift in its core product. Determined to accelerate product development cycles, Quantum's management decided in the early 1990s to create a set of cross-functional product teams to address market needs for key product lines. Teams were made responsible for the definition, development and introduction of new products and for the revenues and gross margins generated by the products. The teams were overlaid on the existing functional structure, with the functional divisions responsible for executing the plans initiated by the teams and for providing effective career paths and skill-building programmes.

The creation of permanent process teams increases the complexity of a company's organizational structure. It also introduces many opportunities for conflicts and misaligned objectives across the functional and team structures, which increases the need for careful delineation of roles and responsibilities between the process teams and functional divisions. This is not always easy to do. At Quantum, for example, functional managers were made responsible for 50% of the employees' performance evaluation, even though most team members spent about 75% of their time on team-related activities (and hence only 25% on their functional job). This situation gave rise to a number of conflicts. Furthermore, functional units were responsible for hiring, career path development and planning skill-building programmes. However, the skills needed to be a successful team member were quite different from that needed within the various functions, to the point that the company was only able to identify 14 out of 450 employees with the right expertise to be effective team members.

To address these potential pitfalls, Kraft's senior management focused on creating appropriate synergy between the process and functional structures through a combination of innovative mechanisms. In particular, the company placed senior functional managers in formal team support roles, where they could act as a liaison between the teams and the functional departments. The company also changed its performance evaluation system to promote functional support for the teams, including implementing joint accountability to evaluate the synergy between the functions and the process teams.

Refining the role of teams . . .

Going yet further in the direction of process integration, some organizations have decided to eliminate functions altogether and instead organize themselves along process teams. The resulting form is often referred to as the 'horizontal organization'. The horizontal organization concept was first described in the literature by Frank Ostroff and Doug Smith.[5] They described their vision of the horizontal organization as one where:

> *work is primarily structured around a small number of business processes or work flows, which link the activities of employees to the needs and*

---

[5] Ostroff, F. and Smith, D. 'The Horizontal Organization'. *McKinsey Quarterly*, **1** (1), 1992, pp. 148–168.

capabilities of suppliers and customers in a way that improves the performance of all three. Work and the management of teams get performed more by teams than by individuals, and these teams assume real managerial responsibilities. . . . Career paths follow work flows: advancement goes to people who master multiple jobs, team skills, and continuous improvement.

The following principles are commonly seen as forming the core of horizontal organizations:

- The organization is structured around processes and not along functional tasks. In practice, this requires identifying key organizational processes and forming cross-functional teams taking responsibility for the end-to-end performance of these key processes. Organizations that have not yet identified their core processes can apply the same principle, by identifying key customer objectives and putting together a team of individuals with the right set of skills to satisfy or exceed those customer objectives.

- Performance objectives and evaluation systems are linked to process objectives. This requires a dedicated focus on understanding the key parameters of process performance and redesigning organizational systems around them. This usually requires shifting the focus of evaluation and reward mechanisms from functional goals and profitability to customer related indicators (e.g. customer satisfaction).

- Managers assume the role of coaches and provide overall leadership, instead of actual day-to-day control and management. In theory, with properly skilled and empowered teams, there is no need for operational process-related decisions to be moved up the managerial hierarchy. If we further develop this logic, the hierarchy of managers can be removed (or at least decreased to a minimum) and managers have to assume new roles and develop new skills.

Implementing the above principles usually calls for a radical redesign of the entire organizational structure and systems. Consider the example of Xerox. For much of the 1970s, Xerox had watched in horror how smaller, cheaper and better machines from Japanese competitors destroyed its near-monopoly in the lucrative office copier market. The 1980s were spent gaining back, yard by yard, the ground that it had lost by the mile. Despite considerable progress, Xerox still was not competitive when compared to its rivals at the end of the 1980s. Paul Allaire, CEO of Xerox, summarized the situation as follows:

. . . If we are to accelerate the rate of change, then we must change our basic approach of managing the company. For decades, we've run the organization as a large functional machine, which is governed by decisions made at the centre. We've created a system that is complex and which prevents people from taking responsibility . . . We need to create discontinuous change; incremental changes will not get us to our vision.

As part of the effort to create discontinuous change, a team of bright young managers from across the world was given the explicit mandate to look at the future 'organizational architecture' for Xerox. They were given no prior constraints. The term 'organizational architecture' included more than simply the organizational structure; it also included recommendations for changes in processes, behaviours and management style. The team decided to adopt an organizational architecture aligned with the core processes of Xerox, and relying extensively on cross-functional teams. Central to the new structure was the concept of business divisions, a set of cross-functional business teams led by a 'business team general manager'. Each business team was in charge of a product offering in an 'end-to-end' manner, with responsibility for the lifetime performance of the offering, including customer satisfaction, market share and profit (ROA).

Complementing the move to the new organizational architecture, process sponsors and owners were identified and made accountable for specific processes. Process ownership and management involved taking full responsibility for a process, including agreeing upon customer requirements, defining measurements and documentation standards, defining and monitoring work processes, and applying the quality tools to ensure that processes were under control and that output met customer requirements.

Aside from Xerox, a few other organizations have moved towards creating horizontal organizations, though perhaps in a less holistic manner. AT&T's Network Systems Division restructured its entire business around processes, and started setting budgets by process parameters and awarding bonuses to employees based on customer evaluation. General Electric's Lighting business scrapped its functional structure in favour of a horizontal design with more than 100 processes and programmes. Motorola's Government Electronics group redesigned its supply management business as a process with external customers at the end.

In practice, few organizations have been able to match up to the ideal of a truly 'horizontal organization': eliminating functional hierarchies to rely on process teams only. This is because functional homes provide a valuable locus for functional expertise and knowledge development, something that few organizations can do without. As a result, very few can be positioned at the top left corner of Figure 6.1. In fact, some organizations, which went all the way to process teams only, have since moved back towards restoring their functional homes, at least partially.

In an effort to further refine and clarify the roles of functions and process teams, some organizations are experimenting with another organizational form, which has been labelled the *knowledge-process model* in Figure 6.1. The knowledge-process model recognizes the need for both functional homes and cross-functional process teams and thus explicitly acknowledges the dual needs of an organization: the need to create knowledge and the need to create value for customers by utilizing this knowledge within customer-oriented processes. Functions are assigned the job of developing knowledge, while process teams are responsible for creating value for customers.

The knowledge-process model goes further than Quantum's process team structure described earlier, which left the functional departments in control of resources and execution of tasks. In contrast, the knowledge-process model moves responsibility for the execution of value-creating tasks solely to the process teams. The functions become almost like 'schools' assigned two major responsibilities:

- summarizing current knowledge, searching for new knowledge, and teaching this to all their members; and

- developing guidelines or best practices for utilizing the knowledge within value creating processes.[6]

The knowledge-process model also has an impact on career paths. When employees are part of a cross-functional process team, they are evaluated on the amount of value they create within the process. Periodically, they move back for some time to a function where they are evaluated on the basis of how well they perform the following two tasks:

- increasing their own knowledge by absorbing knowledge within the functional home; and

- increasing the functional home's knowledge base by bringing to it knowledge from the process team(s).

Organizational forms similar to the knowledge-process model can be found in various sectors. Womack and Jones report how Honda makes its engineers in Japan and North America go through repeated learn-apply-learn cycles. Honda engineers alternate between their 'home engineering function' where they develop complex engineering skills, and 'development teams', where they apply their acquired knowledge in value-creating processes.[7] Many consulting and software companies also operate in such an organizational form where employees spend most of their time working in cross-functional teams with the sole objective of delivering value to customers (e.g. through a consulting report or a software system). At regular intervals, or on an as-needed basis, they return to centres of knowledge (e.g. special areas of consulting expertise such as international tax laws, or emerging technologies, such as Java for software firms), to update their skills before moving on to another cross-functional team.

Finding the balance

Large organizations tend to be complex and hence typically cannot be classified into pigeon-holes as neatly as in Figure 6.1. Balancing the degrees of functional specialization and process integration is a delicate task. Each company has to seek its own unique trade-offs between the right amount of emphasis on the two aspects.

---

[6] Womack, J. and Jones, D. 'From Lean production to the Lean Enterprise', *Harvard Business review*, March-April 1994, pp. 93–103.
[7] Womack, J. and Jones, D. *op. cit.*

Frequently, this results in the organization making constant changes to its structure, as it searches for the 'optimal' organizational form. At other times, it leads to the adoption of multiple structures within the same firm, each adapted to the specific needs of a part of the firm.

Some guidelines can be proposed about factors influencing the choice between functional specialization and process integration:

- *Rate of technology change*: When the rate of change in the fundamental technology underlying the business is very high, so are the benefits of a functional emphasis within the organization. Functional groups provide good homes for the development of new knowledge and help organizations keep up with changes in technology. Upon creating process teams, organizations frequently find that their functional skills are at risk.

- *Rate of market change*: When the firm's market is changing rapidly, process orientation needs to be higher within it. Process teams provide a sharp and unrelenting focus on customer value, an aspect which often gets obscured in functional hierarchies. Cross-functional membership allows the process team to respond flexibly to changing drivers of customer value.

- *Inter-dependence of process activities*: No two processes in an organization are alike. Each process consists of several subtasks, which can be inter-linked with various degrees of complexity. When the process subtasks are stable with low degrees of interdependence, they can be done relatively independently within functional groups. This type of environment encouraged specialization of tasks at the start of the industrial revolution. When the process tasks are not very well defined and/or are highly interdependent, then process integration becomes more necessary. Process teams allow for a high bandwidth of communication across team members and facilitate effective value creation.

In general, then, a rapid rate of change in *markets* would favour organizational forms in the *upper left* corner area of Figure 6.1, while rapid changes in the firm's technologies would be better served by structures positioned in the lower right corner. Other things equal, a high degree of interdependence across process activities would favour organizational structures in the upper left corner .

Getting there . . .

While there is no silver bullet for the structural woes of firms, the emphasis on process teams within organizations is on the rise. Firms are trying to address some of the shortcomings of yesteryear's functional structures by creating process teams to listen to customers. However, the move towards process teams is not simple and comes with several challenges.

First, organizations need to understand the drivers for customer value. Focusing on the true drivers of customer satisfaction can lead to interesting insights about long-held assumptions about customer value. For example, as discussed in Chapter 2, Taco Bell had long assumed (much like other fast-food firms) that customers appreciated large and fancy restaurants. In the late 1980s, John Martin, Taco Bell's CEO, initiated a move to understand the true drivers of customer satisfaction. The result was one simple answer—food. Good quality food, cooked in clean surroundings and served at the right temperature. This discovery was completely at odds with much of Taco Bell's existing strategy. Taco Bell had spent much of the 1980s decreasing the cost of food given to customers so that they could spend more on building fancier restaurants and on advertising.

Secondly, firms need to understand the organizational processes which create true customer value. This can imply the creation of detailed core process maps for the whole organization, much as Xerox did, or the simple identification and mapping of a few key processes, as in Alcatel. As described in Chapter 2, creating process charts and mapping process flows can be a cumbersome and frustrating process, often simply because employees within functional departments lack an understanding of the entire, end-to-end process. Frequently, existing processes have to be redesigned or new processes have to be created from scratch to deliver customer value. The Trustees Savings Bank, for example, had to create a new process to sell products to customers within a branch, something which had never been done within their branches before.

Thirdly, and as mentioned in Chapter 4, organizational structures and systems have to be redesigned in harmony. When process teams are created, performance evaluation and reward systems often have to be reexamined closely. Because it is foolish to reward behaviour A and hope to obtain behaviour B, a push towards process integration will either not work or will not be sustainable without adequate support from redesigned organizational systems. It is important to note that the introduction of process teams typically adds to the complexity of an organization. Hence, the exact delineation of roles and evaluation procedures across process teams and the rest of the organization has to be thought through carefully. Quantum, for example, faced several ambiguities in internal evaluation procedures when team members, while assigned almost full time to process teams, were being evaluated 50% by their functional chiefs.

Finally, a cultural change process has to be initiated at all levels of the organization, starting from senior management to front-line employees. Each member of the firm has to come to terms with, and accept the need for, the new organizational philosophy and the new way of working it calls for. An internal publication within Xerox, circulated after its move to the new process-oriented organizational structure, summed up the challenge succinctly:

> *We are no longer the organization we used to be and we are not yet the*
> *organization we intend to be. We are in a state of transition where former*
> *processes may not work any more. New roles and responsibilities may not*

be totally defined or understood. Key interfaces between organizations may not be solidified. This can be a difficult state in which to operate. Many people are uncomfortable with change. They feel anxiety as power bases shift and familiar working patterns are altered. This is the normal outcome of dramatically reengineering an enterprise.

Case 6.1

This case was written by Soumitra Dutta and Enver Yucesan, Associate Professors at INSEAD. It is intended to be used as a basis for class discussion, rather than to illustrate either effective or ineffective handling of an administrative situation.*

Xerox (France) 2000:
A Race Without a Finish Line

We will need to change Xerox more in the next five years than we have in the past ten.

Paul A. Allaire
Chairman, Xerox
4 February 1992

The strategic vision of Xerox for the 1990s was to be the leader in the global document market by providing document services which enhanced business productivity. After many years as *the* copier company, Xerox had tried unsuccessfully to expand into information technology from its traditional strengths of document handling, printing and copying. With continued analysis of its customers' key requirements and its own strengths, it had become clear that information handling would continue to depend to a great extent upon documents, either on paper or digital media. Thus the strategic vision of the company was updated to be 'The Document Company'. To make this strategic intent explicit, Xerox even changed its names in the mid-1990s to 'Xerox, The Document Company'.

For much of the 1970s, Xerox had watched in horror as smaller, cheaper and better machines from Japanese competitors destroyed its near-monopoly in the lucrative office copier market. The 1980s was spent gaining back yard by yard, the ground that it had lost by the mile. Central to this comeback was drastic cost-cutting in manufacturing and an obsessive dedication to quality which had gradually changed the culture and management of the company. However, Xerox remained less than healthy at the end of the 1980s. Even as it regained precious lost market share, brutal price competition held its equipment revenues flat and its pretax profit margins on business equipment decreased to around 4% from 17% at the start of the decade. Many of its forays into noncopier businesses, such as workstations, also turned out to be wasteful ventures. Even more ominous was the direct attack that Japanese competitors were poised to launch in the 1990s on Xerox's most lucrative product segment—high performance copiers. At the start of the 1990s, Xerox could hardly afford to stand still and rest on its past laurels.

Xerox faced several important challenges in the global document market on its path to achieving the strategic vision, termed as Xerox 2000. First, the businesses making up the document market in the second half of the 1990s were seen to be different from the way businesses had been traditionally defined within the company. Second, there were basic changes in customer needs, in competitive actions and even the nature of competitors which needed to be addressed. Third, the pace of technological innovation was increasing rather than slowing down. Paul Allaire expressed the need for change within Xerox in the following manner:

> . . . *If we are to accelerate the rate of change, then we must change our basic approach of managing the company. For decades, we've run the organization as a large functional machine, which is governed by decisions made at the centre. We've created a system that is complex and which prevents people from taking responsibility . . . We need to create discontinuous change; incremental changes will not get us to our vision.*

COMPANY BACKGROUND

'This is a race without a finish line.'

David Kearns, ex-CEO, Xerox

Xerox was formed in 1961 when the Halold company renamed itself after its highly successful model 914 office copier, the first viable xerographic office copier. Protected by several patents, Xerox grew at a phenomenal rate and completely dominated the US copier market during the 1960s and the early 1970s. After a series of antitrust actions, Xerox agreed to license its technology to competitors, and this allowed IBM and Kodak to enter the copier business in 1970 and 1975 respectively. Japanese companies also entered the market and concentrated upon mass-producing low volume copiers. By the early 1980s, IBM and Kodak had made significant headway in the upper end of the copier market, and Japanese companies had started dominating the lower end of the market.

Xerox's origins in Europe lay in Rank Xerox, which was formed in 1956 as a joint venture between the Rank Organization plc (of the UK) and Xerox to manufacture and market reprographics and other office information equipment throughout the world, excluding the Americas and Japan. Rank Xerox quickly rose to become one of Europe's leading high-technology companies, providing a wide range of products, systems and services to handle customers' document management requirements. In 1997, the Rank Organization sold its stake in Rank Xerox back to Xerox and the company, Rank Xerox, was renamed as simply Xerox.

From the start, Rank Xerox led the European market with a steady stream of improved new products. Starting from the mid-1970s, however, Rank Xerox's position was challenged by Japanese companies with small, high-quality, low-priced copiers. Similar to the developments within the US copier market, by the mid-1980s, 80% of the copiers in use throughout Europe were Japanese supplied. As part of a recovery strategy, Rank Xerox undertook a series of benchmarking studies with Xerox (North America) to determine the underlying reasons for their lack of competitivity. They looked externally at world-class companies and internally at Fuji Xerox, a Xerox joint venture in Japan which had won the Deming Prize in 1980. These studies revealed that Xerox and Rank Xerox were at a disadvantage in terms of both cost and quality, and that they were slower to develop and introduce new products. Following further studies of the success of Fuji Xerox, Rank Xerox joined with Xerox in the development of a quality strategy to recover its leadership position.

Rank Xerox started implementation of its quality strategy in 1984. Initially the focus was on product quality and cost. The emphasis soon moved to incorporate all aspects of its operations. The focus then became changing the culture of the company to one of quality. By 1986, Rank Xerox had contained the erosion of its market share and started to climb back to leadership. Rank Xerox was also helped with the anti-dumping duties levied against Japanese companies in 1986–87. By the late 1980s, Rank Xerox was one of the few European companies which had not just halted, but also reversed Japanese advances in its key markets with the result that in 1993, it was the leader in the European document market in terms of revenue generation.

By 1996, Rank Xerox was one of the top 150 companies in Europe, serving over 500 000 customers, employing 22 000 people and earning revenues of Ffr 32.5 billion per year. In 1996, Xerox in France employed around 3600 people and generated about Ffr 4.7 billion in revenues annually. Rank Xerox had manufacturing factories at Mitcheldean (UK), Venray (Holland) and Lille (France) employing a total of 4500 employees and sourcing materials and services from over 400 local suppliers. Rank Xerox's customers represented every level of commercial, industrial and government organizations, from the small local organization to the largest international ones. Their requirements ranged from small office copiers to sophisticated international networks of computers, copiers and printers. Over the years, Rank Xerox established itself as a leader in European industry as witnessed by its string of awards including the first European Quality Award in 1992 (see Exhibit 6.1.1).

LEADERSHIP THROUGH QUALITY

Xerox is a quality company. Quality is the basic business principle for Xerox. Quality means providing our external and internal customers with innovative products and services that fully satisfy their requirements. Quality improvement is the job of every Xerox employee.

The quality policy stated above has served as the basic business principle for Xerox since the inception of the 'Leadership through Quality' efforts in 1984. Didier Groz, the director of strategy and quality for Xerox France explained:

We were fortunate to have had a very clear vision. We had a 'Green Book' in 1984 describing the different elements of our quality policy [see Exhibit 6.1.2]. This has not changed even today. The different pieces of the puzzle were all there in 1984. Different divisions within Xerox and Xerox France have taken and applied the pieces in their own order, but the end state is the same for everyone—to be the leader in the global document market.

Didier Groz continued:

Of course the emphasis within the application of our 'Leadership Through Quality' principle has evolved and changed over time [see Exhibit 6.1.3]. Initially, it was making our customers satisfied. Next, the emphasis shifted to being recognized externally as a quality company. This was followed by our attempts to integrate quality in all aspects of our business and management processes. Today, our emphasis is on quality in the support of our four priorities: customer satisfaction, employee satisfaction, market share and productivity. Such an emphasis on productivity is important because we want to be a role model for our customers. We believe that we are in the business of helping our customers improve their productivity through the document.

Franz Scherer, the CEO of Xerox France between 1993 and 1996, emphasized that deep impact of the quality movement:

> *We did not discover quality as a management fad. Rather it was a common approach for the fundamental overhaul of the corporation. Thus we took the behavioural aspects of quality very seriously right from the start.*

In 1994, Xerox was sure of the benefits of its quality policy. However, it faced 'a crisis of opportunity' in the words of Paul Allaire:

> *We are facing a crisis of opportunity. On one hand, we see attractive markets, and we have superior technology. On the other hand, we won't be able to take advantage of this situation unless we can overcome cumbersome, functionally driven bureaucracy and use our quality process to become more productive.*

Didier Groz explained further:

> *For one, the company's classical functional structure was not the way the customer perceived the company. The customer did not wish to go from one function to another. Due to the functional silos, all cross-functional decision-making was pushed up the hierarchy within the organization. This reduced our ability to respond quickly and to be closer to the customer. For example, in France we had eight regional organizations responsible for sales. Service reported directly to the managing director through a separate functional hierarchy. Sales staff had a mindset of 'sell and forget'. They had little incentive to be concerned about the relationship with our customer after sales. In the same manner, service staff collected valuable information from the customer during their site visits, but rarely passed them on to the sales staff. In case of any conflicts between sales and service, decision-making had to be pushed up to the managing director.*

Such a crisis of opportunity presented several challenges for Xerox. It had to strengthen the connection between Leadership Through Quality and productivity improvement. In particular, it had to decrease the time to market radically, ahead of competition, with offerings that exceeded customer expectations. It also had to develop process breakthroughs which would achieve quantum improvements in business performance while empowering individuals and work groups to improve their business processes continuously. Thus a major change agenda was required to transform Xerox into the leader in the global document market. The Leadership Through Quality principles were seen as providing the bedrock for this change agenda and for achieving world-class productivity as illustrated below in Figure CS6.1.1.

Figure CS6.1.1 Leadership Through Quality

THE NEW ORGANIZATIONAL ARCHITECTURE

I began to suspect that our organization was getting in the way of achieving the changes we required.

Paul Allaire

Towards the end of 1990, Paul Allaire formed a Quality Improvement Team (QIT) consisting of some of the best and brightest managers a few levels down in the organization from different functions and different geographies to look at the challenges facing Xerox and help to understand the nature of the problems and the alternatives. The team, which renamed itself as the 'Future-tecture' team, was given the explicit mandate to look at the future organizational architecture for Xerox without any constraints. The organizational architecture was seen as more than an organizational structure. It was supposed to include recommendations for changes in processes, in behaviours and in management style.

The Future-tecture team developed four alternative approaches to the future organizational architecture of Xerox: a business division architecture, an improved marketing architecture, a geographical/transnational architecture, and optimizing the current architecture. Each alternative was evaluated against the degree to which it met the following criteria: capitalizing on employee energy, reducing time to market, simplifying the running of the business, increasing customer focus, and improving competitiveness. After a good deal of analysis, the Future-tecture team came to the conclusion that the business division's architecture would provide the best platform for achieving the Xerox 2000 vision.

In the summer of 1991, the top management team of Xerox spent time working with a core group of the Future-tecture team to move their approach to the next level of detail and to resolve the key open questions. Next, another group, the Organizational Transition Board (OTB), comprising 15 key senior managers from different functions and geographies was set up to finish the design of the new Xerox. Between autumn 1991 and spring 1992, the OTB spent hundreds of hours consulting with different Xerox teams across the world and with the top management team to piece together the new organizational structure, design the new processes and describe the required new behaviours.

The organizational structure

The core idea behind the new organizational structure was to create business divisions—units and teams—with complete end-to-end responsibility for a particular Xerox offering (see Exhibit CS6.1.4). There were three points of coordination between the divisions. First, a customer business unit had primary responsibility for direct interactions with Xerox customers and for ensuring customer satisfaction in a particular geographical region (such as the United States, Americas and Europe). Second, a technology management division was responsible for nurturing technological competencies and leveraging them in the different Xerox offerings. Third, a set of strategic service units was formed to support the business divisions in areas such as manufacturing and logistics.

A business division was to have control over the complete value-adding chain including business planning, product planning, development, manufacturing, distribution, marketing, sales (including control over its own sales force/channel) and customer service. Each business division could decide to conclude agreements with other units to provide for services on a contracted basis, for example, manufacturing from a manufacturing unit, sales through a salesforce that is shared with other business divisions, or customer support from the customer operations division.

The core of each business division was a set of cross-functional business teams (See Exhibit CS6.1.4) led by a business team general manager. Each business team was responsible for an offering in an end-to-end manner with responsibility for the lifetime performance of the offering, including customer satisfaction, market share and profit (ROA). The business team general manager was envisioned as the primary general management job at the operations level and as a key developmental position for future senior managers of the corporation.

Customer business units were responsible for the support of all activities which related to the customer, including service, administration, integration of major customer relationships, sales support, and providing local administration for the business division salesforces. Customer business units were responsible for customer satisfaction within a specific geographical region and for integrating sales, service, customer support and administration to ensure that a single consistent face was presented to the customer. The operations of the units were funded through contracts with the business divisions. Business divisions were not forced to use the customer business units and could opt to choose other channels as appropriate.

With major decisions delegated to the business divisions, the corporate office was seen to be responsible for strategic direction-setting, interbusiness division resource allocation, maintenance of the values of the company, and the creation of a context that enhanced that ability of the rest of the corporation to be effective. A senior manager from the corporate office, termed as an 'operations executive', was to act as a coach to ensure the business success of each division. Operations executives had no dedicated staff and were also supposed to act as 'corporate champions' for the division(s) within their responsibility. In addition to the operations executives, two senior executives from the corporate office were each responsible for the technology management process and strategic services.

After the articulation of the new organizational architecture in early 1992, different divisions within Xerox gradually changed their structures during late 1992 and much of 1993 in accordance with the new architecture (see Exhibit CS6.1.5 for the new organizational structure of Xerox France). The changes were dramatic because it implied a move from a classical 'command and control' organization consisting of discreet functions, such as sales, marketing, service, manufacturing, personnel and finance, which were organized on a formal hierarchical basis, to a much more cross-functional and participative organization where team orientation and self-managed work groups became commonplace. Bob Walker, director, business management systems and quality for Xerox UK, explained further:[1]

> In the functional, command and control organization, it is typical to have functional conflict, increased bureaucracy and overresourcing within functions—a functional myopia which causes a limit to adaptability when change is necessary. All of this was present in Xerox. We wanted to shift the paradigm to a much more cross-functional integrated team approach to running the business.

The organizational structure was seen as only one component, the 'hardware' of the new architecture. The other side, the 'software' consisting of processes, people, management leadership, culture and values was seen to be equally and probably more important. To ease the transition to the software of the new organizational architecture and to create a 'common language', a graphical management model was formed (see Exhibit CS6.1.6). The management model had six dimensions with the customer at the centre with bidirectional connections with other major elements of the model. Management leadership and commitment was seen as driving all other activities and connected to the customer via the people. The emphasis on productivity was a common thread flowing through each of the six elements of the model.

THE BUSINESS PROCESS ARCHITECTURE

> Business processes are those processes through which we create, and deliver value to our customers . . . Business processes drive business results. The whole entity is defined by its business processes . . . We must redesign those business processes that were rooted in the functional organization and align them with the cross-functional requirements of the divisional structure.

Xerox 2000 Leadership Through Quality
Strategy Briefing Book, February 1994

Effective and efficient business process management was viewed to be critically important in the new organizational architecture. When Xerox first implemented Leadership Through Quality in 1984, it focused on process identification and improvement within tasks. As employees became familiar with work processes, Xerox applied it to large-scale, functional processes. Later Xerox moved into cross-functional and cross-organizational processes. This progression, and the understanding of its processes that Xerox developed, enabled it to define its new business process architecture. This was an integrated process structure, describing all the process areas which made up Xerox's business, and outlining the 76 subprocesses needed. It was seen as Xerox's template for future process and systems development.

Xerox's key processes at a macrolevel were defined as shown in Exhibit CS6.1.7. Franz Scherer provided a summary description of the architecture:

---

[1] Walker, R. 'Rank Xerox – Management Revolution', *Long Range Planning*, **25**, 1992, p. 11.

The first core process is the time to market. It is the process of bringing new products to market. It involves the design and engineering of a new product, its manufacturing process design, its launch and marketing and, finally, its maintenance. This process was one of the weakest for Rank Xerox in terms of control and development costs, failure rates and time required for introducing a new product (speed to the market).

A second core process is the integrated supply chain. It provides demand driven manufacturing and flow of goods and involves the management of transportation, receiving and handling of materials, packing and shipping equipment, operating warehouses; in general, it provides planning, acquisition, distribution and recycling of all assets.

The third core process is market (engagement) to collection. This process involves identifying new market segments, identifying prospective customers within these market segments (market management subprocess) progressing these potential customers to engagement and ordering (prospect to order subprocess) and invoicing, delivery and collection (order to collection subprocess).

The fourth core process is customer service and involves sales support, after-sales support and product maintenance in order to achieve delighted customers.

In addition, enabling processes, which supported the business internally, were defined company-wide and locally for business management, human resources management, financial management, and information management. Different divisions of Xerox further adapted the generic process architecture of Exhibit CS6.1.7 to suit their local needs. For example, Xerox France adopted the process architecture shown in Exhibit CS6.1.8. The business process architecture served as a common language for the organization, laying the foundations for defining appropriate management roles and responsibilities.

Process sponsors and owners were identified and made accountable for specific processes. Exhibit CS6.1.8 indicates the assignment of sponsorship and ownership for processes within a specific core process of Xerox France. Process ownership and management was based on taking process responsibility, such as agreeing upon customer requirements, defining measurements and documentation standards, and defining and monitoring work processes; and applying the quality tools to ensure that processes were under control and that output met customer requirements. Process owners defined standards of operation against different criteria (such as customer feedback, benchmarking and best-practice sharing), and ensured that any conflicting requirements were addressed and resolved. Exhibit CS6.1.9 lists the specific roles and responsibilities for process sponsors, process owners and other players involved in process management. Interface issues at both core and subprocess levels were managed in regular meetings of process owners. For example, the monthly European Integrated Supply Chain Process Steering Committee set action plans to resolve strategic and tactical issues of inventory management.

The process architecture was also seen as providing the basis for identifying and implementing significant process improvements to achieve the desired goal state of achieving world-class productivity. As described in an internal document:

The New Productivity means changing processes and shifting the emphasis from doing

Case 6.1

things cheaper and faster to the really significant improvements that come from doing things differently or not at all. The architecture forms the basis for understanding the whole and guides management's focus for systematic breakthrough and continuous improvement efforts.

Process reengineering or breakthrough was seen as a desirable approach to transforming the organization and improving its performance. Process reengineering involved clean-sheet design focusing upon critical business processes that created a competitive advantage. The goal of process reengineering was set ambitiously—to seek twofold to tenfold improvements in quality, cost or time. Due to the vast cross-functional scope of process reengineering efforts, the need for adequate senior management involvement and sponsorship and the necessity to integrate it into the strategic planning process was emphasized. The role of the corporate office, division presidents and their first level of management was to identify the potential return from reengineering processes, commit the necessary investments and identify boundary conditions for implementation. The responsibility of management at every level within the organization was to form empowered, cross-functional implementation teams to design the new process.

Once breakthroughs had been implemented and were mature, the focus would shift to continuous improvement which had traditionally been one of the cornerstones of the Leadership Through Quality movement. The use of quality tools to document and measure existing processes in order to understand their current realities along with performance gaps was central to the efforts of continuous improvement. Individuals and teams were expected to attain proficiency in the disciplined use of the quality tools for this purpose. Exhibit CS6.1.10 illustrates the relation between process reengineering and improvement.

The business process architecture would also have a major impact on guiding and forming behaviours within the new vision of the organization. Didier Groz emphasized the impact of the process architecture for managing people:

We do not view the process architecture as a bureaucratic apparatus for the functioning of people. Rather, it forms the fundamental context for empowering people through the creation of autonomous, process-oriented work teams. In our view, processes are not controlling, but empowering.

He elaborated further:

We are slowly but surely moving towards a process-oriented organization. This implies a radical change in not only our structure but also in our behaviours, roles and responsibilities. While the structural change is easier to accomplish, the behavioural change is much harder. Our management processes now reflect our process orientation. For example, we are increasingly managing Xerox affairs at the European level through meetings of appropriate process sponsors and owners.

Franz Scherer emphasized the focus on implementation:

Today it is a different game—a game of rigorously implementing reengineered processes. Before general managers were often defending their head counts. Today, that is no longer a bone of contention. Everyone is aware of the levels of productivity implied by the reengineered processes, and there is competition to achieve or exceed those levels.

MANAGEMENT AND PEOPLE BEHAVIOURS

Our people have been a competitive strength and will continue to be so in the future. At the same time, this new environment is going to require new skills, knowledge and experience. Perhaps even more importantly, it will require new attitudes on the part of many managers . . . Before focusing on what needs to change, we should be clear on what needs to be preserved and strengthened.

Paule Allaire
4 February 1992

There were three things which were not going to change. First, the core values of the corporation (see Exhibit CS6.1.11), which had been stated at the start of the Leadership Through Quality movement in 1984, were to be renewed and strengthened during the transition to Xerox 2000. The corporate core values determined the four business priorities for the organization which were in the following order: customer satisfaction, employee motivation and satisfaction, market share and return on assets. Didier Groz elaborated further on these business priorities:

> *Our first priority is our customer. This is natural because we exist to satisfy and exceed our customer's needs. Employee satisfaction is our next priority because we believe that satisfied employees are critical for customer satisfaction. For example, if our service staff are not motivated and satisfied then it is likely that they will not be able to satisfy our customers when they visit our customer premises. Employee satisfaction is related to customer satisfaction. If our employees are visiting dissatisfied customers, then they, too, become demotivated.*
>
> *If our customers and employees are satisfied, we believe that our market share and return on assets will increase. It is interesting to note that our business priorities have evolved over time. In 1985 we had three priorities in the following order: return on assets, customer satisfaction and market share. Over time, we developed the conviction that customer satisfaction, through motivated and satisfied employees, is the right way to achieve our desired market share and return on assets. We believe this to be a mature ordering of priorities.*

Second, the commitment of the organization to the Leadership Through Quality principles were to be affirmed. Didier Groz explained:

> *Our quality policy and processes have served us well over the last decade. Committing ourselves to our Leadership Through Quality policy implies a number of role model quality behaviours for our managers. For one, it means placing customers as the number one priority through the use of a customer-driven approach. It means viewing empowered people and teams as the primary enablers to delivering business results. Also, it implies managing by fact and conducting root cause analyses to be results oriented. Finally, it means exhibiting personal strength and maturity in leading people.*

Third, the strategic intent of the corporation to be 'The Document Company' was to remain intact. The intent was the same, only the means for achieving the intent was to change. There were five general requirements on the kind of people skills required in the new vision of the corporation: general management experience/skills, ability to understand and deploy technology, market orientation, ability to delegate and empower, and good cross-organizational teamwork. A senior manager commented on the requirements:

> We know that we have few staff of the desired profile today. We will need to invest significantly in our people. We all need to undertake a learning experience—build on our strengths and rectify our weaknesses. Towards this goal, we have identified 23 different specific leadership attributes (see Exhibit CS6.1.11) which we desire in our management and eight different dimensions (see Exhibit CS6.1.12) of the type of culture we wish to create in the new organization. Our employees are very aware of these priorities because the compensation of each one of our people is dependent on the degree to which they have contributed to the achievement of these four priorities.

A management process was devised to drive the desired management model and the business priorities through the organization down to each team and individual employees. A programme of continuous self-assessment called Business Excellence Certification (BEC) was instituted for each operating unit. The Business Excellence Certification programme required each operating unit of Xerox to assess its position relative to the desired state of the six different components of the new management model (see Exhibit CS6.1.6). The self-assessment was conducted quarterly along than 40 different subelements (see Exhibit CS6.1.13) of the new management model and was validated by senior line managers from other units. An internal document described the assessment programme thus:

> Achieving certification under the BEC programme means a Xerox operating unit has demonstrated an operational command of Leadership Through Quality tools and processes, is achieving good business results characterized by continuous improvement, and has demonstrated good prevention-based processes throughout the organization.

The quarterly self-assessments were combined with monthly reports on the four business priorities of customer satisfaction, employee satisfaction, market share and return on assets to yield a consolidated cumulative diagnosis which in turn was validated by line managers from other units. This consolidated diagnosis was used to generate the 'vital few' actions for the unit. A policy deployment process (see Exhibit CS6.1.14) was then used annually to negotiate and mutually decide upon operational plans and goals for each unit, team and individual. A senior manager described the policy deployment process:

> It is a vehicle to make every team and each individual work to support the objectives of the company. It serves as a great tool for focusing the minds and energies of the entire unit on the strategic priorities for the year. The policy deployment process is bidirectional. On one hand, it involves the downward cascade of strategic priorities and vital few actions. On the other hand, it includes an 'upward' negotiation at each level in which each team negotiates its actions to support the strategic objectives. The end result is that each individual has clear personal targets and goals, can relate his or her activities to other people and understand how to contribute to the success of the company.

THE TRANSITION

We are no longer the organization we used to be and we are not yet the organization we intend to be. We are in a state of transition where former processes may not work any more. New roles and responsibilities may not be totally defined or understood. Key interfaces between organizations may not be solidified. This can be a difficult state in which to operate. Many people are uncomfortable with change. They feel anxiety as power bases shift and familiar working patterns are altered. This is the normal outcome of dramatically reengineering an enterprise.

Special skills were required within the organization during the change process. The Leadership Through Quality movement had instituted special skills throughout the workforce. But were these skills going to be enough for employees to prosper in the new organization? Would the cross-functional business teams yield the desired dramatic breakthroughs in productivity and profitability? Was the new management model sufficiently clear and powerful to provide effective leadership through the transition period?

While Xerox in Europe had come a long way from its low days of the mid-1980s, its revenues and net profits (see Exhibit CS6.1.15) were under pressure. The organization was again going through a major transition. Was this change going to lead to the establishment of a new enterprise which is not only profitable, but is also able to motivate and get the best out of its employees? Was this going to lead to the achievement of the Xerox 2000 vision?

Case 6.1

Case 6.1

Exhibits

Exhibit CS6.1.1 Awards won by (Rank) Xerox and its units

Year	Award	Unit
1978–90	UK National Safety Certificate	Mitcheldean
1983	Dutch Quality Award	Venray
1984	British Institute of Logistics Award	International HQ
1984	British Quality Award	Mitcheldean
1985–87	Gold Award from the Royal Society for the Prevention of Accidents	Manufacturing
1986	British Quality Award	Welwyn Garden City
1986	Netherlands Logistics Award	Venray
1987	Prix Industrie et Qualité	Lille
1988	Fax of the Year - Xerox 7020	Norway
1989	Netherlands Co-Partnership Award for Procurement	Venray
1989	Scanstar & Worldstar Award for Packaging	Norway
1989	Major Commendation for Environmental Improvement	Mitcheldean
1989	Best Factory Award for Quality, Productivity, Cost and Customer Satisfaction (British Institute of Management)	Mitcheldean
1990	Grand Prix de l'Accueil Téléphonique	France
1990	Office Equipment Supplier of the Year	Sweden
1990	Sword of Excellence for Safety	Mitcheldean
1992	Irish Quality Award	Ireland
1992	Women in Management Award	UK
1992	Belgium Quality Award	Belgium
1992	European Quality Award	(Rank) Xerox Ltd

Exhibit CS6.1.2 Leadership Through Quality: 'Green Book' elements

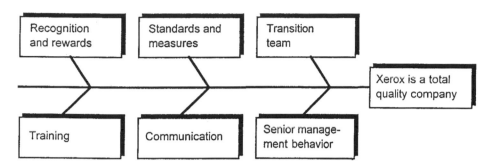

Case 6.1

The Leadership Through Quality movement is like starting a revolution. The first thing you do is make sure you have the right generals (Senior Management Behaviour). Having put your generals in place you begin to take over the radio and television stations, the press and the rest of the media so as to gain control of mass communications (Communication). Then come the schools and universities. You begin to condition the young generation by installing your value systems and ways of thinking and behaving (Training). Those generals, troops and others who are seen to be supporting the cause are decorated and paid handsomely (Reward and Recognition). In parallel with these other activities you throw away the existing laws, redefining the legal system to support your aims (Standards and Measurements). Finally, dove-tailed into these other elements which have built the required infrastructure, your troops continue the conversion of the population to the new regime (Transition Team).[2]

---

[2] Walker, R. 'Rank Xerox – Management Revolution', *Long Range Planning*, **25**, (1) 1992, p. 10.

Case 6.1

Exhibit CS6.1.3 Quality progress agenda for (Rank) Xerox

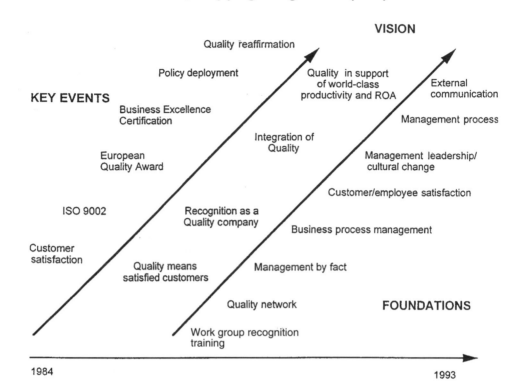

VISION

Quality reaffirmation

Policy deployment

Quality in support
of world-class
productivity and ROA

External
communication

KEY EVENTS

Business Excellence
Certification

Management process

Integration of
Quality

European
Quality Award

Management leadership/
cultural change

ISO 9002

Recognition as a
Quality company

Customer/employee satisfaction

Customer
satisfaction

Business process management

Quality means
satisfied customers

Management by fact

Quality network

FOUNDATIONS

Work group recognition
training

1984

1993

Exhibit CS6.1.4 The new organizational structure

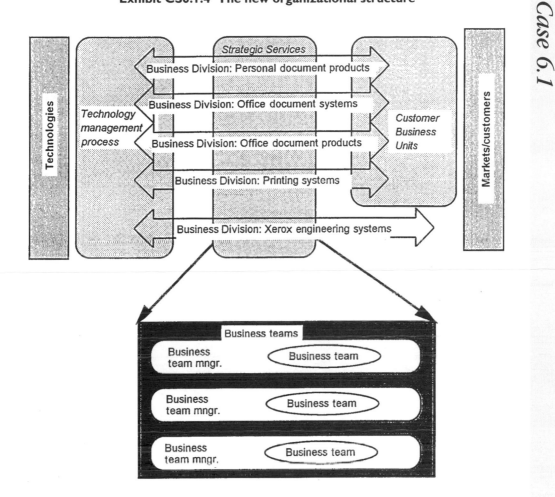

Exhibit CS6.1.5 Xerox (France) organizational structure

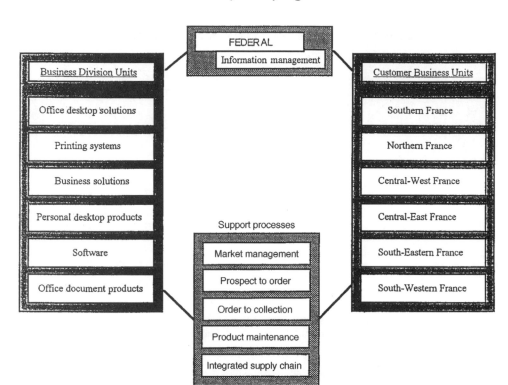

Exhibit CS6.1.6 Xerox 2000 Leadership through Quality: management model
(continues)

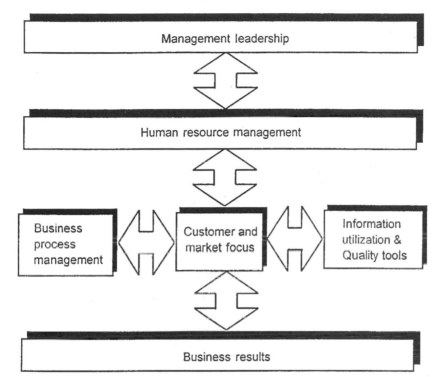

Management leadership	Xerox management displays a customer focus, exhibits 'role model' behaviour, establishes clear long-term goals and annual objectives, establishes strategic boundaries and provides an empowered environment to achieve world-class productivity and business results.
Human resource management	Xerox management leads, motivates, develops and empowers people to realize their full potential. All employees are personally responsible for continuous learning and acquisition of competencies required to creatively achieve business objectives and to continuously improve productivity for customers and Xerox.
Business process management	Business processes are designed to be customer-driven, cross-functional and value-based. They create knowledge, eliminate waste, and abandon unproductive work, yielding world-class productivity and higher perceived service levels for our customers.
Customer and market focus	Current, past and potential customers define our business. We recognize and create markets by seeing patterns in customer requirements. Anticipating and fully satisfying those requirements through the creation of customer value achieves Xerox business results.

Case 6.1

Information
utilisation &
quality tools

Fact-based management is led by line management, is achieved through accurate and timely information, and by the disciplined application and widespread use of Quality tools.

Business results

Xerox, The Document Company, is the largest and one of the most productive and profitable companies in the global document market.

Exhibit CS6.1.7 Core Processes of Xerox

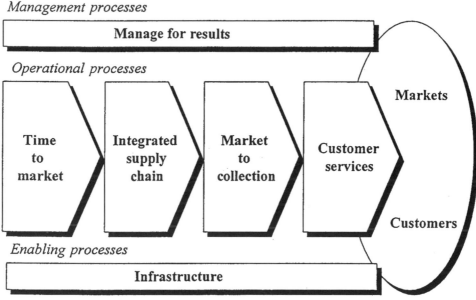

Exhibit CS6.1.8 Business process architecture of Xerox (France)

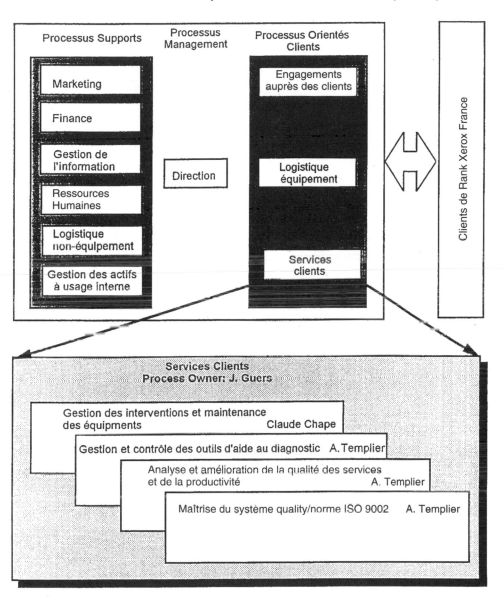

Case 6.1

Exhibit CS6.1.9 Descriptions of roles and responsibilities (*continues*)

Process sponsor

- Name process owners and in consultation with them defines
    - the limits of the different processes
    - the desired states for the processes
    - the priority actions ('vital few') for the processes.
- Start programmes for process improvements and process reengineering, get necessary resources and process owners' involvement in the programme.
- Follow up on the process owners' actions to reach the established desired states
- Inspect process performances compared to the desired states as well as the evolution of the improvement/reengineering programmes and their related impact on productivity.
- Represent his or her process in the Process Management Committee.

Business process owner

- Obtain information regarding the process, set up measures and follow up on benchmarking processes.
- Define in consultation with the process sponsor, features about the process such as:
    - conformity (production in accordance with the expected result)
    - efficiency (optimal output)
    - capacity (undercontrol)
    - measures (performance indicators on quality, efficiency and productivity)
    - management (follow the performance and take necessary corrective measures).
- Inspect process measures and features as described above.
- Manage the cross-functional nature of processes to ensure efficiency and the process integrity.
- Plan and run process improvement/reengineering projects to reach the desired state.
- Retain responsibility for the programme at the company level even if responsibility for certain specialized stages of the process is shared with others.

Case 6.1

Exhibit CS6.1.9 *(continued)*

Process actor

Every person playing a part in the execution of the process is a process actor. The process actor is:

- responsible for his or her part of the process output and conformity relative to the customer's need

- required to define his or her needs with his or her supplier (upstream process actor) and to challenge the supplier to get the right output.

Quality officer

- Provide process sponsors, owners and actors with methodological support to:

  - gather process information and choose appropriate measurement indicators

  - manage zero defects/defaults

  - use quality tools for 'problem resolution' and 'processes improvement'

  - organize benchmarking and best-practice studies.

- Facilitate intraprocess and interprocess interfaces though the Quality network.

Exhibit CS6.1.10 Process improvement and process reengineering

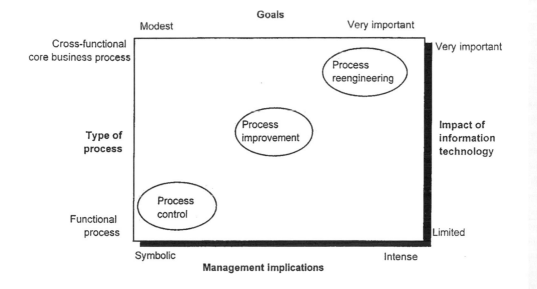

Exhibit CS6.1.11 Corporate Values (*continues*)

- We succeed through satisfied customers
- We aspire to deliver quality and excellence in all we do
- We require premium return on assets
- We use technology to develop market leadership
- We value our employees
- We behave responsibly as a corporate citizen

Leadership Attributes

Strategic leadership	Thinking strategically about major trends affecting the business.
Strategic implementation	Able to translate broad strategic plans into concrete plans and actions.
Customer-focus	Keeping the organization focused on understanding and responding to the needs of customers.
Shared vision	Creating and communicating a clear and inspiring vision for change.
Decision-making	Able to make sound, timely decisions with available information.
Quick study	Able to understand complex business processes and technologies quickly.
Operational performance	Being focused on accomplishing relevant objectives and business goals.
Staffing for high performance	Able to attract and retain talented people.
Organizational talent	Recognizing and developing workforce competencies.
Delegation and empowerment	Developing staff confidence in their ability to lead and define the business.
Teamwork	Managing teams effectively.
Cross-functional teamwork	Ability to maintain working relationships across functions.
Innovation	Producing results through the creation and management of new initiatives.
Business results	Driving both short-term and long-term goals.
Leadership Through Quality	Committed to Leadership Through Quality as the basic business principle.

Exhibit CS6.1.11 *(continued)*

Openness to change	Open to new ideas and willing to experiment.
Interpersonal empathy	Recognizing the impact of one's behaviour on others.
Personal drive	Demonstrating a deep-seated need for achievement and excellence.
Personal maturity	Demonstrating resilience in response to short-term and sustained stress.
Personal consistency	Evoking trust in others by being open and behaving in predictable ways.
Environment perspective	Aware of developments in the larger business, political and social arena.
Business perspective	Understanding the management requirements for running a business.
Technical knowledge	Having in-depth technical knowledge about Xerox's products and services.

Exhibit CS6.1.12 Cultural dimensions

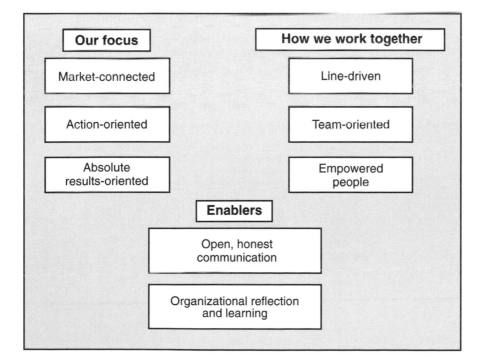

Exhibit CS6.1.13 Business assessment and certification

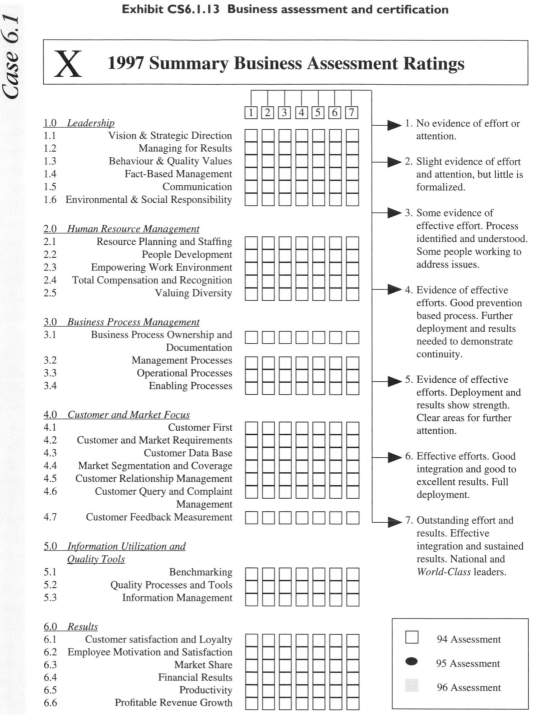

X 1997 Summary Business Assessment Ratings

| | 1 | 2 | 3 | 4 | 5 | 6 | 7 |

1.0 *Leadership*
1.1 Vision & Strategic Direction
1.2 Managing for Results
1.3 Behaviour & Quality Values
1.4 Fact-Based Management
1.5 Communication
1.6 Environmental & Social Responsibility

2.0 *Human Resource Management*
2.1 Resource Planning and Staffing
2.2 People Development
2.3 Empowering Work Environment
2.4 Total Compensation and Recognition
2.5 Valuing Diversity

3.0 *Business Process Management*
3.1 Business Process Ownership and Documentation
3.2 Management Processes
3.3 Operational Processes
3.4 Enabling Processes

4.0 *Customer and Market Focus*
4.1 Customer First
4.2 Customer and Market Requirements
4.3 Customer Data Base
4.4 Market Segmentation and Coverage
4.5 Customer Relationship Management
4.6 Customer Query and Complaint Management
4.7 Customer Feedback Measurement

5.0 *Information Utilization and Quality Tools*
5.1 Benchmarking
5.2 Quality Processes and Tools
5.3 Information Management

6.0 *Results*
6.1 Customer satisfaction and Loyalty
6.2 Employee Motivation and Satisfaction
6.3 Market Share
6.4 Financial Results
6.5 Productivity
6.6 Profitable Revenue Growth

1. No evidence of effort or attention.

2. Slight evidence of effort and attention, but little is formalized.

3. Some evidence of effective effort. Process identified and understood. Some people working to address issues.

4. Evidence of effective efforts. Good prevention based process. Further deployment and results needed to demonstrate continuity.

5. Evidence of effective efforts. Deployment and results show strength. Clear areas for further attention.

6. Effective efforts. Good integration and good to excellent results. Full deployment.

7. Outstanding effort and results. Effective integration and sustained results. National and *World-Class* leaders.

☐ 94 Assessment
● 95 Assessment
▨ 96 Assessment

Exhibit CS6.1.14 Policy deployment process

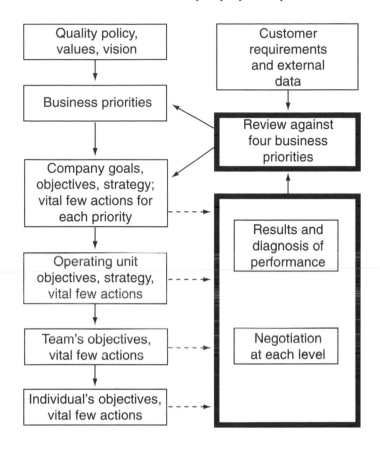

Case 6.1

Exhibit CS6.1.15 Results for Xerox in Europe

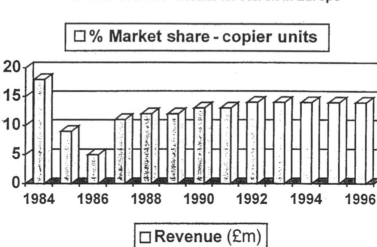

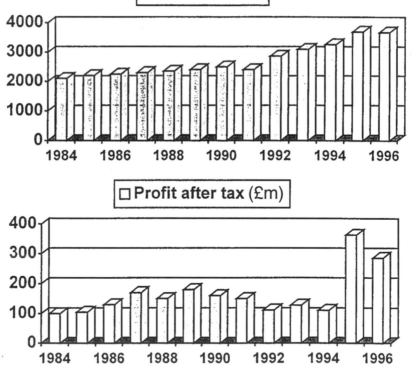

Harvard Business School

9-692-023
Rev. 2/17/92

Case 6.2

Doctoral candidate Clayton Christensen prepared this case as the basis for class discussion rather than to illustrate either effective or ineffective handling of an administrative situation. Some information in this case about Quantum Corporation's products, employees and markets is disguised, in order to protect information which is proprietary to the company.

Quantum Corporation: Business and product teams

The April 1991 announcement of Quantum Corporation's fiscal 1991 financial performance capped a remarkable five years for the Milpitas, California-based manufacturer of rigid disk drives. Quantum had logged $878 million in revenue—up 627% over its 1987 level—and net earnings had shot up from $9 to $74 million, boosting it from ninth to third place among the world's independent disk drive manufacturers. 'I think our decision in 1989 to manage through business teams was a key to this growth,' reflected Dave Brown, Quantum's vice chairman and chief operating officer. 'The major questions now are whether we have enough people in the company who have the talent and training to staff the number of teams we'll need in the future, and what the relationships among the teams ought to be.'

COMPANY BACKGROUND

Quantum Corporation was founded in 1980 to manufacture and market a four-model product line of 10-, 20-, 30-, and 40-megabyte (Mb) 8-inch Winchester disk drives for minicomputer manufacturers such as DEC, and makers of sophisticated multiuser word processor systems such as Xerox and Wang. Quantum's first products were well received, and sales reached $119 million by 1985. But a 1985 decision to integrate the technologies further—especially the disk drive controller and the system interface—led to the launch of a line of lower cost (and subsequently lower priced) drives targeted at the rapidly growing multiuser microcomputer and workstation market. This strategy moved Quantum toward new customers, higher unit volumes, and smaller form-factor, or diameter (5.25-inch) drives, and proved disastrous. Quantum was late to market with its new designs, and its product employed a nonstandard interface not readily accepted by most PC manufacturers. As a result, sales and profits tumbled as its initial 8-inch products aged.

The company was saved from total failure, however, by a subsidiary firm, Plus Development Corporation, whose Hardcard® product—a 3.5-inch hard disk drive (initially with a capacity of 10 MB)—was sold directly to PC owners through computer retailers and direct-mail marketers. Matsushita Kotobuki Electronics Industries, Ltd (MKE), a leading manufacturer of video cassette recorders, produced Hardcard in one of its plants in southern Japan. Plus' rapidly growing revenues in the 1985–1988 period offset the evaporating sales of the original Quantum OEM (original equipment manufacturer) business. (See Exhibit CS6.2.1.)

In 1988, Quantum bought the 20% of Plus's shares it did not already own, and two Plus executives, Steve Berkley and Dave Brown, became Quantum's new chairman/CEO and president/COO, respectively. Plus was kept a wholly owned, independently managed subsidiary. The balance of Quantum's business, consisting of sales to original equipment computer manufacturers, became known internally as 'Quantum OEM.'

Quantum's new management team saw the opportunity to build upon Plus's technological experience in 3.5-inch disk drive design by developing a new family of small form-factor, high-capacity products for the rapidly growing OEM market for engineering workstations and high-end desktop personal computers. In late 1988, Quantum announced it would stop developing and manufacturing all 5.25-inch products and replace them with 3.5-inch disk drives of comparable capacities. Quantum also decided to phase out its Puerto Rico manufacturing operations and source the majority of its drives from MKE. The results were dramatic. Whereas two-thirds of Quantum's 1988 revenues (and all of its OEM sales) had come from 5.25-inch products, in 1989 it derived 100% of its revenues from 3.5-inch products. The company continued this strong momentum in fiscal 1990 and 1991.

PREPARATIONS FOR CONTINUED GROWTH

Between 1988 and 1991, senior management launched three initiatives to reinforce Quantum's OEM momentum in the face of factors which they believed would characterize the disk drive industry in the 1990s: shortened lead times, stiffer competition, and slower market growth. Dave Brown played a leadership role as the organization pursued all three.

The first initiative was to reduce the time required to define, engineer, and produce new products in volume. As strong competitors' product lines overlapped across a broadened range of the capacity spectrum, Brown felt it would be increasingly difficult to sustain an enduring competitive advantage around performance-differentiated products. Although he was confident that the Quantum/MKE design engineering and manufacturing teams could continue to offer the best price/performance products in the industry, the period during which a new Quantum product could retain a performance edge before being leapfrogged by a competitor was shortening dramatically.

Quantum's OEM customers were accelerating their own development cycles, and the time during which they would evaluate and select the disk drive for a new computer system was also becoming increasingly short—opening and closing within a few months. Being even a few weeks late in product development could cost tens of millions of dollars in lost revenues and lower margins on whatever sales Quantum could make on a late product. From the industry's early days, disk drive prices had been falling about 5% per quarter. 'The price will decline on you whether your product is in the market, in the factory getting the production bugs worked out, or in the development lab waiting for a new chip to come in from a vendor,' reflected a Quantum marketing manager. This effectively meant that if Quantum could shave three months off its time-to-market cycle, not only could it garner revenues and profits from each product for an additional three months, but its average price realization over the life of each product would increase 5% as well.

The second initiative was to broaden the product line and provide customers with both follow-on generation product plans and currently competitive products. In addition to expanding the number of models targeted to the single-user PC and workstation markets from 10 to 27, Quantum in 1989 initiated development of a high-capacity family of drives for multiuser workstations and file servers. An additional program to develop smaller, lower-cost 2.5-inch drives targeted at the rapidly emerging notebook computer market was launched in 1990.

The third initiative was to push decision making lower into the company, and to begin managing critical programs through a team approach. Brown could remember sitting in an executive staff[1] meeting in late 1987 and asking why they were not shipping a particular product. To get at the answer, he had to start wading through all the details about whether enough components had been ordered, enough capacity put in place, the right forecasts provided, and so on. After that discussion, he realized management was part of the problem. 'We were trying to manage details we weren't knowledgeable about. We had a bandwidth problem—the executive staff just didn't have enough time or brain capacity to keep making all the key decisions. To maintain our growth, we had to push decision-making down into the organization, to the levels where the most informed people were.'

Quantum had tried team-level coordination before. In 1985, the company promoted an experienced manager and charged him to manage development of a key family of 5.25-inch drives by coordinating the work of functional employees. In retrospect, the attempt failed at least in part because the project was extremely ambitious technically, and in part because the project leader had a hub-and-spoke management style—all information and decision-making flowed through him. In 1988, the company hired from outside a senior manager with program management experience and gave him a similar charge on a development project. As one executive recalled, that attempt failed because, 'The project manager didn't understand the business well enough to be able to make the right decisions. Even though he ostensibly had the authority to make key decisions, he didn't have the knowledge or credibility to make them—so the decisions kept getting bumped up to the executive staff anyway.'

---

[1] The Quantum OEM executive staff consisted of Chief Operating Officer Dave Brown and the vice presidents of each of the company's functions: Engineering, Marketing, Operations, Quality, Finance, and Human Resources.

On Quantum's next major product development program, Impala, two functional managers, Mike Spencer from Marketing and Bill Benson from Engineering, assumed *ad hoc* responsibility for managing the development effort. No one assigned these two managers to work together closely, nor did they ask permission to alter the way their functions interacted; they simply went ahead and did it. Spencer and Benson were widely successful. Quantum announced 40- and 80-Mb Impala models in February 1988, and the line generated revenues of $320 million in the next 18 months.

INITIATING FORMAL PRODUCT TEAMS AT QUANTUM OEM

In summer 1989, Brown established an explicit product team system within Quantum OEM to manage the definition, design, manufacturing, and marketing of all new products. Although the Impala line of 3.5-inch drives was Quantum's major product, a follow-on generation of 60- and 120-Mb drives, code-named Cheetah and built upon the same product architecture as Impala, had been under development since early 1988. A second product family under development, code named Aerostar, was a line of 100- and 150-Mb drives targeted at the multiuser engineering workstation and file server markets. Brown established the Cheetah and Aerostar Teams to oversee these efforts. A third team, code named Wolverine, was created in early 1990 to manage the development and launch of Quantum's 2.5-inch products, in 40- and 80- Mb models, for the notebook computer market. Exhibit CS6.2.2 charts the development and market introduction times for each of these products.

The product teams' charter was to work in a coordinated way to address *market* needs. To ensure that no important details would fall through the cracks in the team management system, Brown carefully delineated which issues were the responsibility of the product teams, the functional organizations, and the executive staff. Teams were responsible for the definition, development, and introduction of new products; for the revenues and gross margins generated by the products; and for the inventories required to support the revenues. Furthermore, product planning was to become a continuous process, rather than an occasional, event-triggered exercise, within the teams. The product teams also were responsible for achieving Quantum OEM's Fast Cycle Time objective: to slash development time from 18 months on average to 12 months. Product team leaders, because they had critical general managerial responsibilities, came to be viewed within the company as key 'heavyweight' employees.

Functional VPs, on the other hand, were charged with managing ongoing functional activities and expenses, providing effective career paths and skill-building programs, and executing the plans and staffing the programs initiated by the teams. For example, the marketing and sales groups represented Quantum's entire product line to customers, hired and developed qualified people, and staffed and supported the product development teams. The objective was to use teams to coordinate internal operations to meet market needs, but to have the teams by largely invisible to customers. Likewise, Quantum OEM's VP-Engineering (and VP-Manufacturing) was responsible for allocating engineering (manufacturing) personnel to the various teams' development projects, to ensure that projects were supported responsively, and to provide specialized services as needed.

Guidelines for teams

The OEM executive staff established several principles for staffing, managing, and evaluating the product teams.

1. Care in selecting team members

The executive staff chose initial team members carefully. Brown used two questions to determine whether someone could successfully serve on a product team: 'Would the executive staff trust this person to run this business?' and 'Can he or she think like a general manager?' Such questions implied that he pool from which team members would be drawn comprised directors or managers in one of the functional organizations, one or two levels below the functional vice presidents on the executive staff. Although Quantum employees were known throughout the industry to have above-average capability, Brown was able to identify only about 14 people in the 450-person organization in 1989 whom he felt were capable of succeeding in the team management structure. Of them, four had to be relieved of team responsibilities within six months because they proved not to have the aptitude required.

Teams originally were to consist of a core group of six members, one from each function. The core team members were *collectively* responsible for the general management of their project; they were explicitly charged *not* merely to represent their functional point of view. In general, the team member from engineering was to be the leader of the team during the initial phases of the program. As the product approached its commercial launch, the marketing member assumed team leadership duties. A member of the executive staff who also assigned to each team as a sponsor and coach, to assist in communicating with and getting decisions from the executive staff, when necessary.

Core team members generally were to spend about 75% of their time on the team's work and the remainder on other functional responsibilities. As the work load demanded it, other members could also be drawn from the functions and added to the teams, on a less-than-full-time basis. In practice, team leaders found it awkward to distinguish between core and other temporary, non-core members, many of whom made invaluable contributions to the team. Over time the 'core' was gradually dropped.

2. Performance evaluations

All Quantum employees were eligible for a profit-sharing bonus that had been running about 5% of salary, but core team members could be awarded significant additional compensation, depending on personal, team, and company performance. Performance was assessed and weighted on three dimensions:

Performance Dimension	Weight	Performance Assessed By:
1. Achievement of team objectives	25%	Executive staff
2. Individual contribution to team	25%	Ratings by team leader and executive staff sponsor with input from peers
3. Competence in dealing with functional responsibilities	50%	Functional manager

3. Co-location

Insofar as possible, the six core team members were to be housed together in work areas called Bump Spaces, where the layout forced frequent informal encounters. Initially, only Wolverine members were colocated because many of the people were new and space was being added. In addition, those who were to be dedicated to the team for several months (usually a dozen or more engineers) were also to be moved to the team's Bump Space. By early 1991, the value of colocation was fully accepted and all new teams were colocated, even if it meant substantial rearrangement to free up the needed space.

4. Process as product

Developing a *process* for managing the development and launch of new products with fast cycle times—and not just the development of the physical products themselves—was an important goal of the team approach. Although Dave Brown conceded that the first team might be relatively inefficient and that each new team would have to go through its own learning cycle, he hoped subsequent teams would benefit from the processes developed by predecessor teams, while not feeling constrained by them. To nurture effective team management processes, a facilitator—either a consultant or a specially-trained Quantum employee—met with each team in its weekly meeting. The facilitator's role was to pull team discussions back from tangential discussions, to ensure that no member's defensiveness blocked the team's progress on an issue, and to keep discussions moving on an agreed-upon timetable. At each meeting's end, the team took 10 minutes to evaluate the meeting's effectiveness in results achieved and process employed, and the members offered ideas to improve future meetings.

PRODUCT TEAMS' PERFORMANCE

Although Quantum's teams ostensibly were to operate along the same guidelines, their performance differed substantially.

Cheetah/Lightning team

Formed just as the Cheetah product was ramping up to mass production. Cheetah was Quantum's most successful team. The Cheetah was a close technological relative of the successful Hardcard/Impala product series, employing the same optically encoded, closed-looped servo architecture for positioning the read-write heads as had been used on the earlier products.[2] Thanks in part to the coordinated leadership of the product team and to customers' booming sales of personal computers and engineering workstations, Cheetah generated over 30% of Quantum's total sales revenues in calendar year 1990, despite Cheetah products' having been on allocation (capacity constrained) for much of the year.

Spencer found that in managing the Cheetah Team after the transition to mass production in early 1990, his team tasks had become very much those of the general operating manager: determining customer priority, forecasting manufacturing requirements, controlling the size and

---

[2] Optical encoders beamed light through grids on glass plates that extended from the head assembly. By measuring the alignment of the grids to a fixed reference point on the base of the drive, this system closed the feedback loop to the actuator that controlled head positioning.

location of inventories, and reevaluating pricing strategies. At the same time, knowing Cheetah's life expectancy probably would not exceed 18 months, Spencer and others saw the need to initiate next-generation product development efforts, targeted at future needs in the same markets Impala and Cheetah had addressed so effectively.

To simultaneously manage the revenues and gross margins of Impala and Cheetah, and yet provide sufficient team depth to staff the Wolverine project, the executive staff restructured and put Mike Spencer in charge of the Lightning *Business* Team. The purpose of the business team was to manage a *family* of product lines, each of which was in a different stage of development. Two new product teams were established to manage the development and launch of follow-on products using the basic Cheetah architecture: products code-named Gazelle and Bobcat. Since Cheetah was now fully in production, the operations management responsibilities of the old Cheetah product team were passed to a Lightning *operations* team. This meant that general management responsibility for the product line as an operating business rested with the *operations team* (and above it, the business team), rather than an individual.

Although it made little sense to Quantum's managers to draw team organization charts when their structures were changing so regularly, this reorganization essentially resulted in a two-tier team structure. Members of the business team were responsible for the success of its product teams and its operations team. The product teams were responsible for products from conception to mass production, at which time responsibility shifted to the operations team.

At the time he was put in charge of the Lightning *Business* Team, Spencer was fresh from successfully, leading the Impala team. He was asked to infuse a similar everybody-take-charge culture into the integrated business effort as well as its subprojects. He was aided in this by the fact that many Cheetah Team members also had previously shared crossfunctional team management experiences on the Impala program.

Recognizing the need to 'seed' Lightning's subsidiary operating and product teams with personnel experienced in team management, Spencer assigned a core member of the original Cheetah product team to serve simultaneously as a member of the Lightning Business Team and as leader of an operations or product team. The operations team had its own crossfunctional core group of six members, but the product teams, especially during their early stages, were staffed by three members, from engineering, marketing, and operations functions only. To make it clear that manufacturing, marketing, and other issues needed to be integrated with product engineering decisions, even at the earliest stages, Spencer avoided labeling the new product teams as development teams.

Even though its task was much more complex than the original Cheetah Team's charter had been, the Lightning Team proved to be a successful manager of the broader effort. Cheetah customers were happy with the product and its pricing, MKE was producing it with high yields, and gross margins were attractive. The first of the follow-on products, Gazelle, was introduced in mid-1991, only four months late from its aggressive 12-month time-to-market plan.

The Aerostar Team

In contrast to the Cheetah/Lightning Team, the Aerostar product team's history was characterized by frustration and conflict. Introduced in September 1990, Aerostar's product line was several months late to market, overran its budget, and was unable to garner the margins management had targeted when initiating the development effort.

The Aerostar Team's product design was technically ambitious. At 100 and 150 Mb, Aerostar was the largest-capacity 3.5-inch drive Quantum had ever attempted; at 18 ms access time, it was by far the fastest. In addition, rather than employ the type of off-disk optical encoder technology used in the Hardcard, Impala, and Cheetah drives, Aerostar had special codes embedded on each disk surface that helped the heads self-adjust to stay precisely over the track on which they were reading or writing data. Whereas IBM had long used this positioning technique in its drives, this technology was new to Quantum. Finally, the Aerostar Team attempted a broader interface[3] development effort than had ever been attempted at Quantum. In addition to the AT interface, which linked the drives to IBM and compatible computers, the Aerostar program called for simultaneous development of an ESDI interface for complex systems employing multiple disk drives. By offering multiple interfaces, Quantum hoped to gain flexibility in addressing the markets for sophisticated file servers and multiuser workstations—segments in which it had not recently competed.

As a conscious strategic move, Quantum management decided to manufacture Aerostar itself. This required significant investment in advanced manufacturing processes and equipment, and start-up of that operating system on a technically demanding product. It also required establishing additional relationships between design engineering and manufacturing so the Aerostar product and process choices would be fully compatible. Building these new capabilities while developing the Aerostar product added another dimension of risk and complexity to the team's tasks.

Many, however, felt the decisions made in staffing the Aerostar Team contributed to the program's woes. 'We learned that we needed three or four people on a team with the personality to energize the team—and that those key people need to respect one another, 'reflected one early member. The team's first leader, in fact, had accepted his team assignment reluctantly. As the functional engineering manager charged with Aerostar's design, he resented the diversions team management required, feeling that resolving product engineering design crises was a more pressing target for his managerial energies.

In response to the team's struggle to meet its time-to-market mandate, the executive staff kept trying to find a combination of people who could solve Aerostar's problems—but to no avail. The Aerostar core team had 4 sponsors, 4 leaders, and 16 members in its first 18 months. But it seemed that the modest commitment, low morale, and lack of interfunctional respect that characterized the team's culture in its early months persisted in the group, regardless of the personalities who occupied the offices in Aerostar's Bump Space.

After months of delay and intense involvement of the executive staff, Aerostar was finally introduced to the market. Because it represented an important strategic thrust into a new

---

[3] The drive's interface was the logic circuitry that allowed the drive to communicate with the computer's operating system. Until about 1984, interface circuitry was installed on a separate circuit board in the computer and typically was provided by a third-party company, other than the drive manufacturer. By the late 1980s, most drives were sold with interfaces embedded within the drive housing. Hence, at the time Quantum initiated interface development for Aerostar, few independent disk drive manufacturers had extensive experience in interface development. The three primary types of embedded interfaces were the Small Computer System Interface (SCSI), used primarily on Apple computers; the AT interface for IBM and compatible equipment; and the Enhanced Small Device Interface (ESDI), a circuit used primarily with large, multiunit systems.

market segment, the executive staff determined not to let Aerostar's belated launch taint the entire initiative. They therefore began development of an additional product built on Aerostar's basic architecture, code named Gemini. To manage the operational issues associated with Aerostar after its introduction and to oversee launch of Gemini and additional products, a Business Team/Product Team structure similar to that of the Lightning Team was instituted, and the umbrella team was labeled the Apollo Business Team.

The Wolverine Team—A more detailed look

Whereas the Cheetah and Aerostar Teams had been formed around products that were well into the development process, the Wolverine Team—first as a product team and later as a business team—was created to manage the development and launch of a completely new product—a 2.5-inch drive for notebook computers—and was to be sold in 40- and 80-Mb versions.

Wolverine, a mid-1989 Bill Benson brainchild started to move toward full product development status in October 1989, when Larry Peterson, a development manager who had worked on typewriter design at SCM, was hired specifically to be the initial leader of the Wolverine Team.[4] After Benson and Peterson worked four months to refine the product concept more thoroughly, a full team was recruited in January 1990 to manage the effort. Although none of the original Wolverine Team had prior crossfunctional team experience, the group coalesced relatively quickly under Peterson's leadership into an effective organization.

The Wolverine core team met every Monday afternoon from 1:30 to 4:30. 'That is where the decisions get made,' noted Peterson. 'If you want to contribute, you've go to be there.' It was not unusual for team members to fly back from Japan to be at a key team meeting.

As one of its first actions, the Wolverine Team assigned priorities to its objectives: (1) minimize time to market; (2) create intrinsically low-cost 40- and 80-Mb drive designs which could also be manufactured at low costs; (3) minimize power consumption (an important factor for battery-powered notebook applications); and (4) achieve the fastest possible access performance (the time required to retrieve and transmit a block of data from the drive to the computer).

The priorities guided the team in making several difficult choices in design technology. For example, developing a unique ASIC[5] controller chip might provide the Wolverine product with an extra performance cushion to compensate for possible performance-limiting compromises made in later design phases. However, the priority given to time-to-market over high performance led the team to select the ASIC that had been developed for the Cheetah product.

The time-to-market priority also encouraged the team to explore ways to regroup and restructure critical development activities. Quantum engineers thought of a typical product design effort as a sequence of Design, Fabrication, Assembly and Test cycles (DFAT), through which the team iterated three or four times before product introduction. Historically, these DFAT iterations had been conducted serially, with each iteration averaging 3 to 4 months:

---

[4] Peterson's newness to the disk drive world was not unique on the Wolverine Team. Gregg James, the Quality representative, joined the team from Freightliner, a manufacturer of Class-8 trucks.

[5] ASIC is an acronym for Application-Specific Integrated Circuit.

$$D_1 \quad F_1 \quad A_1 \quad T_1 \rightarrow D_2 \quad F_2 \quad A_2 \quad T_2 \rightarrow D_3 \quad F_3 \quad A_3 \quad T_3 \rightarrow \text{Mass Production}$$

The design steps were the exclusive responsibility of Quantum. Component fabrication was the responsibility of MKE and of third-party suppliers. Assembly (also called 'pre-production builds') was done at MKE. Units made in the preproduction builds were tested by Quantum and MKE engineers independently, and results were shared and discussed.

To accelerate time to market, Peterson's team attempted to overlap the test and design stages, hoping to reduce the time for each iteration from 3 or 4 months to 2 months, as follows:

$$D_1 \quad F_1 \quad A_1 \quad T_1$$

$$D_2 \quad F_2 \quad A_2 \quad T_2$$

$$D_3 \quad F_3 \quad A_3 \quad T_3 \rightarrow \text{Mass Production}$$

This overlapping plan meant that many of the problems uncovered in the tests of A_1 units were not known to the design team when D_2 began, and the second DFAT cycle's purpose became primarily manufacturing learning, rather than design improvement. The design phase of the third cycle, therefore, needed to catch and correct a larger proportion of the design problems than under the prior serial structure.

Peterson found, as did the original Aerostar team leader, that managing the Wolverine Business Team *and* his 30-person product engineering function was an overpowering task. He was relieved, therefore, when in May 1990 the executive staff decided to install one of its own VP-Marketing Mark Quinn, a full-time Wolverine Business Team leader. Quinn's title was changed to VP-Portable Storage Products, and Mike Spencer became corporate OEM marketing vice president.

The Wolverine Team's overlapping DFAT strategy made it more important than ever that details not be overlooked as critical information passed between functions. To ensure smooth coordination with MKE and component suppliers based in the United States and Japan, Quinn invited Kurt Jackson, an experienced Lightning operations team member with functional responsibility under the VP-Operations for the Quantum-MKE relationship, to join the Wolverine Team. As iterations of drawings for the design emerged from the team, engineers would electronically transfer them to MKE for manufacturability[6] evaluation. MKE would then transmit the drawings that same day to its parts suppliers to permit them to start tooling design and longer-range equipment and capacity planning.

Engineering-marketing interactions within the Wolverine Team often centered around allocation of preproduction models. Marketing had an insatiable appetite for early prototypes to give to key customers for early evaluation and design into their systems. But engineering also needed preproduction models for testing in the DFAT cycle. Wolverine Team meetings provided a forum where, with functional representatives taking a general manager's perspective for revenues

---

[6] Manufacturability evaluation proceeded in two stages at MKE. The first was a 'repeatability' evaluation, which included an analysis of tolerances actually required and achievable on the parts. The second was an analysis of processes and design modifications, which would reduce the cost of fabricating and assembling the parts.

and gross margins, such demands for preproduction models could be granted priorities and managed.

To understand whether the posture taken by the functional members represented the viewpoints of functional or general management when such issues were addressed, the casewriter characterized all comments made a typical team meeting. The results, summarized in Exhibit CS6.2.3, show that the executive staff seemed to have succeeded in selecting members from the functional groups 'who could think like a general manager.' Team members were frequently vocal about issues that had little impact on their functions and, on occasions, advocated positions that were in the team's, but not their function's, interest.

As they approached Wolverine's launch, team members uniformly felt what they had learned the first time around would make their performance on follow-on development programs much stronger. A partial listing of lessons learned, prepared for team discussion by Larry Peterson, is included as Exhibit CS6.2.4. Like the Lightning and Apollo Teams, the Wolverine Team eventually became a business team, establishing product teams to manage specific product lines. Exhibit CS6.2.5 characterizes the composition of these teams in mid-1990.)

THE TEAMS' IMPACT ON QUANTUM'S PERFORMANCE

Although team management had not yet had its full desired impact on Quantum's fast cycle time objective (see Exhibit CS6.2.6), Dave Brown and others felt this would improve as more people became experienced in core team management. Moreover, management felt that the team management system had brought additional valuable benefits to Quantum.

One important benefit was that Quantum's product planning was better integrated into the company's strategy. Whereas historically, product development was often initiated by resource availability (i.e., 'We have four engineers freeing up on this project next week, so we'd better start working on another product'), continuous business team-level product planning has become the trigger for product development initiatives.

The team system also had taught the executive staff to delegate important decisions. 'Team management had given us a model of how to empower people,' Brown reflected. 'It has also taught us the value of specificity—how to delegate big, important pieces of responsibility into the organization. It has taught us a *process* for managing the company, now that things are more complex. Our key people have started thinking like general managers. They care less about whom they work for and about getting done what needs to be done. The boundaries between the functions are beginning to blur.'

Although several executive staff members had wondered whether adopting a division organization might have been a simpler way of organizing, most felt teams gave the company the flexibility it needed to organize and reorganize around shifting markets. In Brown's view, 'In order to maintain our growth we need to compete in a broader range of market segments than we did in the past. Segments appear and disappear rapidly, and customers in each segment have somewhat different needs. Top management continually has to ask, 'What do we focus on?' We have to keep looking for new ways of organizing to get the right decisions made by the people best qualified to make them.'

Individual team members also felt the team experience had augmented their managerial skills. One member's enthusiasm was not unusual: 'Sometimes I leave those team meetings thinking, 'Wow! I can personally affect this business!' It's really exhilarating. Team management gives you a better understanding of the over-all problem set—you see how it all fits together. It is great general management training.'

Kurt Jackson, who had served on the Lightning and Wolverine teams, claimed, 'If I were to leave Quantum (which I'm not planning to do) I'd write "Crossfunctional Team Management Experience" across the top of my resume. It's really been valuable.'

The personal benefits to the *managers* involved in the team system could, however, be somewhat asymmetrical. Mark Quinn, who opted to leave his position as VP-Marketing to assume leadership of the Wolverine Team, reflected that 'It was a tough situation for a while. Previously, I had 60 people in my organization. I had been used to asking people to do things and having it get done. Now I'm responsible for the success of this business, but I don't really have direct authority. It would be a lot easier to get the job done if I had power to 'hire and fire,' rather than just 'influence and persuade.' Initially, other team members had a hard time adjusting to Quinn's assertive leadership style as well. Quinn recalled that the group 'eventually developed a norm that when I go too 'directive,' a team member would yell 'Boss!' I then would back off and try to get the job done through 'influence.'

Another executive staff member mused that 'the team system is great for the team members. It gives them major responsibilities, a great learning opportunity, a visibility—it puts them on the fast track. For the executive staff, however, it means less authority, no additional financial rewards—and you still get your ass kicked when things go wrong.' One functional vice president, in fact, had been reassigned when he proved unable to adjust to the new role the team system demanded.

Problems in Quantum's team management system

Even in the more successful teams, managers found that the team structure itself could exacerbate problems. Mark Quinn described one issue:

> Last week, I learned that a group of engineers (working under the supervision of the functional engineering VP in pursuit of a Wolverine Team objective) was late on a key milestone. I stepped in and resolved the bottleneck. This particular issue had just fallen through the crack. Since the team has primary responsibility for getting Wolverine to market, I think the engineering vice president watched it less closely than he would have if his own neck had been the only one on the block. And since I'm not directly managing the engineering effort, I probably learned about this problem several weeks later than a closely involved functional manager would have.

An Apollo core team member reinforced this point. 'When the team's involved, functional managers seem to feel that the team will take care of it. They feel less responsible for getting the job done. And from the team's perspective, we really don't have the competence to tell whether the functional groups are getting their jobs done well or not.'

When the Wolverine Business Team established an additional product team, called Mongoose, to manage the development and introduction of the next-generation product, all of the core business team members' schedules were already so stretched with the launch of the initial Wolverine product that Quinn could not expect any of them to serve simultaneously as members of the Wolverine Business Team and to lead a new product team, as Mike Spencer had done in the Lightning Business Team. The Wolverine Team therefore staffed the Mongoose Team with the best people it could find, but there was concern that the Mongoose Team members' lack of stature and team experience might make the team less effective. Quinn reflected:

> Folks will sit in those meetings and say, 'I'm only a quality specialist. I can't say what we ought to do.' It's hard getting these product teams to work—to get individual members to drive the decisions and progress of the group. They tend to think of themselves as functional people with circumscribed authority, rather than as general managers with shared responsibility for this business. I think the executives in Quantum understand this team concept very well. But the people farther down in the organization really don't.

'Do you know what the key is?' queried another manager. 'It's the people. Give me the pick of the best five or six people, and we'll have a great team. With the right group of people who understand the business and have the confidence to run it, you don't need a boss to give orders. As a group, they'll just go after the objectives we set for the business. But how do you empower people—I hate to use that word, but it's exactly what we're trying to do—when they don't want power, or aren't experienced in using it?'

In Quantum's rapid growth environment, the shortage of experienced team members was a critical issue for management to resolve. 'It would be great if they could all turn out like the Lightning Team,' one manager reflected. 'But that takes *people*. Where are they going to come from? From my experience on the Aerostar Team, I am sure the project would have gone better in a simple functional organization. I think having no team is better than a having an ineffective team.'

One incident affecting the Cheetah and Aerostar Teams illustrated the people-specific nature of Quantum's experience with teams. Both teams had decided to use the same new ASIC in the controllers embedded in their respective products. The ASIC vendor informed an Aerostar Team member that it would be 14 weeks late in delivering its first prototypes. Because procuring and testing the ASIC was on the team's critical time path, the chip's delayed delivery forced the team to adjust its entire schedule, pushing its time-to-market back by a couple of months. A few weeks later, the Cheetah Team also became aware (from the vendor, not the Aerostar Team) of its intention to deliver the ASIC chip 14 weeks behind schedule. Within three days, the Cheetah Team member charged with component procurement had renegotiated the ASIC delivery delay down from 14 to 2 weeks, for the Cheetah *and* the Aerostar project.

Reflecting on this incident, VP-Operations Claude Quichaud commented, 'Fast cycle time is not a system. It requires fundamental behavioral change at the individual level, everywhere in the company. It's also a simplification to think of time-to-market as only an issue of product design. You can have very fast development programs feeding into a very slow company.'

Key issues

As Dave Brown reviewed the experience and concerns regarding the impact team management was having at Quantum, three issues stood out:

1. *Team management*: Would the business and product team management process *really* be instrumental in moulding Quantum into the fast cycle time leader that Brown felt was imperative? In his more skeptical moments, he worried that all that had been demonstrated to date was that teams worked well when the technical challenges were not excessive and when team members were experienced and took initiative. Many of the products Quantum was contemplating for the future were technically more complex than those in the Impala-Cheetah sequence, and Brown felt they had already promoted all of the employees who had the requisite experience to be part of a successful team. Under such conditions, could they forge ahead with team management with the confidence that it would succeed?

2. *How to train*: How should Quantum train more people to be strong team contributors? In the past, new employees typically had been brought into the functions, and then the best people in the functions had been selected to work on the teams. Should Quantum continue this process, possibly asking functionally trained employees first to cut their team teeth on small projects before being assigned to larger business teams? Or should it hire people specifically for their team management capability and make a product or operations team their initial assignment, as had been done with Larry Peterson and the initial Wolverine Team? And had he and the executive staff been right in establishing team membership as the 'fast track' at Quantum OEM?

3. *How to operate next time*: It seemed, as Brown examined the information summarized in Exhibit CS6.2.4 and CS6.2.5, that team management as a *process* was working well. People seemed to be thinking as general, rather than functional, managers. But did he, or anyone else at Quantum, have the right model in mind for how teams ought to work the next time around? How could management define an improvement path for team performance and the company's team capabilities?

As competitive time-to-market pressures intensified, as new market segments in the computer industry continually redefined the boundaries of Quantum's markets and the sources of future growth, and as disk drive technology accelerated, Brown could see no good alternative to managing through Quantum's kind of team structure. Whether this team structure could *succeed*, however, given the limited number of employees with the depth in multiple functions and the general management breadth and experience needed to staff teams faced with increasingly complex tasks, was the critical question.

Exhibits

**Exhibit CS6.2.1 Selected financial data, fiscal years ending 31 March, 1982 to 1991
($ millions)**

	1982	1983	1984	1985	1986	1987	1988	1989	1990	1991
Summary Income Statement:										
Quantum Revenues	13.7	41.8	67.1	120.3	121.2	47.8	120.5	—	—	—
Plus Development Revenues						73.0	66.0	—	—	—
Total Reported Revenues	13.7	41.8	67.1	120.3	121.2	120.8	186.5	208.0	446.3	877.7
Gross Margins	3.3	18.3	26.0	46.6	48.3	41.5	38.5	59.8	131.7	215.9
Operating Profit	0.2	11.9	15.8	28.8	21.5	6.8	(11.9)	12.4	67.8	108.3
Profit After Tax	0.2	7.8	10.7	21.0	22.2	8.8	(3.2)	12.9	47.2	73.9
Summary Balance Sheet										
Cash		20.8	18.7	47.0	78.0	65.6	59.1	50.9	79.3	139.6
Total Assets		56.6	74.8	99.5	107.2	137.3	141.7	157.0	243.2	497.5
Long-term Debt		0	0	0	0	0	0	0	0	0
Total Liabilities		8.1	14.5	17.1	20.3	29.2	27.7	56.9	89.6	259.2
Equity		48.5	57.9	82.4	86.9	108.1	114.0	100.1	153.6	238.3

Case 6.2

Exhibit CS6.2.2 Product development (1988–1992) – Teams, products, and market

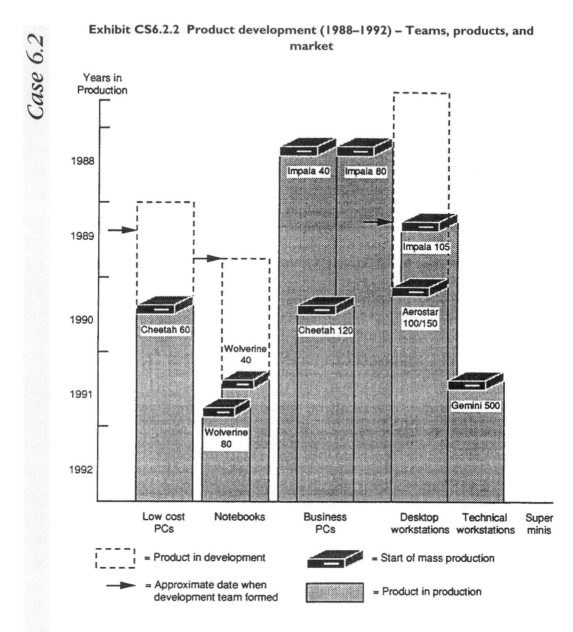

Exhibit CS6.2.3 Sources of input on issues discussed at a representative Wolverine Team business meeting

Importance	Topic/Subject Addressed (A) or Decision Made (D)	Engineering	Marketing	Mfg./MKE[a]	Quality	Finance	Human Res.	Team Leader
Recurrent	Get to market quickly (A)	X		X	×			×
2	Get good preproduction samples into key customers' hands quickly (D)	X	×	X				
1	Design Wolverine's next product to the specification of Quantum's largest corporate customer, even though it was not the largest potential customer in the market segment where Wolverine was targeted. (D)	X	O	X	O	O		O
2	Allocate fewer (mainly defective) preproduction samples to engineering for testing (D)	×	×					
2	Pressure MKE to use a new thin-film head supplier who seems more capable than MKE's preferred long-standing supplier (A)	×		X	×			
3	Determine priorities of potential customers (A)		X	×				
4	Define format and agenda for off-site strategic planning meeting (D)	×	X	×	X	×		X
3	How to get test results from preproduction builds faster (D)	X		×	X			
2	Discuss design/engineering progress (A)	X		×	×			
4	Discuss schedule conflict at MKE for preproduction builds between Wolverine and Gazelle Teams (A)			×			×	X

Legend: Importance refers to the amount of time spent in the meeting considering the topic and the sense of urgency members conveyed about the issue. 1 = most important; 4 = least important. Those issues that required decisions to be made in the meeting are denoted by (D), and those that were addressed without requiring a decision are denoted by (A).

X indicates that an individual strongly supported a position that was consistent with the team's final decision or sense of direction.

x indicates that an individual weakly supported a position or direction that the team eventually decided upon.

O indicates than an individual strongly opposed the position or direction that the team finally decided upon.

o indicates that an individual weakly opposed the team's final decision or sense of direction.

[a] The manufacturing representative in the meeting was a Quantum employee, who was responsible for the Quantum-MKE interface.

Exhibit CS6.2.4 Wolverine Team lessons learned: Key points of fast cycle leverage

Do Again:

Spend time on up-four architecting; *no specification changes.*

Actions for next go-around:

Get critical mass of resources available for early focus on architecture.

Define architecture and get functional and team buy-in.

Implement formal process on spec changes.

Get MKE input on mechanical design early.

Actions for next go-around:

Get MKE to send mechanical engineers to Quantum for a residence period.

Ensure connection of MKE mechanical engineers with MKE's vendor base to get early design feedback prior to tooling start.

Co-locate offices and development labs; heads/media engineering groups; test, process and continuous engineering groups, etc.

Actions for next go-around:

Ensure building layout plans are optimized to preserve as much as possible the present environment.

Brainstorm ways of adapting new facilities to maximize co-location benefits.

Do Differently:

By a dedicated spin stand with a *drive's* read channel already installed.

Actions for next go-around:

Reverse decision to cancel purchase of spin stand. Add to the approved capital budget.

Make better use of modeling; bring finite element analysis capability in-house.

Actions for next go-around:

Hire consultant to review needs and recommend software.

Purchase software.

Train engineers and run sample test case for practice.

Use designer/draftspeople for CAD input and operation. Free mechanical engineers to do more testing.

Actions for next go-around:

Screen, interview, & hire.

Influence work done in advanced product engineering so that they develop more 'building blocks,' which can leverage future development projects.

Exhibit CS6.2.5 Team management structure, June 1991

Case 6.2

Functional areas

	Engineering	Operations	Marketing	Quality	Finance	Human Resources	Strategy & Planning
Lightning Business Team	X	X	X	X	X	X	
Cheetah Operations Team	X	X	X	X	X	X	
Gazelle Product Team	X	X	X				
Bobcat Product Team	X	X	X				
Apollo Business Team	X	X	X	X	X	X	
Aerostar Operations Team	X	X	X	X	X	X	
Gemini Product Team	X	X	X				
Wolverine Business Team	X	X	X	X	X	X	
Wolverine Product Team	X	X	X	X	X	X	
Mongoose Product Team	X	X	X				

(Left axis label: Executive Staff Sponsors)

Support	Director, Advanced Engineering						Director, Strategy & Planning
	Other Functional Directors	Other Functional Directors	Other Functional Directors	Other Functional Directors	Other Functional Directors	Other Functional Directors	
	Other Functional Managers	Other Functional Managers	Other Functional Managers	Other Functional Managers	Other Functional Managers	Other Functional Managers	

Note: X indicates where a member of a function serves as a member of a team.

Case 6.2

Exhibit CS6.2.6 Time to market for Quantum's major new product lines

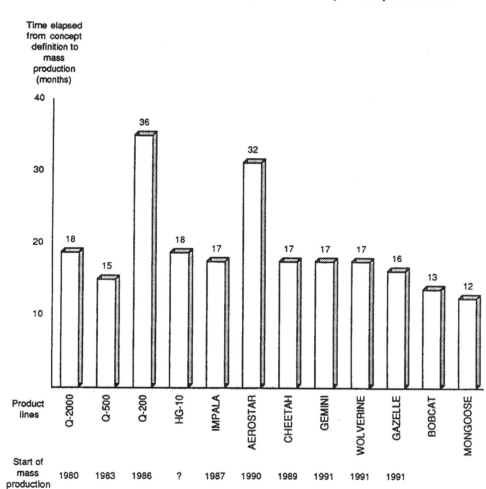

Time elapsed
from concept
definition to
mass
production
(months)

Product lines: Q-2000, Q-500, Q-200, HG-10, IMPALA, AEROSTAR, CHEETAH, GEMINI, WOLVERINE, GAZELLE, BOBCAT, MONGOOSE

Values: 18, 15, 36, 18, 17, 32, 17, 17, 17, 16, 13, 12

Start of mass production: 1980, 1983, 1986, ?, 1987, 1990, 1989, 1991, 1991, 1991

OB 235
18.03.98

Case 6.3

Case 6.3

*This case was prepared by Dr R. Morgan Gould as a basis for class
discussion rather than to illustrate either effective or ineffective handling of
a business situation. Research Associate Michael Stanford, IMD, and
Research Office, Kate Blackmon, London Business School, contributed to
the development of this case.*
*It was developed within the research scope of Manufacturing 2000, a development
project conducted with global manufacturing enterprises.*
*Revolution at Oticon A/S (A) and (B) received the EFMD 1994 European Case Writing
Competition Award for Best Overall Case: Excellence in Management, sponsored by IBM
International Education Centre, Belgium, and for Best in the category: Innovations in
Organizational Structures, Systems, and Processes, sponsored by Centre for
Organizational Studies, Barcelona, Spain.**

Revolution at Oticon A/S: The Spaghetti Organization (condensed)

*We told them we were going to take all existing departments away. Nobody could
hide anymore as everything would be out in the open. We would be able to look at
what they were doing, and they could see what we were doing. This was a shock to a
lot of people. They asked, 'How are we going to cope with this? Where are we going
to sit? Is everybody going to look at us all the time? What about people like me who
are managers, how are we supposed to talk to our employees privately? And where?'*

Torben Groth, former middle manager at the 'old' Oticon

Case 6.3

Never mind that CNN, BBC and other international news bureau were rushing to Oticon A/S, the Danish manufacturer of high-quality hearing aids in Hellerup, Denmark. The real story was not only the innovative new structure at Oticon, but also the revolutionary new assumptions of what it meant to work and how one worked. At Oticon, management had given employees the power to drive change and the opportunities for ceaselessly pursuing new challenges. Employees would be the ones responsible for setting the pace for change. The question was—would it work?

Lars Kolind, Oticon's new CEO, was convinced that the best strategy for achieving long-term competitive advantage would be to create a work environment that unleashed individual ability and to design a company proficient in the management of change. Aware of the gamble, he had personally invested over DKK 26 million in this strategy.

Oticon A/S was a niche company with product lines devoted exclusively to systems for the improvement of hearing. Unlike several of its chief competitors—huge, diversified multinational companies—it manufactured and distributed only three main product lines:

1. Behind-the-ear (BTE) hearing aids, which were produced in large series as standard products and used primarily by people with relatively severe hearing loss;

2. In-the-ear (ITE) hearing aids, which were produced for individual users, primarily people with mild or moderate hearing loss; and

3. Systems which eased communication at home, work and in various public situations, including loudspeaker systems and loop amplifiers for schools, churches, etc.

OTICON HOLDING A/S

Oticon Holding A/S, which had previously served exclusively as a financial function, was restructured in 1990 into a group management company to gain more optimal business opportunities, including possible acquisitions in an industry undergoing rapid consolidation. (Refer to Exhibit CS6.3.1). The change provided Oticon A/S with a stronger market focus by giving the subsidiaries greater visibility and supporting distributors for its hearing aids and accessories in just over 100 countries. Oticon had sales companies in 13 countries whose main task was to build and maintain close cooperation with the professional hearing aid dealers; Oticon Export A/S served this same function through independent distributors in the remaining 85 countries. The main functions—research, development, marketing, purchasing, and production—were realigned to meet the new market focus.

THE HEARING AID INDUSTRY[1]

The market

The main component of the worldwide audiology market was the manufacture and sales of hearing aids accounting for 89.8% of the $1.13 billion total audiology market in 1993, and was

---

[1] Market data from Market Intelligence Research Company, 1993.

projected to rise to 90.4% by 1998. The market was mature, with compound annual growth projections from 1990 to 1998 estimated at 5% per annum (refer to Exhibit CS6.3.2); some experts, however, estimated growth as high as 6.8% per annum. Industry experts anticipated that the industry structure would remain stable, with a host of established companies participating. As well, a considerable number of mergers and joint venture activities were expected to take place in an effort to expand product lines and technology.

The US historically had been the dominant market for audiology, with Europe and the Pacific Rim comprising the second and third largest markets, respectively. (Refer to Exhibit CS6.3.2). Industry experts believed that this market structure would not change much, though higher growth rates were likely to occur in Eastern Europe, Asia and especially China; the US market was expected to remain flat to the end of the century. While world revenues grew by 9.9% in 1987, they fell by 3.3% in 1988—chiefly because of the decline in sales of custom hearing aids in the huge US market. Oticon estimated that sales figures for 1993 would show a 7% decline in the US market.[2]

Experts projected that prices would remain steady and price competition would continue to be an important dynamic in the market. Differentiation of the product, or the addition of value-added features, were among the strategies available to manufacturers in this mature market. Sales correlated directly with manufacturers' efforts to educate end-users and to provide technical support to dispensers.

Technological development

The major function of a hearing aid was to increase the volume of sound heard by the wearer through amplification of sound. Miniaturization of hearing aids made possible the development of the first behind-the-ear hearing aids, replacing technology from the 1950s. Recent technological advances included remote volume control devices, sophisticated sound filtering, and multichannel digital programming for different levels of hearing loss and variations in hearing environments.

With the introduction of digitally programmable hearing aids, the industry was greatly transformed overnight. Programmable systems allowed the dispenser to customize the hearing aid and to adjust it as the user's hearing values changed. Multichannel hearing aids treated low-, medium- and high-frequency sounds differently, using non-linear amplification.

Hearing aid technology was moving towards the creation of a single hearing aid which could be adjusted to many types of hearing loss through programming, causing many current products to become obsolete. Oticon was at the leading edge of developing hearing aids with non-linear amplification, and sought to become a leader in programmable, fully-digitized hearing aids.

Oticon's product line

The hearing aid market was segmented into behind-the-ear (BTE) and in-the-ear (ITE) hearing aids. BTE hearing aids, used for relatively severe hearing loss, were more standardized. They were fitted to the contour of the outer ear and then worn behind the ear so that a relatively small

---

[2] Personal correspondence with Leif Sorensen.

number of sizes were required to fit most people. ITE hearing aids, for mild to moderate hearing loss, were individually designed by taking an impression of the client's ear and then fabricating a customized shell. The market for BTE hearing aids was stagnant, while the market for ITE hearing aids was considered the higher-growth segment.

In 1991, Oticon introduced MultiFocus, the world's first fully automatic hearing aid with no user controls. This device, which could treat 70% of people with hearing loss, exceeded sales projections by more than 100%. The PerSonic series was a range of six new BTE hearing aids. As technology moved toward digital signal processing, Oticon added an integrated circuit development team, in one of the largest design centres in Denmark, to develop new chip technology. The creation of new products had shorter and shorter life-cycles (currently 4–7 years), but most of them were upgrading existing Oticon products rather than creating new market segments.

Competition

Siemens Audiologische Technik (Erlangen, Germany) a division of Siemens A.G., and Starkey (Minneapolis, USA) were the world's two leading hearing aid manufacturers, in both volume and market share. Starkey was the leader in ITE hearing aids. Oticon A/S was the third largest hearing aid manufacturer in the world. Oticon had cooperated for many years with the other two Danish hearing aid manufacturers, GN Danavox, and Widex A/S, on technical matters— especially concerning issues on standardization. Oticon exported 90% of their production. Other competitors included Phillips Hearing Instruments, Dahlberg, and Phonak.

The hearing aid industry was becoming more competitive, with over 100 companies in the market. Oticon was targeting the high-priced segment of the market, where audiology expertise and reputation differentiated them from low-cost manufacturers. Lars Kolind sought to enhance Oticon's market focus by giving greater attention to Oticon's customers, who were the nearly 5000 key hearing aid dispensers and hearing clinics most committed to end-user satisfaction. The industry practice of close collaboration with physicians and medical facilities would continue, but again only with those more professionally focused.

Distribution

Hearing impairment had to be diagnosed by a professional audiologist. Then, the hearing aid device would be purchased from a dispenser—i.e. the audiologist, a physician, or a licensed independent hearing aid fitter—who would buy directly from the manufacturer. Increasingly, retail outlets were also springing up (similar to those for vision wear), making it more difficult for small independent dispensers to compete with large chains. Nevertheless, the number of independents continued to grow, with a near tenfold increase of audiologists in the US alone over the last decade.

Access to distribution networks was also made through acquisitions. Bausch & Lomb, for example, had recently purchased Dahlberg, the manufacturer of 'Miracle Ear', a hearing aid with high brandname recognition in the US, for its network of 800 franchises and 200 Sears Roebuck Miracle Ear hearing aid outlets. Oticon similarly sold to chains of hearing aid dealers who were responsible for increasing volume in Europe, the US and the Far East. Club Hearing Instruments A/S. a project headed by Soren Holst, was launched in 1992 to service the dealer chains. These chains had their own service departments, marketing, stock control and distribution systems, eliminating the need for the hearing aid supplier to perform those functions.

OTICON'S HISTORY: A STEADY COURSE

Oticon was founded in 1904 by Hans Demant, whose wife was hearing-impaired. When he returned from a visit to the US, he brought his wife one of the first electronic hearing aids, and soon others were asking for this new product. Mr Demant started importing hearing aids for sale in Europe, operating as essentially a trading company.

Oticon began its own production of hearing aids during World War II, remaining a family-owned business until the mid-fifties when new management took the company into mass production. Under this same management, the company rose to the number one position in the world market by the end of the 1970s. With 15% of the world market, and sales in over 100 countries, Oticon had established itself as a leader in miniaturizaton, the technology use in mass production of behind-the-ear hearing aids. 'What counted then was miniaturization, and we were very good at that', said Lars Kolind in looking back at Oticon's 'golden age'.

Although Oticon's second management had proved its effectiveness by becoming number one in the hearing aid market, the company was 'conservative' like many others of its era. The functional departments—marketing and sales, finance, manufacturing and operations—were headed up by directors who, in turn, made up the top executive group responsible for all strategic decisions. 'Basically, Oticon had been an extremely conservative company for many years and was still the same when I joined in 1984. We had lots of departments', reported Torben Groth, who had been a middle manager in those days. The hierarchical structure worked well for the mass production of hearing aids, providing the necessary coordination and control for a manufacturing company.

Crisis at Oticon

'What we didn't know was that hearing aids would move two centimetres, from behind the ear to right into the ear, and that's a very long distance', recalled Lars Kolind. With the advent of in-the-ear (ITE) products, Oticon was faced with a rapidly declining market share as competitors reaped the benefits of the technological breakthrough and, with no ITE product of its own to offer in this changing market, Oticon was not a player at that time. By 1987, in-the-ear (ITE) hearing aids already accounted for just under 50% of the world market, while behind-the-ear (BTE) represented only slightly more than 50%.

Oticon's troubles had actually begun in 1979, but the favourable exchange rate between US dollars, the source of most of Oticon's income, and Danish crowns, where its major costs were incurred, enabled the company to continually show improved financial performance through 1985. In reality, during the period 1979–1985, Oticon lost a tremendous amount of competitive power. By 1985, a reversal of the favourable exchange rate—the declining value of the dollar against most European currencies, coupled with competitors' introduction of the ITE products put Oticon in a highly threatened position. Oticon's market share tumbled from 15% to 7%; the world champion lost its position as market leader and fell to third place. Some industry watchers seriously questioned whether Oticon could even survive this disastrous development. Following losses of DKK4 million in 1986 and DKK41 million in 1987, Oticon's Foundation Board decided that new management was needed to overcome the crisis.

Oticon's third management

Lars Kolind was an unexpected choice for CEO as he had had no previous experience in the industry. He had, however, come from a company producing scientific instruments (Radiometer) that had gained first place in another niche market. Furthermore, his values corresponded to those of the Foundation Board members for the most part. Thus, Lars Kolind and his team were only the third management of Oticon since its founding in 1904.

The previous management, however, remained fully involved in Oticon's operations. The former CEO, Bent Simonsen, assumed a seat on Oticon's Foundation Board. Torben Nielsen, the former technical director, took a position in manufacturing operations. The previous sales and marketing director, Henning Monsted Sorensen, went to Paris as General Manager of the Oticon subsidiary in France. The previous finance director, Bengt Danielsen, retired. Throughout Oticon's history, management continuity had been a constant, and the arrival of Lars Kolind would do little to change that tradition.

Crisis management

Lars Kolind immediately introduced drastic cost-saving measures and refocused the business on specific key segments. He moved quickly and mercilessly to bring down overhead costs and to cut unprofitable product lines. Some 10–15% of the employees at headquarters lost their jobs. The turnaround was dramatic. Within six months, Oticon returned to profitability, despite heavy losses in the first two quarters of 1988. By 1989, Oticon reported a profit of DKK22 million. (Refer to Exhibit CS6.3.3).

Lars Kolind immediately saw the need to change the market focus. Since quality had become a requirement for participation, nearly all competitors were meeting quality standards. Thus, being a high-quality, high-cost producer no longer assured Oticon a competitive advantage. Lars Kolind decided that Oticon should become the preferred partner with the most professional hearing clinics and hearing aid dealers in the world.

Lars Kolind promoted several younger managers who had been middle managers before the change to higher levels, where they would assume responsibility for making the necessary changes for survival. But, although the financial turnaround was successful, the feeling remained that nothing had really changed. Lars Kolind commented 'I didn't think that I had managed to establish a long-term stable change at a significantly better level. Although I had cut costs and loss-making product lines and activities, I didn't believe that would significantly improve our competitive position in the long term.'

COGITATE INCOGNITA: THINK THE UNTHINKABLE

You are not alone when you fight, there are people who want to work together with you. But I saw that certain major competitors were driving new technologies and were moving fast. I was concerned that we had not established a solid base for the long term. And, I realized that this concern was not generally accepted at all inside the company. I was really alone in wanting to take the company significantly further. Imagine my situation. Everybody from outside, including the Board, was saying that we had a really fantastic management. They were satisfied and wanted to leave things the way they were. I felt more alone than I had ever been.

Lars Kolind

On New Year's Day, 1 January 1990, Kolind wrote a four-page memo where he described his dream for the kind of organization he believed would achieve a sustainable competitive advantage for the future: an organization that would lead in creativity, innovation and flexibility. Lars Kolind asked all Oticon employees to 'think the unthinkable'. Tossing out all assumptions about work and workplaces, Kolind challenged his managers to being again with a clean slate. All paper would go. All jobs would go. All walls would go. The foundation upon which Oticon would build its future would rest on twin concepts: dialogue and action. Everything at Oticon would be designed to support these two ideas as the means for making breakthrough accomplishments in creativity, speed and productivity. 'We really wanted to be innovators in this industry', said Kolind, 'but we were known for being exactly the opposite. We wanted to combine innovation with new records in productivity levels.' Kolind set ambitious goals: a 30% increase in productivity in three years, an objective which became known as 'Project 330'.

Lars Kolind's vision

What Lars Kolind had in mind was a completely different kind of company; it would have just one team—of 150 employees at headquarters—all continuously developing. This company, the new Oticon, would not just run faster, but better.

Lars Kolind asked each employee to examine his own job and focus on what he did well. Each employee should be able to do several tasks: those he did very well, and those where new skills could be learned. Lars Kolind called on everyone to assume several jobs to maximize individual contributions. He was strongly convinced that paper hindered efficiency. If people were expected to work with other employees on several projects, then each person should be able to move about freely. Paper inhibited that capability. The other barrier to working together, in Lars Kolind's opinion, was a physical one: walls. If people were to become one big team, then the walls had to go. Lars argued in his memo:

> *We need an open work environment, one that is interesting and exciting, where employees become part of a bigger entity and can more easily understand how their job relates to Oticon's strategies and goals. The time needed for the many controls in a traditional organization could be devoted instead to serving clients better.*

First reactions

The memorandum was meant to initiate a dialogue for a new way of doing things. But, by February 1990, Lars Kolind had had enough discussions with his management to know that the older, more traditional managers were against his proposal. Torben Groth recalled that some managers were not only bewildered by the concept, they were strongly opposed to it. Many of the younger people, however, agreed that the company needed the very changes that Lars Kolind was prescribing. Loyalty had always been a strong value at Oticon and, when Lars indicated his commitment to this new way of working, everyone extended their support—at least on the surface. The document 'Think the Unthinkable' was drafted for presentation to the staff and the Foundation Board.

There had been little resistance when Lars Kolind first presented his vision, but also not much confidence that it would ever materialize. However, momentum picked up quickly when discussions began in earnest about relocating Oticon's headquarters and the new kind of organization Lars had in mind.

THE CHANGE PROCESS

Merging two cultures

Lars Kolind felt he had to be bold when considering the problem of where to locate Oticon's new headquarters. Previously located in two sites—marketing and product development at one site, corporate management, administration and distribution at the other—two distinct and incompatible cultures had emerged. Moving one to the other might invite disaster. Instead, thought Lars, why not create a new *third* culture and move them both into that? Oticon was in a strong position financially after weathering the crisis. But, Oticon was still a high-cost, high-overhead company, despite the many significant reductions. This situation, together with a house divided within itself, led Lars Kolind to consider radical solutions.

From March 1990—when 'Think the Unthinkable' was first introduced—through the remainder of the year, Lars Kolind devoted himself to preparing Oticon's employees for the move to the new headquarters. In November 1990, he hired Sten Davidsen to manage the change process. Davidsen came from Den Danske Bank, where he had been working on a bank merger project. Davidsen first prepared a one-page 'map' of the change process (refer to Exhibit CS6.3.4), so that all employees would be informed about how they fit into the plan. A statement of corporate values was drafted and circulated (refer to Exhibit CS6.3.5).

Small working groups were formed to carry out the actual tasks related to the organization changes. One group worked directly with the architects and engineers as they designed the new headquarters building. Another group edited and published 'Project 330', the newsletter that kept all employees up-to-date on progress. A group of 13 people called 'Superusers' was created to train people how to use the software for the new information system. 'We got a lot of people involved in the process—as many as possible—and gave them a free hand in deciding for themselves what responsibilities they had in their area', remembered Niels Jorgen Toxvaerd.

Monthly meetings were held, and a number of academics and consultants were invited to present the latest thinking on organization design, communication, and team work. Lars Kolind's new management group was similarly involved in re-examining some of Oticon's fundamental business assumptions, renewing product lines, reviewing the company's strategy regarding markets, products and competitive moves.

Resistance becomes revolt

As the change process progressed, and it became apparent that Lars Kolind was intent on implementing his vision, resistance among Oticon's management intensified. Open resistance finally emerged when considering a relocation to Thisted in Jutland, a remote part of western

Denmark where Oticon's manufacturing site was located. Managers who had been covertly resisting became completely open in their opposition to the move; they began to work actively against Lars Kolind. Despite all of Lars' efforts to prepare everyone for the move to the proposed new headquarters in Jutland, he was suddenly facing a full-scale revolt. Should he proceed with his vision? Had he perhaps already gone far enough? Was it possible that Oticon had indeed achieved sufficient fiscal health and that such radical restructuring was no longer really necessary? Lars Kolind recalled:

> I knew what I wanted to do and felt powerful enough to handle the opposition. They honestly felt that I was on the wrong track. They believed that resistance was their only way to prevent me from ruining the whole company; therefore, they openly tried to obstruct what I was doing. I believe they showed loyalty to the company that I think one should respect. If the boss is making the mistake of the century, then it's right to try and stop it.

Losing to win

Lars Kolind abandoned his decision to move to Jutland very reluctantly. He had, in fact, already pursuaded the Board that moving the headquarters to Jutland was the right thing to do, but he had known that getting the employees to agree would be difficult. To help persuade them, Lars had described ambitious plans for creating appealing working conditions and an attractive environment. But, when Oticon's employees emphatically turned his proposal down, Lars Kolind relented. There was no way that Lars could know that he had, in fact, gained ground:

> I gained their support by stepping back and saying, 'Okay, you won that one. I respect you'. There was never a negative word from me. I accepted the decision, that was it. I decided to forget about the past and look, instead, toward the future. From one week to the next, there was a complete change.

Lars Kolind's commitment to the future

Lars Kolind persisted unrelentingly in order to ensure that his vision of the future would be realized. 'I gave in on where Oticon is going to relocate, but I insisted on carrying through with this reorganization plan. I decided that was the way it had to be. Full stop. 'The day of the move, in August 1991, was a watershed event for Oticon employees. Within the first week, a complete change in perception began to emerge. The things that they had discussed, and could then only imagine, had become a reality. What they were experiencing bore little resemblance to the old Oticon, but they encountered very few surprises; after all, they had played a central and critical role in designing the new Oticon.

Continuity and change: leaving the past behind

Returning to the past was never an option for Oticon employees; the past was finished and there was no way back. Lars Kolind admitted that this strategy had been deliberate and carefully thought out: once Oticon employees were settled in their new headquarters in Hellerup, Lars invited the press to witness the revolution at Oticon. There was extensive press coverage—television, radio, newspapers and magazines—as the management experiment rapidly caught the imagination of the business press. Journalists were given complete freedom to walk around unaccompanied, so that the employees became the spokesmen for this radically new way of working. Lars Kolind recalled:

> I made no secret of the fact that I was using the press coverage as a tool, as a way of burning the bridges behind us so that it was obvious to everybody that there was no return. People's perception of what they are doing depends very strongly on their alternatives and, if there are no alternatives, they are less likely to carry on about 'how wonderful it was in the old days'.

On continuity: Oticon's strategy and competences

Although Lars Kolind's intent was to market the revolution as a significant change from the past, the fundamental business strategy remained largely intact. Oticon would shift its market focus to those dispensers who were most concerned about end-user satisfaction. This move towards the more professional dispensers would be accompanied by a new product strategy: to provide the dispenser with a full (complete) line of hearing aids.

Oticon would maintain its past standards, focusing on high quality and technical excellence, and would remain at the high-price end of the market. At the same time, the related concerns of efficiency, coordination and control would need attention. Lars Kolind believed that these high objectives could be achieved after the transformation, but he also knew the risks involved.

THE SPAGHETTI ORGANIZATION

> We used to have these marvellous offices for management on the third floor of the old building, as well as Jaguars in the garage. But, we realized that our company functioned not because of our organization, but in spite of it. This clearly had to change, and I saw no other solution: except to discontinue the concept of a formal organization. So, we threw it all away, and we introduced something new—which has been named 'the spaghetti organization'.
>
> Lars Kolind

The spaghetti organization was not a matrix organization. The only structure was provided by the projects, with the additional dimension that the staff were multi-skilled. In a matrix organization, a chip designer would work on chip design—possibly on, say, three projects simultaneously, but at Oticon he would also be doing marketing or finance on another project. The new organization could also be called a 'chaos' organization, i.e. one that was most emphatically unhierarchical, chaotic, always changing, and with no organization diagram.

Finding a job inside the new Oticon

When employees walked into Oticon's new headquarters on 8 August 1991, they no longer had their old jobs, but they were expected to create new ones. Torben Groth explained:

> You had to position yourself in the new organization, creating your own job in a new context, with new relationships. Middle managers would become nearly obsolete as the company changed from a management-driven to a project-driven organization, i.e.

the task would be the driving force, not those managing the tasks. Some managers felt lost for a time as nobody attempted to tell them what their job would be—that was part of the process. Lars purposely left us alone—to sort out who was capable of working in this new flexible and dynamic environment.

Lars Kolind decided that traditional job descriptions and formal positions were inhibiting. Typical Oticon employees had a range of capabilities, and thus no one ever fitted a job profile exactly. So why not design each job to accommodate the particular person? But, as it was not usually possible to have a unique position for each employee, why not make better use of people by enabling them to apply their skills in several different positions? Lars Kolind recalled:

We quickly agreed that all employees would have a portfolio of jobs, and we were tough; we said at least three jobs, with the main one in their profession or using their greatest competence and the other two in outside areas. This concept really expands an organization's resources: engineers are doing marketing, marketing people manage development projects, and financial people help with product development.

Oticon's non-offices

Oticon not only eliminated the organization chart, its employees also no longer had 'traditional' offices, i.e. there were no walls. Lars Kolind decided to have absolute transparency in the work environment, where 'intentional disturbance' was taken for granted. Everybody had a desk and a computer, but that desk could be moved—ideally within five minutes. There were no files, just a caddy where certain essential paper information was kept. No partitions—i.e. no offices.

Project teams would physically sit together, thus enabling decisions to be taken instantly rather than waiting for a meeting to be arranged. Marketing people would be aware of what advertising was doing next door. Coffee bars were strategically located to stimulate and encourage discussion. Elevators were replaced by a central spiral staircase that was wide enough to permit chance encounters and dialogue.

Communication

Kolind also believed in the advantages of a paperless office. He explained: 'We shifted from formal communication to dialogue and action. To have this kind of flexible workplace, we obviously had to get rid of the paper.'

Lars asked Torben Petersen, hired in the spring of 1990, to set up an electronic information technology system that would support informal communication. Petersen remembered the process:

I initially looked for ways to gain efficiency through the elimination of paper, but the whole emphasis changed a bit in the process. It became more and more clear that there were two things to be achieved: flexibility and sharing knowledge. Flexibility was necessary so that desk stations could move around easily—without mountains of paper. Working on multiple projects required that the team members be able to access information quickly without having to find all the documents.

Torben Petersen designed a retrieval and archive system where all documents, reports, memos, letters, etc. could be accessed easily. Everyone received this new electronic archives and e-mail system, as well as the software for word processing and financial spreadsheets. Other material, such as information about competitors and relevant journal articles, were also available on the system.

All incoming mail was scanned into the system and could be read by everybody. After being scanned, it was immediately destroyed by a paper shredder (except legal documents, of course) and then discarded through a transparent chute that was visible in the company dining area. Being able to literally watch the morning mail disappear confirmed the process for everyone in a uniquely entertaining way.

The system was designed to enable communication within and across projects, thereby supporting work processes. Petersen observed:

> *Successfully creating a framework for a cross-organizational corporation was really the most important accomplishment. The difference between now and before is in the way people interact—they solve problems differently. Everyone in the company, not just in the same department, is a potential co-worker. What you notice most is that the different skill levels in the organization are becoming more and more invisible.*

MANAGING THE SPAGHETTI ORGANIZATION1[3]

> *The changes we will be making from 1991 to 1996 are as radical as the sum of all the changes we made from 1904 to 1991.*

Lars Kolind

A Management Committee made up of the 'functional' directors handled all decisions relating to the strategic direction of the company, similar to Oticon's traditional approach. A subcommittee of this Management Committee dealt with new projects being proposed, while the 'owner' of a particular project—also a member of the committee—was required to find and recruit a project leader. The project leader, in turn, was responsible for assembling his team using available employees and for finding the necessary resources to support them. In recruiting team members, persuasion and direct negotiation were the only management tools used.

In the spaghetti organization, members rotated from project to project, almost always working on two or more projects simultaneously. A sophisticated computer program kept track of all current commitments; thus a project leader could quickly scan the list and learn who was available. People soon discovered who made many commitments but never followed through, as well as whose contribution was substantial and critical to the team's success. Such knowledge would inevitably influence a person's future demand. Similar performance 'measurements' were made for project leaders, i.e. a good reputation would attract talented team members.

The pace of technological change made it necessary to maintain knowledge and competence to remain competitive. Circuitry design and digital signal processing were only two recent examples.

---

[3] Appendix CS6.3.A gives a summary of critical events at Oticon from 1990–1993.

However, a project-managed organization tended to overlook professional development, and risked the possibility that valuable gains in knowledge would be lost or not shared. Oticon therefore appointed coordinators as 'guardians', frequently past functional managers, to oversee the professional development of functional skills—i.e. marketing, finance, technology, etc. These coordinators were expected to ensure that Oticon's competences met strategic requirements in each functional area. Skill coordinators were given their own budgets to finance additional training and other needs that were not covered by the project budgets. They were also responsible for conducting salary and personal development reviews, and for recruiting new people with potential growth in areas outside their original area of expertise. Each headquarters employee reported to both a skill manager and one or more project leaders.

One pervasive problem at the new Oticon was finding a visible means for rewarding people as they developed. By abolishing the concept of 'career', professional development could not be rewarded by promotion to a higher level—to an office or a title.

Product development cycles

Competitors' advances in technological development had seriously affected Oticon's product development process. The development of miniaturization technology had led to the introduction of (ITE) hearing aids, where Oticon had been left behind, and subsequent developments in programmable hearing aids were transforming the industry overnight. The race to get new, technologically-sophisticated products to market first was fast becoming the competitive criterion for success.

In November 1990, Torben Groth, manager of technological development, created Oticon's first integrated circuit design centre. Digitally programmable hearing aids also required greater sophistication in software design. More importantly, these technological developments were also transforming how dispensers served their customers. With the new programmable hearing aid, Oticon had to consider supplying its dispensers not only with the new software and perhaps the PCs to run that software, but also the training needed by the dispensers to use it. The situation and the practices for the dispensers was likewise radically altered.

Product development at Oticon required a very sophisticated planning system. Because team members worked on multiple projects, the activities of the projects were tightly linked. Slippage or reassigning resources on one project, then, inevitably had an impact on all other projects. Efficiency suffered when such slippage occurred. Hiring additional people, however, in order to ensure that resources were always available was also inefficient.

Time to market with new products was estimated to take a little longer than previously, according to one senior manager at Oticon. He attributed the increase to resource constraints, the additional time required to 'sell' or negotiate inside Oticon, and the lack of a formal coordinating function across projects. Lars Kolind disagreed:

> Time to market is very difficult to measure because it depends on how much innovation is involved. The level of innovation today cannot be compared with 3 to 5 years ago—there was much less innovation then, only minor changes. My estimate is that time to market for comparable products has been cut in half, or even more than that. Some major products that will be introduced in the next few years will have been

Case 6.3

under development for 3 to 5 years—that is, for innovations such as going from a fully analog to a fully digital signal processing hearing aid.

As to coordination, I believe there is a very high level of coordination taking place because so many people know what other people are doing. I am aware of some criticism that our formal coordination is fairly weak, but I honestly don't care. Because I believe that there is no better coordination than informal action.

THE SPAGHETTI ORGANIZATION AT WORK

In the spaghetti organization, product development and manufacturing practices were greatly transformed (refer to Appendices CS6.3.B and CS6.3.C). Years earlier, Lars Kirk had initiated the new philosophy at Oticon's manufacturing sites, building competences among his people across all product lines and engaging them in direct problem-solving on the line. Former supervisors either left or took on substantially different roles. Peter Finnerup had spearheaded the development of a new product by creating a strategic alliance with three of Oticon's competitors, thus creating less visible physical boundaries for Oticon as an organization as well.

The development of MultiFocus was another example of where new work methods were used to bring a new product to market. The MultiFocus hearing aid provided superior performance by automatically adjusting the hearing aid's amplification to the immediate environment. This eliminated the need for a remote control device—the most recent innovation—and ensured a very high quality of sound reproduction. MultiFocus—the result of 12 years of R&D at Oticon's Eriksholm research centre located near Elsinore—was a big success, exceeding sales projections by 100%. Perhaps no other development at Oticon had contributed so greatly to improved financial performance. Moreover, three new product extensions of MultiFocus would be introduced within a similar 18-month period, nearly half the time required for earlier product introductions.

Action of a different kind: the critical success factors

There is absolutely no way that works better than having transparency and fostering dialogue. Maybe I make it sound like a kind of religion with me, but I am a very strong believer in the value of transparency.

Lars Kolind

Activity at Oticon had reached ever-higher levels, perhaps the most meaningful single indicator that Oticon's employees were realizing greater potential. These activities were not driven by a boss, but by an organization fully engaged in doing its business. The kinds of activities had also changed: instead of reading memos, scheduling and holding meetings, Oticon's employees were taking action, and the results were evident in the new level of communication at Oticon. Being a manager meant working in an entirely different way. Soren Holst observed, 'The projects have become the driving force for resources and direction now. Our message to the people is 'do not do as you wish, do as the task requires'.

Niels-Jorgen Toxvaerd added:

The actual management task is different now from what it used to be when you used your power to push things through instead of negotiating things. Before, you could get things done by using a certain group of people, now you have to do it in a different way. I believe that our ability to react to new problems, to new opportunities, is much faster now than it was before.

Marketing and R&D linkages had been strengthened. Soren Holst explained:

It's a chain which starts at the audiological research centre in Eriksholm. Removing and lowering many internal barriers, as well as those between Oticon and its customers, has improved the whole process. We have learned to adjust our strategy to support the new products and processes.

Inge Christoffersen also commented:

The fact that the process was unpredicatble made it so everyone involved would be able to influence the changes. Even Lars didn't know where it would end. The result would be a combination of the endeavours of the entire company. We were able to influence the project and the end results.

Changing work practices

The staff found that, while many of their tasks remained the same, they gained a broader knowledge of what was happening in the organization and how their particular contribution fitted into the overall picture. Being able to observe what others were doing made people aware that their colleagues were also busy. The sense of being a 'team' provided greater confidence—a willingness to ask questions when necessary and to be less afraid of making mistakes. Ida Raun Petersen reflected 'The new Oticon develops your sense of self, if you work here for a while.' Helle Thorup-Witt stated 'We feel that everyone is equally important. Because we are all contributing. As we gain in competence, we also feel better about ourselves. In the past, your importance was related to your position, now it comes from your contribution.'

STUMBLING TOWARD SUCCESS

After the new Oticon was introduced in August 1991, the company suffered millions of DKK in losses in three quarters—from the second quarter of 1991 through the first quarter of 1992. Profitability resumed at the beginning of the second quarter of 1992, but dark days persisted for some time. Just when investments in the new Oticon headquarters building were being made, a product introduction in the US failed completely. Then, negative cash flow from the investments in the new organization increased, which prompted the Board to accelerate the search for a financial director to complement Lars Kolind.

Lars Kolind and the board had been planning to hire a financial director after the change process was completed sometime in 1992. However, by late 1991, they agreed that Lars should start searching immediately in order to appoint a financial director as soon as possible.

In January 1992, Niels Jacobsen was appointed Executive Vice President of Oticon. He and Lars agreed not to divide the functional responsibilities formally. Inevitably, however, their respective

Case 6.3

roles did evolve: Lars Kolind assumed responsibility for overall management, and the research, development, sales and marketing, while Niels Jacobsen focused on day-to-day operational issues. Niels, who was a very traditional kind of manager, introduced the controls and cost measures needed to balance the creative, flexible new organization. Lars remembered 'He was a great help in managing the business more effectively, but it was not easy. Having just loosened everything a lot, it was difficult to tighten up again.' Sten Davidsen observed:

> He liked manuals and following procedures. He was constantly reminding us about the bottom line, and, he wouldn't accept any outpayment—all investments went through his hands. Financial systems were installed, not only for accounting purposes, but also for inventory and for planning in production. Then, he integrated them; before, they had been separate systems. He had the determination and motivation to put everything together.

Lars Kolind and Niels Jacobsen both realized the need for constant dialogue and for having joint responsibility for managing the company. Otherwise, the new flexible organization could have been jeopardized by a regression to the previous traditional structure because of the concern about short-term cash flow and profits.

Performance: results

Before the 1991 crisis, Oticon typically had run at a profit level of DKK18 million a year, or approximately DKK9 million each six months. Following the introduction of the new structure, Oticon quickly began to recover. By the second half of 1992, Oticon's profits were nearly DKK18 million for that period alone and, by the first half of 1993, were over DKK36 million. Within a two-year period, Oticon's profitability level had quadrupled. Turnover had also previously been stagnate, at about a 2% per year annual growth rate. But, by 1992, Oticon enjoyed a growth of 13% in a very flat market and increased another 23% in 1993. (Refer to Appendix CS6.3.A).

SUSTAINING CHANGE

Lars Kolind's vision had been realized: Oticon had managed to achieve a breakthrough in creativity—to become a trendsetter. He elaborated:

> By taking people out of their traditional way of thinking, we got them to work together successfully. We were able to reduce our product development cycle, and our time to market by 50%. Out best competitor requires a couple of years to produce a new product, but we can do it in less than 12 months. Our new approach—being flexible, having multi-jobs, using know-how intelligently, and being open and transparent—was what enabled our organization to have a sustainable competitive advantage. Is it sustainable? It not, then we will invent something else. But, I think it is sustainable because it is very, very difficult to change from a traditional hierarchical, departmental, slow-moving organization to what we have now.

It seemed increasingly clear that Lars Kolind's vision was the right one for Oticon. In any case, one thing was certain: there could be no turning back. Not only did Lars Kolind have his own personal funds staked on the future, but many of Oticon's employees had also bought company shares. Lars believed that they were not merely 'lucky'; he was convinced that they had truly found a strategic advantage that competitors would find difficult to replicate.

Inge Christoffersen shared her thoughts about Oticon's future:

> Oticon will have changed again in five years, but we are used to change processes now. I think that companies today will have to find ways to change very quickly and be willing to adapt. My advice to companies that want to know how to gain competence in making changes is: You have to abandon your usual way of thinking, forget old habits, and leave a lot of traditional thoughts behind.

Will they be able to do it? If not, then *Oticon's competitors, beware!*

Case 6.3

Case 6.3

Exhibits

Exhibit CS6.3.1 The Oticon Group

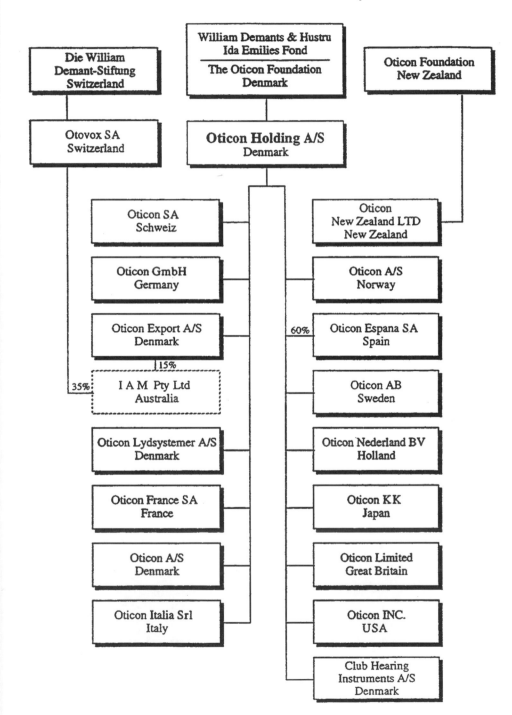

Exhibit CS6.3.2 Worldwide Audiological Market: Revenue & Forecast by Product Type[4]

	Total revenues ($ millions)	Revenue growth rate	% Revenue hearing aids	US	Europe	Pacific rim	Rest of world
1988	893.22	–	89.0	41.8	33.5	12.8	11.9
1989	936.04	4.8	89.2	41.7	33.6	12.9	11.8
1990	981.33	4.8	89.4	41.6	33.8	12.9	11.8
1991	1029.55	4.9	89.5	41.4	33.9	13.0	11.7
1992	1080.20	4.9	89.7	41.3	34.0	13.1	11.6
1993	1133.71	5.0	89.8	41.2	34.1	13.2	11.5
1994	1190.27	5.0	90.0	41.2	34.2	13.2	11.4
1995	1250.05	5.0	90.1	41.1	34.3	13.3	11.3
1996	1313.36	5.1	90.2	41.1	34.4	13.3	11.2
1997	1380.26	5.1	90.3	41.1	34.5	13.4	11.1
1998	1451.19	5.1	90.4	41.1	34.5	13.4	11.0

	Sales by region (%)			
	1990	1991	1992	1993
Scandinavia	15	15	15	15
Western Europe	35	36	40	38
North America	25	23	21	19
Asia	15	15	13	16
Rest of world	10	11	11	12

[4] From *Market Intelligence*, 30 October 1992. 'World Audiology Products Market: Market Size/Forecasts by Region'.

Case 6.3

Exhibit CS6.3.3 Oticon Financial Performance: 1988–1993. Oticon Holding Group (DKK1000)

Principal figures	1988	1989	1990	1991	1992	1993	30 June 1994
Net Turnover	432 756	449 601	455 931	476 531	538 816	661 284	362 100
Gross Profit	197 125	212 910	195 069	214 264	257 891	322 074	NA
R&D	11 534	13 782	15 822	30 422	39 528	44 068	NA
Profit on primary operations	6 893	36 105	16 870	8 731	31 249	85 538	63 500
Profit before tax	−48	22 298	13 127	5 465	18 154	83 738	62 400
Net profit for the year	−5 110	16 944	10 425	−287	8 822	62 305	39 400
Net cash flow	10 836	−19 015	70 392	−73 909	52 393	94 339	35 500
Shareholders' equity	124 033	137 399	172 694	155 805	146 020	191 490	234 500
Total assets, year-end	353 357	378 479	370 853	411 108	395 749	491 355	498 000
Number of employees	1 064	1 087	1 049	1 086	1 069	1 073	1 138
Key figures							
Return on equity	−3.9%	13.0%	6.7%	−0.2%	5.8%	36.9%	21%
Share capital	50 000	50 000	65 000	66 957	66 957	66 957	66 358
Book value/ 100 DKK share	248	275	266	233	218	286	352

Exhibit CS6.3.4 Some milestones

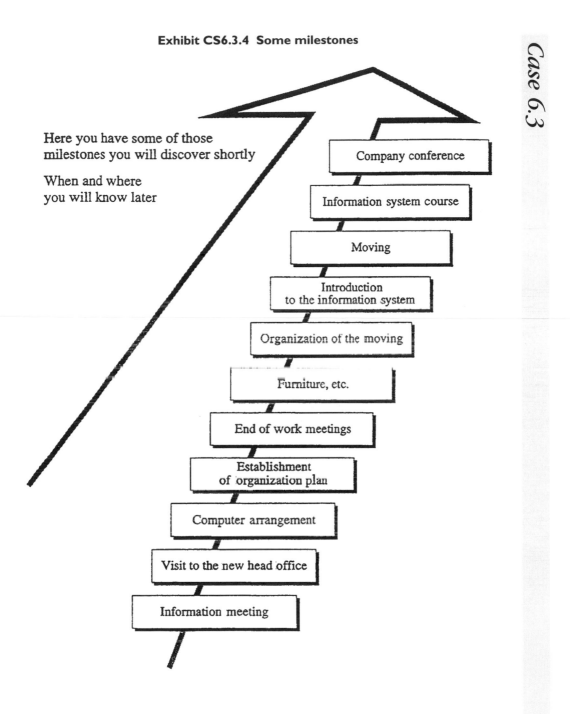

Here you have some of those
milestones you will discover shortly

When and where
you will know later

Company conference

Information system course

Moving

Introduction
to the information system

Organization of the moving

Furniture, etc.

End of work meetings

Establishment
of organization plan

Computer arrangement

Visit to the new head office

Information meeting

Exhibit CS6.3.5 Statement of values at oticon by Lars Kolind

1. All human beings like to take responsibility if they are given responsibility.

2. The people we are willing to trust will return that trust.

3. People innately wish to develop and make progress. People, then, prefer to be challenged and, while they may be afraid of change, don't like things to be always the same.

4. People want a clear understanding of the structure and overall objectives of their work environment, but also like the freedom to fulfill these objectives. People like to have influence over their daily work situation.

5. People wish to be paid according to their performance and their results in a way that is fair and justified.

6. People prefer to be partners in their companies, in both good and bad times.

7. Job security is best achieved through developing individual competence, so that competitors will compete for your services.

8. Every person should be treated as an individual and assisted in his career development.

9. People are interested in knowing how their work relates to the strategy and goals of the company, and wish to fully understand the company's general situation and development.

Appendix CS6.3.A Oticon Holding Group: key events and results 1990–1993
(*continues*)

1990	1991	1992	1993
Profit before tax fell from DKK22 million to DKK13 million due to a decreasing gross profit margin. Return on equity was 6.7%. Sales volume remained flat in 1990, therefore production staff were reduced. Support functions for all distributors were gathered under one function, Distributor Support, managed by Vice President Niels-Jorgen Toxvaerd. New factory section was built in Thisted in preparation for the production of digital and digitally programmable hearing aids. Design centre was established for the development of integrated circuits. Oticon's research and development costs increased by 37% over the last two years.	Sales increased by 5% and gross profit by 10%. The new product introduction in the US was not successful and, coupled with a doubling of research and development costs, the profit before tax fell from DKK13 million to DKK5 million. Oticon's head office was relocated from two separate buildings into one building in Copenhagen, including a total change of organization and work methods seeking greater flexibility, creativity, speed and productivity. The new factory in Thisted was constructed, along with the new Oticon World Distribution Centre, and all Oticon locations were linked on a nationwide network for data and telephone communications.	Sales grew by 13% in a stagnant market. Pre-tax profits jumped from DKK5 million to DKK18 million. Operating profits soared from DKK9 million to DKK31 million, up 85% over 1990. Gross profits increased by DKK44 million, from 45% to 48% of sales, up 20% over the 1991 level. Turbulent currency exchange rates brought about sharp increases in financial expenses, from DKK7 million to DKK12 million. Severe cutbacks among national health services on the costs of health service, and fluctuating exchange rates forced further consolidation in the industry, with two companies merging, another closing, and a third acquired.	Oticon enjoyed its best financial year ever. ■ Net sales increased by 23% in a continuously stagnating market. ■ Primary operating results increased from DKK31 million to DKK86 million, or 13% of net turnover. ■ Profits before tax increased from DKK18 million to DKK84 million. ■ Return on equity rose to 37%. Oticon entered into a strategic alliance with Phonic Ear, an American Company developing and producing wireless communication equipment for possible use as a supplement to Oticon's hearing aids for educational purposes.

Case 6.3

Appendix CS6.3.A *(continued)*

1990	1991	1992	1993
Oticon A/S began work on ISO 9001 quality certification. Lars Kolind's new vision and Oticon Project 330 was introduced. Since 1987, Oticon maintained substantial growth in spite of a contracting market, therefore received the King Frederik IX's Award for Excellence in Export. A complete series of custom in-the-ear hearing aids were introduced on the American market. Oticon held 10% of the world market for hearing aids.	The MultiFocus hearing aid, the first automatically adjustable hearing aid, was introduced in October. CLUB Hearing Instruments A/S was created to serve a very narrow customer segment: large hearing aid dispensing chains operating in Europe, the US and the Fast East. Because these chains had their own service departments, along with marketing, distribution, etc., Oticon products were offered at a price free from these overheads. In November, 19 565 shares were purchased by Oticon employees, bringing the two-year 1991–1992 total to 43 629 employee shares. The market for hearing instruments expanded only slightly.	Research and development costs increased another 30%, after doubling between 1990 and 1991, and tripling— from DKK12 million to DKK40 million— since 1988. R&D costs accounted for 7% of total Group sales, and 11% of Oticon A/S's sales. The PerSonic series—six new hearing aids which, together, covered 90% of all hearing losses—was introduced. The new MultiFocus, and other knowledge-based products, enabled the company to penetrate the most developed markets in Western Europe and Japan. Further production staff reductions were made in salaried production and head office staff.	In June 1993, Oticon A/S received ISO 9001 certification, confirming its quality assurance system was of the highest international standard. Oticon established licenced production in China. Oticon Holding A/S acquired Oticon New Zealand Ltd., as a wholly owned subsidiary. Oticon Holding A/S acquired the hearing care division and joint venture production company from their previous distributor in Australia. These activities will be carried out by the newly established Oticon Australia Pty. Ltd.

Appendix CS6.3.B The NOAH Project (continues)

Peter Finnerup had left Oticon in 1985 after working ten years for the company. He returned in 1991 at the specific request of Lars Kolind, who valued Peter's expertise in software. Peter had been at Oticon when it held the number one market position; a position the company lost when the in-the-ear (ITE) hearing aid was introduced by its major competitor, Starkey. The ITE required an impression of the client's ear from which a customized shell was fabricated; Oticon simply lacked the organizational set-up. Further, Oticon ignored the ITE product because its acoustical performance was inferior to its BTE product. The customer, however, responded differently. Starkey produced twice the volume of ITE hearing aids than Oticon's production of all types of hearing aids. 'Someone was sleeping profoundly during those years', reflected Peter.

Between 1985 and 1991, Peter had been an R&D manager for a large electronics company, then had left to become a partner in a software consulting company. When Lars Kolind approached Peter about returning to Oticon, he was sceptical. But, after 15 minutes of talking with Lars, Peter realized that 'there was something extremely exciting happening and it was very different from what I had seen before: here was a new company, a new era, possibilities for employees to develop themselves, to create your own job, set your own borders, no bosses, no limits, you could define your job. It was like being your own boss in your own company. 'The new Oticon was exactly the kind of work environment Peter Finnerup sought.

Lars Kolind especially wanted Peter to return to Oticon because of his experience in software development. And he gave Peter a free hand in achieving his goals. Peter formed his own development group with a mix of skills: marketing, audiological experts, and software programmers. Customers were invited in to give their views and opinions about the products being planned. The customers identified the functionality they sought in the product, and these suggestions were taken into consideration by the development group.

Competitors had introduced digitally programmable hearing aids, which required computers to gather and then program data about the client into the hearing aid. End users were thus able to have a device that was customized to meet their specific hearing needs without having to make physical adjustments themselves. But, competitors used different programming boxes and, with no standards determined, each new brand used a new 'box'. A common box, the HIPRO (hearing instrument programmer) needed to be defined and developed, so that the dispensing audiologist would not have to specify different data for each patient for each brand.

Together with two other Danish companies, Danavox and Widex, Oticon formed a project called NOAH 'because everyone was in the same box', explained Peter.[5] The goal would be to develop a common software interface between each manufacturer's particular fitting program, using the same audiological information and a standardized HIPRO box.

The alliance was approved in December 1992. Peter and a marketing colleague Frank Henrichsen presented the alliance to Lars Kolind as a *fait accompli*. Lars felt that they were moving perhaps a bit too fast for Oticon, that normally more discussion would be required, but

---

[5] The NOAH group currently consists of five leading manufacturers of hearing aids worldwide: Starkey, Phonak, Widex, Danavox and Oticon. Manufacturers of audiological systems and dispenser-oriented office systems are queuing up to participate as well.

Case 6.3

he nevertheless gave the development group the full go ahead. Later, Peter commented on Lars Kolind's trust in him as a project leader:

> *Lars Kolind lets you be independent and grow, to be a success or a failure. He is willing to give you a great deal of leeway. He realizes that it is a good strategy, after setting a certain target for a group, to lift that group out of the environment in which they function, isolate them and tell them to come up with such and such a product. We have seen the positive effects of this approach inside Oticon, where we can work almost separately, yet have an interaction with the company at the same time. There is no departmental barrier that slows us down.*

> *Being a project leader is a vertically integrated job where you take everything from the bottom and bring it upward, as well as from above down. You write everything yourself. In the old system, milestones were accomplished sequentially, from A to B, then B to C. Now, you have the freedom to go from point B to point C. You can ask, 'Why am I doing this? Can't we do it some other way? Why don't we do it this way instead?'*

Peter Finnerup summed up his feelings about this new product and new market in a 3-word statement: 'It's extremely exciting.'

Case 6.3

Appendix CS6.3.C Manufacturing (*continues*)

Manufacturing of components was done in Copenhagen, and production, assembly and quality control took place in Thisted, Jutland. Research was located in Eriksholm in Elsinore (where Hamlet's castle was located). In 1991, Lars Kolind asked Lars Kirk to assume responsibility for production in both Copenhagen and Thisted, because of the radical new developments that Lars Kirk had undertaken at Thisted. Lars Kirk recalled:

> *Some years before Lars Kolind arrived, we had the feeling that production should be organized differently. Making use of human resources in production was limited, which was typical at that time. Yet we knew that there was a lot of power out there, a lot of knowledge that was not being used.*

Lars Kirk began by sending people to a simulation exercise called 'the production game' that gave them a simple but 'perfect' understanding of how a factory operates. Within two years, nearly eveyone at Thisted had attended this first course. Other courses followed: in quality, problem-solving and communication.

But, Lars Kirk still felt that their operations were not well organized. There were several departments, typically 30–50 people in each one, broken down into groups of 8–9 people. Walls separated the different groups. The way the process worked, the product started—in serial fashion—at one end, moving from group to group until it reached the other end. Each person handled one or two operations. Along the way, complaints about the other groups could be heard over the walls, which was upsetting 'because everyone could hear them but no one could see who was speaking'. If something wrong was discovered by someone from one group, he had to report it in the foreman's office; the foreman had to go to the group where the problem had occurred and then back to the first group. Such incidents wasted a lot of time and capacity.

To reduce this problem, Lars Kirk cut big holes in the walls so that, when complaints were voiced, it was possible to see who was making them. As well, people began talking to each other. The next step was to initiate cross-training: first, within each group so that, one year later, each person could do all the operations within his own group; and then between groups, so that everyone knew what happened before and after the product reached his particular group.

In this way, serial production was replaced by parallel production. In one department, a product family would typically be made up of, say, five different products, each of which could have 10–15 options—i.e., colours or switches. In each department, every group would make all the products, with everyone in each group performing 8–9 operations. The five different products would still be assembled sequentially: Product A would pass from someone in Group 1 to the person in Group 2 responsible for that product. But, all five products would be in production simultaneously, with only 4–5 people working on each product compared to some 25 people under the serial production system.

Production meetings used to take place every second week to discuss problems that had occurred during the previous two weeks. Lars Kirk abolished them. 'The foreman would always claim that next time things would be done differently, then two weeks would elapse with no change', Lars recalled. Therefore, a new production monitoring system was installed: an electronic bar code system that could locate at any moment where each component was in the

Appendix CS6.3.C (*continued*)

production process. 'We gave them very good tools, so that they could calculate how many people were needed for each group as well as for the complete department', he added.

Lars Kirk began moving people between departments. At first, people were shifted once a month because production was on a one-month schedule. But, after everyone had received the cross-training, they could be shifted every week or even every day.

In 1990, Lars Kirk closed the factory for one day and took the entire production staff to a holiday camp to discuss strategies and goals. He asked people to suggest improvements for their particular work situation. They came up with more than 100 ideas. Each department was then broken down into small groups and asked to discuss—without any management involvement—how they could make the department function better. Lars Kirk recalled management's trepidation:

> *Of course, management was worried. Although it may seem difficult to take on responsibility and more power, it is even harder to give it away. I was afraid, as a manager, that people might run over me.*

For nearly a year, middle management remained recalcitrant. 'Typically in their mid-50s, they had always had the right to decide everything, and they could not accept that people below them were now able to decide things for themselves, were taking responsibility', Lars Kirk commented. 'The managers who honestly believed in the idea were quickly revealed, as well as those who didn't. The ones who couldn't accept it had to leave. There was no way back.' And, at this point, things really began to happen.

If someone had a idea, it was his responsibility to go to the people who would be affected; if the idea involved planning, then planning people had to be consulted; if batch sizes, then the foreman should be involved. But, each person was expected to take initiatives and then implement his own recommended solutions. 'This policy applied to the entire factory of 280 people—everyone was to take direct action on his recommendations by going in person to the players involved.' This idea took off like a trajectory, has been heading skyward ever since and has yet to level off.

From 1988 to 1993, management decreased from 11% to 4.5% of the work force, while people earning hourly wages increased from 200 to 280. 'I believe that, in three years, we will have half again as many managers', Lars Kirk predicted. Employee turnover dropped from 11% to 6% over that three-year period, because hourly workers were no longer leaving. Productivity from 1988 to 1992 increased by 18%. Hours spent on rework dropped from 5.6% in 1990 to 3.5%. Rejected batches dropped from 4% to 1%. And material usage dropped from 16 million in 1988 to 4.5 million. Lars Kirk elaborated:

> *Flexibility is the chief gain. We started a new product three weeks ago, the new MultiFocus Compact. Only three weeks ago, production was zero. Now, next week, we are making 1300 per week, equal to 6000 a month. We can move people around in less than one hour, and they have nearly the same efficiency in other departments as in the one where they work regularly. Our ability to supply product is nearly 100% within the normal schedule time. Next, the computers for the production control*

Appendix CS6.3.C (continued)

system will leave the foreman's office and will be installed right inside the production departments, so that people will be able to answer their own questions.

The new bonus system reflects how initiative is rewarded: all the member of the group receive a bonus, not a specific individual. If someone changes groups, he will still receive the bonus given to the 'home' group. No group is either punished or unduly rewarded because of the performance of one individual. We want to build up respect based on what we are doing rather than on power. When people are working and see something is wrong, they will be willing to handle it if they know they can explain the problem, discuss it with no fear of being looked down upon.

The greatest success? Lars Kirk was gone for three months to an education programme in Japan. On his return he found that the plants had run themselves very well in his absence. As well, he realized he could begin to undertake other tasks besides those of the production manager. He went to Lars Kolind with a request for additional responsibilities and was placed on a product team.

A Race Without a Finish Line

Chapter

7

Introduction

Only a few years ago, the 'management of change' was considered a separate section within 'management' as a whole. The world was not standing still, but we could still distinguish times of relative stability amidst more turbulent periods. A winning strategy did not guarantee eternal success as competition would try to emulate it, but companies often benefited from their competitive advantage(s) for a number of years. Customers did change allegiance, but information on products and services was not as easily available and quickly communicated as it is today. New technologies were emerging, but at a pace that still allowed serious companies to keep up.

Over the last decade or so, the world has been changing so rapidly that one can no longer imagine managing in a 'steady state'; the need to continuously adapt and change has become a given in managerial life. As recently as 15 years ago, tools such as word processors, faxes and e-mail were only starting to emerge. Five years ago, the Internet and cellular phones were still very much 'leading edge'! Today, complete industries are challenged by these new media and capabilities, and we observe more and more 'winner takes all' competitive struggles, as compared to the more oligopolistic arrangements typical in the 1980s.

This book has described a number change efforts, some more successful than others. The cases we have presented show that, when properly managed, process-based change efforts can have a very significant impact on customer satisfaction, employee morale and company performance. But the war is never won even for those companies that operated very successful change programmes or even corporate transformations at the time the case was written; competitors do not abandon the field, customers' expectations do not stand still and technology does not stop evolving.

This is a well-known phenomenon in sports, where it is often said that it is easier to gain the lead than to maintain oneself in first place. In bicycle racing, the rider wearing the leader's jersey often finds it much more difficult to escape the pack as other riders carefully monitor the former's every move. In football, every team plays the leader as if its entire season depended on that game; the leading team coming to town is the other teams' equivalent of European cup games!

This phenomenon has been expressed with a catchy summary of 'Nothing fails like success'! In particular, companies that have undergone successful change efforts are bound to confront the following pitfalls.

Losing the 'competitive edge'

Success tends to generate self-confidence which, on one hand, helps instil pride and high morale. The sports world provides regular illustrations of this phenomenon as close games tend to be won by teams and players who are in the habit of winning. The downside, however, is that success also tends to breed complacency and arrogance; the line between self-confidence and arrogance can sometimes be quite subtle. When such traits permeate an organizational culture, it becomes much harder to question existing practices and prevailing orthodoxies, or to propose new approaches.

Take the example of Volkswagen. The enormous success of the popular Beetle, with its air-cooled, rear-engine design, prevented the company from switching to a water-cooled engine and front-wheel drive, until the company was literally in bankruptcy in the early 1970s. R&D and new product development had been forced to concentrate on refining a dead-end technology until it was nearly too late because of the unwillingness of top management to recognize that the old recipe no longer worked.

Or again consider the more recent example of IBM. In the early 1970s, five of its software engineers put forward a proposal to sell a comprehensive set of software programs for accounting, manufacturing and sales. Big Blue had a system of its own, which was pursuing a different strategy, so the five employees left to set up their own company, SAP. This German company is now one of the largest and fastest growing software houses in the world.

Success can create a climate where, as the 'new wisdom' seems to work it becomes the new 'conventional wisdom', and opposition to this new conventional wisdom is not only frowned upon, but actively suppressed. Companies and, in fact, probably most human beings find it easier to accept challenge and to question themselves when they are struggling than when they are leading. It is also easier to innovate and take risks when one does not have too much to lose, than when one has a dominating position to defend.

Over time, companies may also lose their recipe for success. If companies could precisely assess the drivers of success, they could perpetuate them. The problem is that companies have difficulty in identifying those factors, which can give rise to two sorts of dysfunctions: Some companies fear losing the winning recipe, and try to sustain all the conditions that happened to exist when success manifested itself. Progressively some of these factors—staff, systems, structures—become a liability rather then an asset. In other cases, the company modifies an element that happened to be a key ingredient of its past success.

From squeezing to squeezing too hard

When the company's cry for help is well-founded (e.g. in times of crisis) and effectively communicated, most employees can understand that their employer needs maximum effort on their part. An excellent illustration of this phenomenon was provided in Case 5.3 'Pulling Air France Out of Its Dive: Air France under Christian Blanc', where 80% of employees ended up voting for a restructuring plan that required more sacrifices from them than the plan put forward four months earlier by Bernard Attali, Blanc's predecessor, and which they had forcefully rejected.

Leaders tend to find it more difficult to keep pressing employees for more sacrifices—more wage concession, more productivity improvements, voluntary layoffs, relocations—when the company is doing well. There are probably a number of reasons for this. Most human beings are simply not capable of producing maximum effort one hundred per cent of the time; they need to decompress between peak periods. Employees may also have more difficulty identifying possible improvements; the first rounds of change typically eliminate most of the visible slack, the next rounds often require more ingenuity and effort. Finally, while employees in most Western countries tend to accept that shareholders will receive a significant proportion of the financial implications of success, they also tend to expect to receive some portion of these benefits themselves.

This issue takes us back to the pain–gain equilibrium and the concept of the 'balance diamond', (see Figure 7.1) both discussed in Chapter 5.

Leaders of successful companies, even more so that leaders of companies experiencing difficulties, must be able to articulate 'what is in it for the employees'. Financial and commercial difficulties create constraints for companies and employees, but they provide a reasonably natural legitimacy to sacrifices demanded by top management. When most (or all) indicators are positive, top management concerns for the notions of fair process and procedural justice (discussed in Chapter 5) become all the more essential.

Visions and their limitations

Strong visions can constitute a powerful stimulus for employees. They can also strongly support the alignment of efforts within the firm. On the other hand, powerful and strongly enforced visions can also induce selective myopia and/or become ill-adapted over time—both of which can lead to disaster. For example, back in 1977, based on his vision of computing, DEC president Ken Olson asserted: 'There is no reason for any individual to have a computer in their home'. Unsurprisingly in this context, DEC largely missed the personal computer boom.

In the same vein, the old question, 'What business are we in?', can lead to excessive abstraction in terms of vision. For example, an airline can be variously

Figure 7.1 Balanced diamond

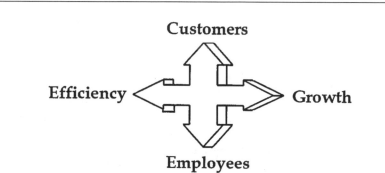

considered to be in the air travel business, in the transport business or in the customer service business. Yet increasing levels of abstraction can lead to a loss of touch with customer needs—to fly safely, quickly, comfortably and inexpensively from one place to another. Similarly, it may be helpful for a company to think in terms of striving to become a 'virtual business' when determining what it is possible to outsource and what is core activity, but it is unhelpful as a vision of what the company can offer customers.

On the same note, some people believe that Bernard Attali's strong expansionist vision for Air France (documented in Case 5.2 'Resistance to Change at Air France: Bernard Attali's Experience'), contributed to the airline's slow recognition of the drastic changes in the environment around 1989–90. Air France's problems were exclusively blamed on external factors—the Gulf crisis and industry overcapacity—rather than internal ones. Meanwhile, the success of rival British Airways was attributed to external factors—the bilateral agreement with the US and low social costs—rather than internal ones (such as state-of-the-art yield management and use of customer information). This superficial view of BA's success prevented Air France from scrutinizing more carefully the real drivers of BA's performance.

More generally, a very strong vision (whether oriented towards growth or cost-cutting) can hamper a company's reading of its situation, as well as making it more difficult to adapt or change course. When top management is fully convinced it is going in the right direction, environmental scanning tends to become selective and the company may become oblivious to disconfirming evidence.

Leadership successions are always tricky

The effectiveness of a leader is based on a complex set of interactions between the leader's characteristics and the circumstances he or she is operating in. Even setting aside the complexity introduced by the fit between the person and the circumstances, it is not always easy to identify all the dimensions that make a leader successful. Some leaders seem to get away with more than other leaders; they get more commitment from employees, they can sell tougher measures and set more demanding targets.

The sources of their ability are not always clear and it is easy to underestimate the impact of the leader's personal characteristics (e.g. a sense of humour, a certain personality or a certain sensitivity), as well as the weight of past events; employees might accept more from a leader who contributed to restoring the prosperity of the company than they would from a newcomer. The importance of that combination of ingredients may only become apparent when the boss leaves—and the successor seems to lack some of these qualities.

At the other extreme, the leader may embody a constellation of attributes that helped the company succeed, which may lead it to appoint a successor in the same image. Yet in a fast-changing context, the future is likely to be substantially different from the past, both in terms of the development of the company and in terms of the evolution of the context. Thus the qualities required of a successor may differ significantly.

For example, the qualities required to initiate a transformation may be quite different from those needed to sustain a transformation; just as those needed to start up a company differ from those needed to bring it to maturity. The company's needs are also influenced by the evolution of the industry, which may be becoming far more international or multidisciplinary (meaning a meeting point for what were previously separate industries) than it was in the past. Thus, any succession decision must weigh up a host of factors in deciding where to promote continuity and where to look for renewal of effective leadership attributes.

Not knowing when to leave

Strong leadership does not just lead to succession problems. Sometimes the problem is that the leader's effectiveness declines over time. There are three basic ways in which this may happen. First, a leader may have associated him-or herself so strongly with a vision or strategy that it is problematic for them to try to envisage, and/or then 'sell' a change in direction. Going back to Air France under Bernard Attali, for example, Attali's forceful growth-oriented message during the first two years of his tenure probably made it more difficult for him to recognize the change in industry conditions, and then weakened both the credibility and legitimacy of his message that the company needed to make a sudden shift to an efficiency-driven approach.

Secondly, power and 'life at the top' take a toll on people, both in terms of patience and in terms of humility. On the patience side, it is much easier for a newly appointed leader to understand and accept, for example, that employees feel a communication gap with management. It is easier because 'I was not in charge up until now, so I am quite open to the possibility that my predecessor might not have communicated very well and I will do a better job'. Two years later, when the same complaint is directed at the same leader now two years into his or her tenure, the problem is now more personal and harder to accept. The prestige and honour bestowed on leaders of large corporations also contributes to making it harder for them to receive criticism and challenge.

Thirdly, a leader's effectiveness is altered by modifications in the company and industry environment they face. In an ideal world, successful leaders would change and adapt their style to different conditions. Yet, it has been argued that most people actually find it very difficult to modify their styles or 'renew themselves'. For example, there is evidence that people tend to reproduce strategic actions over time—both in terms of content and style. Kets de Vries[1] observed a psychological tendency on the part of leaders to start repeating over time strategic choices or themes such as customer service, cost-cutting or reengineering. He attributes this propensity to what he calls the chief executive's 'inner theatre'—the subconscious assumptions and preoccupations, and to the 'scripts' that people acquire over time. These 'scripts' are similar to what Peter Senge calls 'mental maps'.[2]

---

[1] De Vries, M. K. 'CEO's also Have the Blues', *European Management Journal*, **12**(3), September 1994, pp. 259–264.
[2] Senge, P. M. *The Fifth Discipline: The Art and Practice of the Learning Organization*, Doubleday, 1990.

Anecdotal support for this view can be found in the singular behaviour of Colin Marshall, whose actions at British Airways can be traced back to his time at Avis, the car rental firm. While there, Marshall transformed Avis's service image, giving male staff red jackets (which he wore, too) and badges saying 'We Try Harder', which helped Avis dislodge Hertz from the number one position in Europe. He also already displayed the quasi-obsession with customer service that he brought to BA.

Similarly, Air France's boss Christian Blanc repeated tactics at Air France which he had successfully used in previous assignments, including the threat of resignation (which he had carried through in his last organization when the government refused to support him on an issue he considered essential); the use of referendum to push through tough change measures (a ploy he had utilized as a government envoy to negotiate peace with militant separatists in New Caledonia); and the decentralization of operations into profit centres (an initiative he previously used to reform the RATP which runs the Paris metro and bus system). These measures produced divergent results when repeated at Air France: the first and second were successful; the third was probably inappropriate for an airline and was later partially reversed.

The difficulty of genuine self-renewal means that leaders may need to recognize their limitations and remove themselves in order to allow their organization to move forward. While the leader in question may have a lot to contribute to another organization, such lucid acts remain rare.

Sustaining success

The two cases that accompany this chapter are extensions of the very successful stories—Taco Bell and British Airways—presented in previous chapters. The follow-up cases raise different issues, such as "What comes next?" and "How do we top that?" They illustrate the difficulty of sustaining success.

Transformation models typically culminate in a state known as renewal. Companies that reach this stage would have institutionalized learning, know why they have succeeded, know how to perpetuate growth and rarely experience downturns. Unfortunately it is not clear that any company has reached this phase, which has led Chakravarthy to label it 'Nirvana'.[3]

The follow-up case on British Airways, in particular, provides a useful opportunity to reflect on Chakravarthy's contention. Its sustained success—becoming the world's most profitable carrier in 1996—makes it a good test case for determining the accessibility of the 'renewal' phase.

Of course, the difficulty of sustaining success is not limited to companies. The same can be said of once dominant industries, such as shipping, steel or mining, which

---

[3] Chakravarthy, B. 'The Process of Transformation: In Search of Nirvana', *European Management Journal*, December 1996, pp. 529–539.

have been victims of technological change—the advent of alternative forms of transport, materials and energy. Other industries have been revolutionized by innovation from unrelated sectors.

Taking a historical perspective, whole nations and indeed civilizations have struggled with the twin challenges of preserving internal momentum and adapting to external change. The loss of internal drive (complacency or conceit) is perhaps best illustrated by the case of the Roman Empire, whose fall is often attributed to widespread decadence. The failure to adapt to the external environment, on the other hand, is better illustrated by the cases of the Ming and Ottoman Empires, which declined when they stopped taking an interest in what was happening outside.[4]

For example, the Ming Empire in China had a sophisticated canal system and a unified hierarchical administration run by well-educated civil servants. It was technologically precocious with huge libraries, printing by movable type and large numbers of books. Trade and industry, stimulated by the canal-building, paper money and population pressures were also advanced. The output of their iron industry was greater than that of Britain in the early stages of the industrial revolution seven centuries later. They had invented magnetic compasses and had a massive fleet of ships designed for long-range cruising.

That progress came to a halt in the mid-1400s when an imperial edict banned construction of seagoing ships, and China started to pay more attention to preserving and recreating the past than creating a brighter future based on overseas expansion and commerce. The canals and ironworks were left to decay, printing was restricted to scholarly works and not used to disseminate practical knowledge, much less for social criticism, and the use of paper currency was discontinued. The ironworks did not resume production until the twentieth century.

A similar wave of 'introversion' signalled the end of the Ottoman empire. Formerly leaders in the fields of mathematics, cartography, medicine and many other aspects of science and industry—in mills, gun-casting, lighthouses and horse-breeding—the system as a whole increasingly suffered from the defects of being centralized, despotic and severely orthodox in its attitude towards initiative. A culture which had previously thrived on tolerance, attracting talented Greeks, Jews and Gentiles into the sultan's service, became increasingly intolerant towards all forms of free thought. The downfall of the Ottoman empire can be ascribed to cultural and technological conservatism.

Clearly success cannot be sustained indefinitely. But by adapting, it can be prolonged. Consider the simple example of a sports star, someone whose time at the top is inevitably limited. In 1986, disgruntled at never having won a major championship, the golfer Nick Faldo decided to rework his swing. This involved deconstructing the action and discarding the elements that could not be relied upon. From 1987 to 1992 he dominated the game, winning five majors. The next two years

---

[4] As described in the fascinating work of historian Paul Kennedy, *The Rise and Fall of the Great Powers*, Unwin and Hyman, 1988.

proved less fruitful, so once again he withdrew from competition, this time to rework his putting action. By 1996, he had become statistically, the world's best putter. This capacity to self-monitor and willingness to unlearn skills that were painstakingly developed demands both insight and courage, even sacrifice. Faldo's strength did not stem from sheer talent, but from an ability to keep his game together when those around him were crumbling. He internalized every working part of his new swing so that it stood up to the heat of the moment.

Whether at the level of the individual or whole civilizations, the willingness to keep learning may well be the main driver of prolonged success.

Case 7.1

This case was written by Professor Soumitra Dutta, at INSEAD and Niklas Moe, MBA student, at the Haas School of Business at the University of California at Berkeley. It is intended to be used as a basis for class discussion, rather than to illustrate either the effective or ineffective handling of an administrative situation.*

Taco Bell: Reengineering the Business (B) (1995–97)

INTRODUCTION

'In order to be a long-term champion of value you've got to keep on changing & changing & changing. By change I don't mean incremental modifications to an existing set of rules. I mean change that revolutionizes the business. Change which creates a whole new game.'[1]

John Martin had built Taco Bell's transformation over the last decade on the concept of value; but continued stagnant sales and decreasing same-store sales were leading many industry analysts to question the value concept in 1995. Even John Martin had conceded that 'value alone is no longer a competitive advantage'.[2]

Taco Bell could boast of the strongest long-term growth records in the industry, with annual revenue growth averaging 15% a year over the last several years. However, the once invincible Taco Bell and its innovative chief executive, John Martin, were suddenly looking vulnerable in 1995. But Martin was confident of bouncing back:

'I don't know of any company that is better positioned to weather a bump in the road. Maybe I'm a war horse, but I feel good.'[3]

A STRATEGY FOR RECOVERY (1995)

John Martin realized that the 18–24-year-olds who bought 59-cent tacos were too narrow a target to propel sales growth into the year 2000. The company needed a hit to take the

*Copyright © 1997 INSEAD, Fontainebleau, France.
Financial support from 'the Arthur D. Little Fund for the Enrichment of the Learning Experience' is gratefully acknowledged.
[1] 'Mastering Change', Business Forum, University of California Irvine, 18 December 1992, p. 1.
[2] Martin, R. 'Taco Bell's "Lights" menu seeks fix for setback', *Nation's Restaurant News*, 20 February 1995.
[3] 'Taco Bell ready to try new strategy', *Orange County Register*, 23 March 1996.

mainstream business to the next level. Martin once again drew upon innovation to stimulate sales. His next bold stroke was to introduce the health-oriented Border Lights line. Martin reasoned that ageing baby boomers like him would want a low-fat, fast-food option. Consequently, the target audience included older, health-conscious diners, late-teens and twenty-something people who didn't cook at home.

The introduction of Border Lights became very timely as a reply to a 1994 study by the Center for Science in the Public Interest (CSPI) that blasted Mexican food as being very high in fat. Martin conceded that the study had had an impact on Taco Bell's recently softening sales. Response initially was positive. Even nutritionists lauded the company's efforts. Taco Bell seemed, once again, to brand into consumers' minds its name and image as a fast-food innovator.

John Martin predicted that the new line would develop a $5 million new market by the decade's end, and that Border Light items could eventually replace the traditional items. Michael Jacobson of CSPI predicted that the menu makeover would set a new standard for healthier fast-food that should have a ripple effect throughout the industry for years to come.

In line with its strategy to expand points of access (PoA) to a broader audience as well as leveraging the Taco Bell brand, the company expanded its grocery store focus by rolling out a plan to supply about 13 000 or about 40% of grocery stores in the US with a variety of new products, such as the Taco Bell Soft Taco Dinner Kit, aimed at capturing a larger market share. The goal was to generate more business from families. Taco Bell hoped to capitalize on two growing trends: the popularity of ethnic foods and the desire among consumers to eat meals at home. The grocery sales strategy was aligned with Taco Bell's aim to sell food in typical locations and expand its client base to include families. Another innovation was a dual brand venture between Taco Bell and Kentucky Fried Chicken (KFC), where approximately 800 KFC units were offering both KFC and Taco Bell products.

THE FAILURE OF BORDER LIGHTS

Despite strong backing from John Martin and a significant $75 million advertisement campaign, Border Lights never found its audience. Not only did the line not appeal to Taco Bell's core, young male audience, it failed to attract new customers. Restaurant consultant Bob Sandelman commented that most fast-food customers simply didn't care about eating what was good for them:

> People talk a good story when they talk about nutrition, but when they go out and eat, they want to eat what tastes good. And when people eat fast-food, they throw nutrition to the wind.[4]

Jerry Gramaglia, senior vice president of marketing at Taco Bell, commented:

> We allocated 100% of our spending to drive Border Lights, but not to be talking to the 80% of customers who are interested in core products and value. I think there was some tradeoff.[5]

---

[4] 'Taco Bell ready to try new strategy', *Orange County Register*, 23 March 1996.
[5] Whalen, J. and Jensen, J. 'Taco Bell hearing call of the border', *Advertising Age*, 10 July 1995, p. 6.

Bill Campbell, a Taco Bell franchisee in Orange County gave his view:

> What happened was we didn't get the new customers and we didn't get the sales increase.[6]

Despite the failure of Border Lights to find its audience, John Martin was still proud of Border Lights, and insisted that at nearly $600 million in sales in its first year, the low-fat menu had been a successful new product introduction. He acknowledged that the expectations for Border Lights may have been too great. He also spotted flaws in the low-fat offerings, saying items paralleled other Taco Bell items closely, leading consumers to cannibalize the chain's menu. He noted that the company's advertising should have emphasized Border Lights' good taste rather than nutritional nuances. And while Taco Bell was busy pushing Border Lights, it had neglected the young male customers who had propelled its sales all along.

ORGANIZATIONAL CHANGES

John Martin felt that in order for the company to turn around he had to take a more micromanagement approach to leading the operations. He had been burned by the failure of Border Lights and the fact that he pushed the idea so hard, but he knew that his position within Taco Bell was still strong. After having spent a lot of his time and thoughts working on Chevys, he wanted to regain direct control of Taco Bell. As a first step he changed the flow of information so that all Taco Bell vice presidents reported directly to him. The top marketing executive, Jerry Gramaglia, was replaced after barely a year on the job by Peter Waller. John Martin also started to take a more hands-on approach to Taco Bell's marketing strategy. Taco Bell's president, Kenneth Stevens, decided to leave the company.

Taco Bell's employees were expecting John Martin to be able to boost sales by using some of the same magic as he had for the last 10 years. A manager commented:

> 'We all experienced the same glitch in sales and the loss of focus on the core customer. Martin has got everybody focused again. He's a leader. He's rallying the troops behind him.'

Even Martin's new boss and PepsiCo CEO, Roger Enrico, said that Taco Bell was regaining its marketing and product focus and that he had confidence in John Martin, but he also made it clear that PepsiCo wanted to see Taco Bell back on track quickly.[7]

BACK TO BASICS

Rather than trying to revolutionize the fast-food industry as Taco Bell had done in the past, John Martin started to use traditional back-to-basic fast-food techniques such as heavy advertising and refocusing on mainstream products. In an attempt to attract its core customers again, the company introduced fat-laden products such as the Bacon Cheeseburger Burrito, breakfast items, and kid's meals targeted at heavy fast-food users that traditionally had been Taco Bell's most loyal

---

[6] 'Taco Bell ready to try new strategy' *Orange County Register*, 23 March 1996.
[7] Barron, K. 'PepsiCo forms alliance with Lucas Film for use of Star Wars characters', *Orange County Register*, 16 May 1996.

Case 7.1

customers. To back up the new products, Taco Bell launched a $200 million advertising campaign, the largest in the chain's history, in an effort to highlight competitive points of difference in Taco Bell's food, value and service. Richard Reese, a Taco Bell franchisee commented:

> 'We weren't talking to our core consumer in 1995. What's exciting about the current campaign is that it's really gotten us back on target.[8]

Allan Hickok, an industry analyst added:

> 'What they are doing now is back to basics. Trying to drive traffic on value can be very difficult. You can always cut prices. It's so competitive now, it's like airline fare wars. It's just a zero sum game. What operators are looking at right now is operations. What can they do that is a competitive point of distinction, that is different from the next guy on the block?[9]

DECLINING TRENDS (1996)

During 1995, Taco Bell's performance continued to drop. Taco Bell sustained same-store sales declines of 4% for the year and 5% for the fourth quarter. The poor results came despite the highly promoted late, November introductions of several bacon-based menu items. Taco Bell took a $53 million operating loss in the fourth quarter of 1995 and the division's annual US operating profits fell 62% to $38 million. Taco Bell also took a charge of about $80 million related to the division's other chains: Hot-n-Now and Chevys. At least 50 units of Hot-n-Now were closed during 1995 and a significant number of the remaining 150 branches were shuttered. Charges for Chevys stemmed from its slower-than-anticipated rate of growth.[10]

The battle for market dominance in the fast-service restaurant industry was suffering from consumers who demanded 'value' with every purchase. What made the value trend different in 1996 was the customer's definition of the term. It was no longer enough to sell to customers on price. Value now equalled big portions, big flavours and high quality—all for low prices that customers had come to expect.[11]

As a result of not being able to improve sales, Taco Bell felt that they could not continue to support TMUs as daily operational decision-makers. Instead Taco Bell put back store-level management to improve operations. Furthermore, to amplify the focus on core products, Taco Bell sealed an agreement to sell its line of retail (store) products to Kraft.

Having used nearly all possible gains from traditional value strategies in combination with minimum wage increases and higher prices for beef, fast-food chains were forced into 'meal bundlings'. Taco Bell introduced four $1.99 Extreme Value Combos, bundling tacos or burritos and drinks, to compete against the 'value' menus being introduced by other fast-food chains. The

---

[8] Kramer, L. 'Reese, Ghareeb capitalize on Taco Bell corporate-unit sale', *Nation's Restaurant News*, 6 May 1996, p. 15.
[9] 'Taco Bell ready to try new strategy', *Orange County Register*, 23 March 1996.
[10] *Nation's Restaurant News*, 19 February 1996, p.4
[11] 'Consumer Demand For "Value" Is Eating Fast-Food Profits', *Research Alert*, 5 June 1996, p. 1.

price point was approximately $1 less than competitive offerings. However, reversing the trend was not proving to be easy. An industry expert commented:

> The quick-service restaurant Mexican category is stalled, and Taco Bell has to fix that. It is a younger food that some baby boomers are moving away from as they age. Maybe the category has to be broadened to be Southwestern, which has a more of an upscale connotation.

A NEW CEO

The back-to-basic strategy initiated by Martin to boost sales was not able to stop the same-store sales decline. After nearly two years of same-store sales decline, PepsiCo announced in November 1996 that John F. Antioco had been named the new president and chief executive of Taco Bell. Antioco was known to be a turnaround specialist and came fresh from his successful turnaround of the 2500 unit convenience-store chain, Circle K Corp. John Martin still remained in the PepsiCo organization as chairman and chief executive of PepsiCo Casual Dining, a newly formed division comprising the California Pizza Kitchen, Chevys Mexican restaurants and East Side Mario's restaurant chains.[12] PepsiCo announced at the same time that it was reviewing its entire casual-dining business for a possible sale, divestiture, restructuring or other possible realignment in the wake of weak results in the company's overall restaurant group.

Taco Bell decided to review its current advertising agency, Bozell, for a possible change. Peter Waller commented on the change:

> Though Bozell has done some very good work given today's highly competitive, quick service restaurant environment, we believe it is important to explore every option to build our sales and transactions.[13]

During the summer of 1996, McDonald's had introduced the Arch Deluxe line. The line was targeted to a more quality-concerned audience. During the first half of 1997, Taco Bell was still working on its Project Gold. Project Gold was supposed to raise consumers' quality perceptions by upgrading the existing ingredients resulting in a new line of premium products such as steak and chicken fajita wraps. By introducing Project Gold, Taco Bell hoped to move consumers to a higher price level and partially escape the price-driven environment.[14] Food service analysts and Taco Bell insiders thought that Antioco would try to jump-start the chain's sales through a reversal of its value-price positioning.[15] Randall Hiatt, President of the restaurant consulting firm, Fessel International, agreed that Taco Bell needed to create an image move towards quality.[16]

THE END GAME

In January 1997, PepsiCo decided, after being in the fast-food business for 20 years, to spin off its

---

[12] Their combined annual sales of about $400 million were only about 3.7% of PepsiCo's revenues from its KFC, Pizza Hut and Taco Bell divisions.
[13] Benezra, K. *Brandweek*, 27 January 1997.
[14] *Advertising Age*, 10 February 1997.
[15] *Nation's Restaurant News*, 21 October 1996.
[16] *Advertising Age*, 10 February 1997.

$11 billion restaurant business comprising Taco Bell, Kentucky Fried Chicken and Pizza Hut. The move was intended by Enrico to sharpen the Pepsi brand by allowing the company to concentrate on its faster growing Pepsi soft drink and Frito-Lay snack food businesses. He believed that PepsiCo had been weakened by the constant struggle between the packaged goods and restaurant units for funding and managerial talent. He thought the future for the restaurants were bright but that they would do better if separated from PepsiCo's beverage and snack-food operation.

Case 7.2

Case 7.2

This case was written by Jean-Louis Barsoux, Research Fellow, and Jean-François Manzoni, Assistant Professor, both at INSEAD. It is intended to be used as a basis for class discussion rather than to illustrate either effective or ineffective handling of an administrative situation. It is based on publicly available sources (notably those listed in the references).*

Remaining the 'World's Favourite Airline': British Airways 1993–97

INTRODUCTION

In February 1993, Lord King, British Airways' combative chairman was about to retire. For a period of 12 years, he had overseen the airline's remarkable turnaround from slothful state enterprise to industry exemplar. He could withdraw knowing that he had secured his place in aviation history and BA's place as a leading carrier.

Between 1981 and 1993, British Airways evolved from a loss-making, state-owned national carrier with a shabby reputation for customer service—and *the* most unpunctual airline in Europe out of its own home base[1]—into a customer-focused, publicly listed and consistently profitable airline. By 1993, it had been voted world's best airline for five years running in the annual *Business Traveller* survey. This transformation, a testament to the clear vision and strong leadership of Lord King and the chief executive Sir Colin Marshall, was achieved in an industry known for its low margins and its sensitivity to economic cycles. There was little doubt that it ranked as one of the greatest turnarounds in business history.

In terms of the future, King knew he was leaving the airline in safe hands, with the 10-year chief executive Sir Colin Marshall taking over as chairman. Robert Ayling, then marketing and operations director, stepped up to take the role of group managing director.[2] Ayling was charged with the day-to-day running of the company while Marshall concentrated on strategy, overseas investments and external relations. Ayling's promotion merely confirmed the internal speculation that he was Marshall's heir apparent.

*Copyright © 1997 INSEAD-CEDEP, Fontainebleau, France.
Financial support from 'the Arthur D. Little Fund for the Enrichment of the Learning Experience' is gratefully acknowledged.
[1] Corke, A. *British Airways: The Path to Profitability*, Frances Pinter, 1996, p. 120.
[2] The previous head of marketing and operations, Liam Strong, had been tipped to succeed Marshall, before he was poached by Sears in 1991. Unexpectedly, Robert Ayling made the jump from head of human resources to marketing and operations.

Taking over as operational head of one of the industry's star performers (along with Singapore Airlines) looked like the dream job for the 46-year-old Ayling. He had been a key player in BA's transformation and had held a succession of senior posts within the airline. Yet some wondered whether this assignment might not prove something of a poisoned chalice. The strong personalities of King and Marshall had marked the airline, and their business results had set new standards for European carriers. Theirs would be a hard act to follow.

BA'S MID-AIR TURNAROUND (1981-93)

At the start of the 1980s, British Airways was significantly overmanned and losing money. In early 1981, Prime Minister Margaret Thatcher brought in Sir John King with the specific mandate to prepare BA for privatization. In the 18 months before he appointed Marshall as his number two, King cleaned up the balance sheet dramatically, selling off properties and superfluous aeroplanes and cutting loss-making routes. These actions helped to finance some £200 million worth of special severance pay, offered to volunteers willing to leave the airline soon. Within three years, staff numbers had dropped by nearly 20 000 and stood at 36 000 and the operating profits in 1984 exceeded the cumulative total for the period 1970–82[3]. (See Exhibit CS7.2.1 for evolution of BA's financial health and workforce numbers.) King also commissioned the advertising agency which coined the slogan 'The World's Favourite Airline', at the time a quantitative claim based on the number of international passengers carried rather than a seal of passenger satisfaction.

By the time Marshall was appointed, much of the painful downsizing had been effected under Lord King. So although Marshall faced a difficult task in trying to rebuild morale, he did so with 'a clean slate'.

Prior to Marshall's arrival, the airline had been very much about technology and scheduling, and customers had to fit in around those constraints. Marshall instigated a complete change of mindset, which placed the customer at the centre of the airline's preoccupations and was driven by intensive customer service training for all employees. The radical change soon generated considerable enthusiasm and excitement among employees. To reinforce this new orientation, Marshall awarded swift promotion to managers who demonstrated customer service values and activated training programmes that would help managers to support those on the front lines.

Marshall's previous experience in the consumer goods and car rental industries had instilled in him a keen sense of marketing and a virtual obsession with customer service. In executive meetings, the main focus was on service issues—including signposting, smells in 747 toilets, and how to get peanuts out of seat cushions—rather than profitability or aircraft loading.[4]

He restructured the airline twice in his first four years, first to increase accountability, and later to put more emphasis on marketing and IT. He overhauled the human resource systems so that policies for hiring, compensating, appraising and promoting people were aligned with the new strategy and the training received. More generally, the human resources department developed a strong theme of 'caring for the carers' which encouraged a twin focus on customers and on front-line employees.

---

[3] Campbell-Smith, D. *The Struggle for Take-Off*. Hodder & Stoughton, 1986, p. 150.
[4] Young, D. (1990) 'British Airways: Putting the Customer First', in *A Sense of Mission*, eds. A. Campbell, M. Devine and D. Young. Economist Books, (1990) pp. 113–135, 118.

Case 7.2

BA brought in brand specialists from consumer goods companies to explore ways of developing each class of service as brands in their own right. This enabled the airline to maximize yields—the average amount paid by each passenger for each kilometre flown—helped by sophisticated computer programs which juggled ticket prices and discounts to fill the maximum number of seats with full-fare passengers. All this translated into higher profits per seat than most other carriers.

In terms of strategy, Marshall gambled that in the late 1980s aviation would be dominated by two or three carriers in each continental block, and devised a strategy for turning BA into a global network. BA bought 25% stakes in USAir and Australia's Qantas respectively, and took 49% shares in smaller operations in France (TAT) and Germany (Deutsche BA). This enabled BA to funnel more passengers on to its lucrative long-haul flights.

The combination of strategic and marketing manoeuvres, together with systematic cost-cutting efforts and improved asset use, allowed BA to withstand the effects the Gulf Crisis and the ensuing recession.[5] While the airline industry as a whole lost over £10 billion during 1990–93 period—more than its total combined profits since commercial aviation began in 1947—British Airways remained profitable. By 1993, there was real substance to its self-billing as 'The World's Favourite Airline'.

ON-GOING CHALLENGES

Without detracting from BA's impressive turnaround, critical reviews of BA in the early 1990s maintained that several internal challenges still confronted the company,[6] alongside the external challenges of establishing meaningful alliances and marketing arrangements with other airlines worldwide and lobbying various authorities. The commentators suggested that addressing these internal tensions was key to sustaining BA's momentum.

One source of tension was the increasing explicitness of BA's corporate values (notably in mission statements and training programmes) which made it easier to identify and attack contradictions between what the airline said and what it did.[7] As one manager put it:

> The messages coming out of Managing People First [an innovative management training programme] conflicted with the style of the organizational changes that had taken place. MPF was about trust, leadership, motivation, and visioning. In parallel with this, there was a machete environment of hacking out organizations. It produced a conflict in the minds of the managers and staff. I am not saying that the organizational changes were not necessary, but the way that they were made needed to be explained and reconciled with the values portrayed in MPF.[8]

Another question concerned the true extent of the much publicized cultural overhaul. There was little doubt that the airline had fostered a significant change in the values and behaviours of those

---

[5] At the start of 1991, BA announced 4600 voluntary redundancies as a result of the dramatic drop in passenger traffic occasioned by the Gulf Crisis.
[6] In particular, see Hampden-Turner, C. (1992) *Creating Corporate Culture*. Addison-Wesley, 1992, pp. 78–98.
[7] Höpfl, H., Smith, S. and Spencer, S. (1992) 'Values and Valuations: The Conflicts Between Culture Change and Job Cuts', *Personnel Review*, **21**(1), p. 35.
[8] Young, *op cit.*, p. 134.

on the front lines whose job was providing customer service. Yet it was not clear that a similar change in values had occurred in the way back-office people related to one another. One manager confirmed:

> We are inconsistent. You cannot go to one part of the organization and know the whole. It all depends on where you are. It is still quite formal. There is a formality and an informality. It is structured and not structured. It is everything: old-fashioned and modern. Again, it depends on where you are in the company.[9]

There was, therefore, a risk that the profound change in approach to customers would remain separate and compartmentalized rather than serving as a model for improving work relations companywide. One senior manager observed: 'There are still sections of the marshmallow to be tackled. Cabin crews, reservations, check in, and marketing have all changed, but there are other areas that have not.'[10]

The airline also continued to suffer from functional rivalries, especially at middle management level. BA was having trouble integrating the culture of operations—with its emphasis on safety, punctuality and rules—and the culture of service which stressed caring and customized responses.[11] Achieving a more cooperative ethic between functions and individuals was vital in an organization dealing with complex interactive systems, yet this remained problematic. As a senior development projects manager explained: 'One of Marshall's strengths has been to give people individual responsibility, but this has a downside. There are occasions where a team approach is more appropriate. Team-working is still a problem.'[12]

A further tension stemmed from efforts to reconcile continued cost-cutting efforts and customer service, especially at a time when the company was the only Western airline making consistent profits. One of BA's regional general managers alluded to these contradictory pressures:

> Managing costs at the same time as maintaining service to the customer is not easy. Inevitably, cost-cutting impacts the internal part of the service-profit chain. Recruitment freezes, capital-spending freezes, and pay freezes all send a chill down the chain. It is too easy for the customer to feel the icy blast. A staff needs to understand clearly the business rationale for cost reduction and have an opportunity to participate in the tough decisions if the service-profit chain is to remain strong.[13]

The dilemma was particularly acute among middle managers whose ranks were periodically thinned out. Repeated waves of voluntary redundancies were leaving middle managers feeling rather abandoned. As one manager put it: 'Managers are being asked to care for their people, but they have difficulty in doing that if they don't feel cared for at a personal level themselves. Pressure on performance suggests to managers that this is more important than people.'[14] Again, there appeared to be a growing contrast between the people-related values espoused by BA and the getting-it-done values dictated by an increasingly harsh business environment.

---

[9] Salama, A. "The Culture Change Process: British Airways Case Study," Cranfield School of Management, Case Study, 495-00201, 1994, p. 11.
[10] Young, *op cit.*, p. 133.
[11] Hampden-Turner, *op cit.*, p. 92.
[12] Young, *op cit.*, p. 134.
[13] Gurassa, C. 'BA Stood for Bloody Awful', *Across the Board*, January 1995, p. 56.
[14] Young, *op cit.*, p. 133.

A final concern regarding the likelihood of sustaining momentum had to do with the company's habitual approach to change. As one human resources manager observed:

> Colin Marshall's been good at pulling timely crises out of his hat. It was 'heads down' over privatization and our appalling reputation with customers; then 'heads down' again in the crash merger with British Caledonian; then 'heads down' again over joining Galileo (the multiairline Customer Reservation System); and now security is the latest focus. . . . What concerns me is that we have difficulty in harmonizing our efforts and behaving as a team, without a crisis. . . . We've been outstandingly successful, but we are still not a reflective organization, and I'm not convinced we yet know why we succeeded and how we can repeat that experience.[15]

AYLING COMES IN FROM THE WINGS (1993–95)

Ayling's initial background as a civil servant did not obviously destine him to lead the company. Yet in his eight years at BA, Ayling made remarkable progress swiftly accumulating experience in a diversity of functions (see Exhibit CS7.2.2 for details of Ayling's career).

In May 1993, just a few months after his appointment as managing director, Ayling was paying tribute to the achievements of his predecessors and announcing where he thought his own contribution might lie:

> John King gave BA a new sense of commercialism. He brought that with Colin who brought the necessary customer service and disciplines of management to the company . . . My challenge is to take that on to the next stage . . . to achieve as much participation and sense of belonging as possible . . . to create a climate of enthusiasm and competitiveness within the bounds of the acceptable.[16]

At the end of Ayling's first full year in the job, BA was able to announce impressive pretax profits, up 63% from £185 million in 1992–93, to £301 million in 1993–94. Once again BA's performance was in stark contrast with the rest of the airline industry which in 1993 lost $4.1 billion (£2.7 billion) on international scheduled services. BA's progress reflected a cost-cutting drive and increased revenues—and overcame an estimated loss of around £10 million owing to IRA terrorist attacks on Heathrow, London's major airport. BA employees would receive a share of the profits equal to just over a week's basic pay in cash or shares—following a year without profit-sharing.

More importantly, the results silenced critics who thought that Ayling, with his civil service background and scant operational experience, might not have the resolve to pursue the war on costs engaged in by his predecessors. In fact, Ayling went one better, announcing a rigorous target of £150 million of cost-savings for the coming year (ending March 1995).

Ayling was also starting to impress outsiders who had dealings with him. Journalists and analysts considered him a better communicator than either King or Marshall, finding him relaxed, friendly and media-wise. One journalist even commented that, outwardly, Ayling bore more resemblance

---

[15] Hampden-Turner, *op cit.*, pp. 89–90.
[16] *Financial Times*, 18 May 1993, p. 17.

to Richard Branson, the Virgin tycoon, than to his illustrious BA predecessors. In meetings, Marshall increasingly deferred to Ayling's superior knowledge of technical details.

Besides securing impressive business results for 1993–94, Ayling had also launched a novel 'trade fair' where BA employees offered their help as mentors to 40 voluntary organizations, including Age Concern and the Red Cross. The initiative was triggered by a survey in 1993 in which half of the staff voiced their desire to be more involved in community projects. BA costed the manpower value of the time volunteered at £2.7 million.

On the strategy front, BA's international alliances showed mixed results. There was heavy press coverage of BA's involvement with USAir which continued to sustain losses in spite of restructuring efforts. The 24.6% equity stake in USAir was the centrepiece of BA's network of global alliances and brought benefits through a ticket code-sharing agreement, joint frequent-flyer programme and shared facilities at some airports. Pulling out would leave BA 'with a gaping hole in its overall global airline strategy', with the US market accounting for some 40% of world air travel.[17] Of BA's other alliances, only Qantas, the Australian airline, was making any money, though Ayling said that improvements were expected at the French affiliate TAT, following the launch of a restructuring and job-cutting programme, and at the German affiliate Deutsche BA.

By May 1995, Ayling was able to announce pretax profits of £327 million, helped by the fact that BA had taken £160 million costs out of the business during the year. The airline industry remained fragile. In 1994, the world's airlines made aggregate $1.8 billion (£1.2 billion) net profit, but this figure represented only 1.6% of the airlines' aggregate revenues.

BA was also pursuing ambitious fleet expansion plans with the airline phasing out its older, more noisy aircraft by the beginning of the next century. By 2010, BA's long-haul fleet would consist of Boeing 747-400s (enlarged versions of the 747). The airline had also 15 firm orders and 15 options for the new Boeing 777.

In September 1995, BA was ranked fourth in a survey of Europe's most respected companies— and as the top airline. Ayling was pleased, but felt that BA needed to reach for higher standards. He noted that customer approval ratings which had risen sharply in the years after privatization, had started to tail off.[18] He could also see a number of internal challenges emerging. He told journalists that BA was 'more hierarchical than I remember the civil service being and certainly more than some of the better managed British companies of today.'[19] He wanted to change that and, at the same time, give managers a greater sense of professionalism about their work. He also wanted to boost BA's caring side, finding its image is too masculine. Ayling's goal was that BA should be regarded as the *best-managed company in the UK* within five years, conceding that a few companies, such as Marks & Spencer, were still seen as better managed. 'My ambition', he told journalists, 'is to manage BA for the next few years, improve its reputation still further and maintain our financial performance.'[20]

BA launched a new programme called *Leadership 2000* aimed at improving its managers' skills and putting responsibility for keeping customers happy in the hands of those who dealt directly

---

[17] *Financial Times*, 24 May 1994, p. 23.
[18] *Financial Times*, 19 September 1995, p. IV.
[19] *Management Today*, October 1995, p. 56.
[20] *Management Today*, October 1995, p. 58.

with the travelling public. Like Marshall before him, Ayling made considerable efforts to talk to programme participants on a regular basis and, indeed, to as many groups of staff as possible whenever he could.

While acknowledging Marshall's influence on him, Ayling was adamant that they had different approaches:

> When there is a complex issue, for example, I would be quite happy to work through a meeting to try and figure out the best way of resolving it. Colin, on the other hand, has a very accurate and perceptive mind and very often sees straight through to an answer and will do it without needing a great meeting.[21]

In early November 1995, 62-year-old Marshall caused a stir in the City when he announced that he would remain as chairman, but would become non-executive—thus devoting only half of his time to BA—from January 1996. Ayling would become chief executive, taking control of finance and corporate strategy, and preserving his existing responsibilities for operations and marketing. Marshall told the press he had been 'preparing for some time' to hand over to his successor as chief executive, and that the time seemed right with the airline 'running relatively well'.[22] That month's *Harvard Business Review* carried an interview with Marshall where he restated his twin focus on achieving excellence in customer service and the need to care about the problems encountered by staff.[23]

In late November, Ayling made his first pronouncements about the internal challenges facing BA, and what he intended to do about them. He told employees that, while BA was already a well-run company, it needed to be better. He added: 'To be better than we are requires a change in attitude. It requires that we throw off for all time the attributes and attitudes of the public sector'.[24]

As Ayling saw it, these public sector traditions lived on in the refusal of many managers to take responsibility, but not because they feared punishment: 'The paradox here is that people are not punished. Not only are they not blamed, they are not praised either'.[25] Another problem with BA was the profusion of meetings. Ayling felt that these were stimulated by the joint desire to avoid responsibility and to find out what was happening, which highlighted a communication problem.

He illustrated the point by describing an incident from the previous week. It involved the staff at Heathrow's terminal one briefly walking out when they discovered students with foreign language skills had been brought in to help with passengers. Ayling said that staff members had been told six months previously that the students would be employed, the idea being to speed up check-in procedures. 'But,' he concluded, 'the message was still not very convincing for people on the front line. That can only be because the system of communication is not what it should be.'[26]

---

[21] *Management Today*, October 1995, p. 56.
[22] *Financial Times*, 11 November 1995, p. 8.
[23] Prokesch, S. E. 'Competing on customer service: an interview with British Airways' Sir Colin Marshall'. *Harvard Business Review*, November–December, pp. 100–12.
[24] *Financial Times*, 7 November 1995, p. 21.
[25] *Financial Times*, 13 November 1995, p. 12.
[26] *Financial Times*, 13 November 1995, p. 12.

Ayling considered that the best way round this problem—to help employees get BA's message—was to talk to them directly: 'We should not use the written word. Nor should we have managers cascading information to one another, which inevitably means that the message communicated is not the one you started with.'[27] Instead, he proposed making better use of television and satellite technology which made it possible for all employees to receive news directly from the executive at the top.

On 1 December 1995, executives at BA were said to be 'bracing themselves for a wide-ranging management reshuffle' with Ayling seeking board approval to streamline top management, with likely changes in structure as well as personnel.

TAKING THE PILOT'S SEAT (1996–97)

By the time he took over as chief executive, Ayling's streamlined executive team was already in place. The structure of responsibilities had been completely overhauled and the number of direct reports had been slashed from 25 to 11, three of whom were newly named directors (see Exhibit CS7.2.3 for comparison of the executive team before and after).

Within a week of being appointed chief executive, Ayling was named one of *Business Week's* 25 top managers for 1995, based on a poll of the magazine's 220 editors around the world.[28] Only two months before, BA had been voted airline of the year for the eighth consecutive year in the independent survey by *Business Traveller* magazine.[29] Moreover, with global passenger traffic expected to rise by 5–6% a year, and BA's occupancy rates at all an all-time high (with 73.6% of available seats sold), the future looked fairly rosy (see Exhibit CS7.2.4 for evolution of key operating statistics).

This left Ayling with a predicament which he readily acknowledged:

> *It's probably more difficult to take over a successful business than one that needs something done to improve it. If you take over a successful organization it's quite inappropriate to make radical change just for the sake of it. So the change that has to be made is developmental.*[30]

In fact, making a convincing case for change of any kind would not be easy:

> *Although I might see a change as necessary because I've analysed the figures and it's my job to consider the long term, most people in most jobs in the company don't think like that. They think about the day-to-day things—that's as it should be—so it's not at all obvious to them why the changes I think have to be made, have to be made.*[31]

---

[27] *Financial Times*, 13 November 1995, p. 12.
[28] The editors screened the financial figures, questioned analysts and talked to corporate leaders to come up with their selection (*Business Week*, 8 January 1996, pp. 50–60).
[29] This poll is based on replies from 1000 subscribers that live in the UK, mainland Europe and the rest of the world. The replies are reweighted in proportion to the residential distribution of the readers, that is 80% to the UK, 20% to mainland Europe and the rest of the world combined.
[30] *Director*, August 1996, p. 37.
[31] *Director*, August 1996, p. 37.

Soon after taking over as chief executive he presented the board with four objectives for building on BA's existing success, objectives he later reiterated in the company's annual report: first, to sustain BA as 'The World's Favourite Airline'; second, to continue to improve customer service in a more demanding environment; third, to extend BA's reach through alliances and marketing agreements; and, fourth, to 'improve further our management; to be the best managed company in the UK by the year 2000'.

The last objective was the one Ayling felt most strongly about. Better managed, as he saw it, meant fostering an environment that was healthy and democratic: 'People have got to be open with each other, they've got to be challenging and professional, they've got to do what they say'.[32]

Early that spring, Ayling launched a two-week-long business fair, at which the management team presented its ideas for developing the company to some 6000 staff. The presentation was relayed to many more employees via a daily satellite programme. So successful did this prove that BA subsequently launched an internal global satellite channel.

So, in March 1996, BA became the first company in the world to make daily TV broadcasts to its staff. The initiative, costing BA £2–3 million a year, had both communication and management objectives. The broadcasts would include senior managers being 'doorstepped' and interviews with staff involved in industrial disputes. This was part of Ayling's drive to give company news to employees directly, providing 'an improvement on the terrible rumour mill we have at the moment'. At the same time, with BA 'reporters' requiring managers to account for their decisions, it would 'expose those managers who are not performing', explained Ayling, adding, 'we probably have too many managers'.[33] 'I want us to be a modern company, a young company. I want to get rid of hierarchies and deference to seniority',[34] Ayling asserted, reiterating his commitment to ridding BA of its last vestiges of its nationalized industry culture.

In May 1996, the company announced profits of £585 million, up 29% on the previous year, which propelled BA past Singapore Airlines as the world's most profitable carrier. In turn, this produced the biggest bonus payout ever announced by a UK company (totalling £94 million), with each eligible employee receiving a bonus equivalent to just under four weeks' pay. Moreover, the scheme was enhanced, so that employees taking their payment in BA shares instead of cash would receive an extra 20%, the aim being to raise employee ownership from the current 4% to around 10% (66% of staff were already shareholders).

A key contribution to the profit figures were the £150 million in costs taken out of the airline in the previous year. Yet Ayling was determined that BA should not rest on its laurels. So, on the same day that the company announced its record breaking profits, Ayling told employees that BA would have to find £1 billion of savings over the next three years to maintain its competitiveness, with every aspect of the group's operation coming under review to achieve this goal.

Even for a company which had grown use to cutting costs, with a cumulative total of £900 million over five years, Ayling's objective meant stepping up a gear. Managers were set targets on matters such as costs, aircraft punctuality and better use of assets and told to find ways of

---

[32] *Director*, August 1996, pp. 37–40.
[33] *Financial Times*, 6 March 1996, p. 9.
[34] *Financial Times*, 6 March 1996, p. 9.

achieving them. With activities like cleaning and security already having been contracted out, other areas such as accounting, cargo and baggage-handling, and ticket-processing came under pressure to match the costs of outside suppliers.

According to BA, these efforts were needed to prepare for a future in which competition would increase following full liberalization of the European aviation market (from April 1997) and where improved computer systems would allow customers greater freedom to shop around.

In the annual report, Ayling commented on the progress made towards improving management standards:

> Our Leadership 2000 programme has introduced a range of initiatives to encourage better decision-making and greater accountability, rewarding by result and performance and moving away from hierarchical structure. A poster campaign within the company, designed to encourage people to be brave—but not reckless—in taking initiative, became a talking point.[35]

In terms of furthering his customer service aims, BA continued to launch innovations: relaunching their Club Europe and Club World business brands; introducing a 'raid the larder' system which allowed passengers to help themselves whenever they felt hungry; and providing fast-track services for premium passengers through passport and security checks.

On the strategic front, Ayling's appointment had also opened up new possibilities, particularly regarding the unsatisfactory alliance with USAir. The airline's recurrent financial and labour problems were proving costly to BA both in money and management time.[36] Marshall had been closely associated with the strategy of forming a partnership with USAir in 1993, but with Ayling in charge, it would make it easier for BA to sell its 24.6% stake—particularly given that USAir did not provide BA with access to the central and southwestern states, nor with the kind of critical mass BA wanted.

In June 1996, Ayling pulled off his biggest strategic coup, announcing the largest code-sharing agreement ever, with American Airlines (AA). With the two carriers jointly controlling 60% of flights between the United Kingdom and the United States, this would provide a significant boost to the second of Ayling's stated objectives: to extend the airline's reach. Provided the deal was cleared by the US authorities, which might involve renegotiating the agreement governing the allocation of Heathrow slots, it would secure BA's improved access to the world's largest aviation market. By the same token, this far-reaching cooperation and revenue-sharing agreement signalled the end of its relationship with USAir.

Before Ayling took charge, any partnership between BA and AA was regarded as inconceivable by analysts given the 'prickly personalities heading the two carriers'.[37] In particular, the press considered that it had taken enormous skill to win over AA's chairman Bob Crandall, widely regarded as the industry's rottweiler and formerly a fervent opponent of code-sharing

---

[35] Company Annual Report, 1995–96, p. 6.
[36] BA had three (part time) directors on the board of USAir and had been forced to write-down almost half of its investment (£125m) in 1995. BA said, that over the past three years, it had gained $100m annually in extra revenues from the tie up.
[37] Financial Times, 12 June 1996, p. 23.

agreements. As the *Financial Times* observed: 'Although Mr Ayling has proved a tough manager, he is seen as more approachable and personable than his predecessors'.[38]

Ayling agreed that his ability to establish a rapport with Crandall was a significant consideration: 'Personally, I get on with him very well. . . . It's always a factor. If people don't get on, it's difficult to reach an agreement.'[39] As he later observed: 'I do believe in strong leadership, but I think it's important not to get into the cult of personality. Airlines have traditionally been theatrical and high profile, but we have to remember there are 300 000 shareholders out there'.[40]

A few months after the announcement of this mega-deal, BA won the fight to take over Air Liberté, taking a 67% stake in the independent French carrier. The company would be run by the chairman of TAT, BA's French affiliate, and would be kept separate 'for the time being'. This deal doubled BA's share of the French market, raising it to over 20%.

Ayling was also exploring new ways of expanding the company's reach. In particular, BA was pushing back the industry boundaries in terms of contracting out by getting other airlines to fly under its name. The first franchise agreement dated back to mid-1993, but in 1996, BA had nine nonequity franchise agreements, mostly with UK-based carriers, but also including a Danish franchisee and one in South Africa. These small carriers remained independent, but their aircraft were painted in BA colours and their staff wore BA uniforms, while paying BA a fee.

By awarding its less profitable routes to smaller airlines to operate as franchises, BA had generated £50 million in revenues for the year ending March 1996—and Ayling said he wanted to see the figure doubled as this strategy boosted BA's presence in certain parts of the world and fed long-haul traffic. As Ayling observed: 'The days when we could operate 747s to Hong Kong and [still] fly to Benbecula, Scotland, are over'.[41]

In August 1996, questioned about what he saw as his contribution so far, Ayling observed: 'I hope I've created an atmosphere of open dialogue and discussion, in which people will tell me if they think some idea I've got is rather half-baked'.[42] As the interviewer observed, the response was telling in itself. One could scarcely imagine either King or Marshall, from whom Ayling professed to have learned everything he knew about management, making such a remark.

IMPLEMENTING THE BUSINESS EFFICIENCY PLAN (BEP)

First announced in May 1996, the target of £1 billion in efficiencies, had been the subject of consultation and reflection with employees throughout the summer.

On 18 September 1996, Ayling assembled BA's top 300 managers at the Ramada Hotel just off the busy runways of London's Heathrow Airport. Although that morning's *Financial Times* reported that BA had been voted second most admired company in Europe, the CEO's message

---

[38] *Financial Times*, 12 June 1996, p. 23.
[39] *Financial Times*, 12 June 1996, p. 23.
[40] *Sunday Times*, 19 January 1997, p. 3, 7.
[41] *Financial Times*, 19 September 1996, p. 30.
[42] *Director*, August 1996, pp. 37–40.

was far from complacent: 'We must do better because our competitors have caught up, and we are slipping back to the middle of the pack,'[43] he warned. Then Ayling announced that BA would be asking for 5000 volunteers to leave the company over the next 18 months, from November onwards. Over three years, these employees would be replaced by a similar number of new recruits with greater flexibility and more appropriate skills, including languages and the ability to deal with customers.

While they liked his speech, some managers seemed shocked by the projected job losses. 'They may have to think about that', said one. 'Such issues will have to be handled very sensitively.'[44] For others, the plans actually turned out to be less dramatic than had been rumoured. There had been reports that the airline was planning to cut employee numbers by 10 000 and that the company would spend £60 million designing a new logo.

Ayling confirmed that the company was thinking about a new logo, but was not planning to spend £60 million on it: 'I can't think of anything more insensitive than asking for 5000 redundancies then spending £60 million.'[45] Moreover, Ayling emphasized that the company was increasing employee numbers by 2000 a year and pledged that the airline had no plans to reduce headcount in the foreseeable future.

As the press saw it, BA was tackling its labour costs before it was forced to do so by competitive pressures. The *Financial Times* commented approvingly: 'Unlike companies which cut costs when financial disaster strikes, BA has decided to begin the process when its aircraft are full and it is making record profits.'[46]

According to Ayling, BA had in the past been criticized for being short term in its outlook. It was not a mistake he planned to repeat: 'Today marks the start of the next stage on our long journey from privatization . . . In ten years we have transformed ourselves into an airline which has few equals. To be a success in the next century, BA will have to do it all over again.'[47] Ayling also reiterated BA's plans to concentrate on its core business—flying aircraft—and to dispose of activities that could be run more cheaply by outside contractors. Later that month, BA won its ninth consecutive Airline of the Year award from *Business Traveller* magazine.

From the start of 1997, BA offered some modest units for sale, notably its ground-fleet services, its in-flight catering operations and its landing-gear overhaul unit. BA transferred part of its accounting functions to Bombay where it could recruit 200 staff at considerably lower salaries than in the United Kingdom. BA also considered divesting its 9300-strong engineering department, or bringing in outside shareholders, to turn engineering into a stand-alone unit that would offer services to other airlines as well as BA. A few months later, BA shelved this proposal, on the basis that 'market conditions in the aircraft servicing industry "had not moved far enough" to make the plan realistic.'[48]

Some units were scrapped altogether—such as the marketplace performance unit charged with perceiving the company from the customer's point of view (mainly through flying on other

[43] *Business Week*, 30 September 1996, pp. 20–21.
[44] *Business Week*, 30 September 1996, p. 21.
[45] *Financial Times*, 19 September 1996, p. 30.
[46] *Financial Times*, 19 September 1996, p. 30.
[47] *Financial Times*, 19 September 1996, p. 30.
[48] *Financial Times*, 1 May 1997, p. 8.

airlines and writing about it). This unit had been one of Colin Marshall's first innovations at BA, and had played an important role over the years in terms of boosting customer service, but it was considered to have lost its edge.

BA's management also started negotiating with various areas on how they could achieve the cost-cutting targets set. For example, flight operations had been set a target of £46 million over three years which would require 'robust negotiations' with the pilots. As Ayling put it: 'At the moment, I'm not dictating to my management and telling them what decisions to take. We have said: "This is the need, these are the kinds of ideas we have. You know your business, you know there's a redundancy programme, you have your performance targets."'[49]

Over the first few months of 1997, BA secured a number of important concessions from various areas. For example, ground-services staff at Heathrow airport accepted a two-year pay freeze and a reduction in starting pay for new recruits in return for a promise from the airline to keep their 2800 jobs. Baggage and cargo staff agreed to a two-year pay freeze, and to the loss of 400 of 1400 jobs in the cargo area following a £150 million investment in new buildings and automatic equipment.

This was not achieved without some resistance. For BA, which had a 20 year record of voluntary redundancies, at least below management level, hiving-off activities to contractors was tantamount to forcing employees out of the company against their will. The unions made a big deal of an alleged reference by Ayling to a 'virtual airline'. They seemed obsessed with it. This, in spite of the fact that BA had been increasing employee numbers at a rate of about 2000 a year, and in spite of Ayling's repeated commitment that there would be more people working for the airline by the year 2000.

At one stage, anonymous employees had started a vicious poster campaign which involved sending out crude posters to journalists comparing Robert Ayling to Adolf Hitler. As the *Financial Times* commented in January, 'It could be highly damaging to BA if this dissatisfaction were translated into industrial action.'[50] The mere threat of a strike by pilots the previous summer, although averted, had reduced operating profits by £15 million.[51]

In May 1997, there was a 'go-slow' movement by the catering staff whose unit had been offered for sale. This caused embarrassment to BA, with 30 long-haul flights leaving Heathrow with meals for first-class passengers only. More worrying, though, was the threat of strike by cabin crew. The British Airline Stewards and Stewardesses Association (BASSA) announced that it would ballot its 8500 members over strike action. It accused the company of attempting to impose new terms and conditions, including a revised pay structure, set to save £42 million over three years.[52]

---

[49] *Financial Times*, 19 September 1996, p. 30.
[50] *Financial Times*, 25 January 1997, p. 1.
[51] In July 1996, BA flight crews voted to strike on pay and pensions. They disliked the new lower pay rates being introduced for flight crew who joined BA. Nine days later, the strike was averted and both sides were claiming victory. BA had made a number of concessions, but very few which had not been on offer before. The pilots' union maintained that 'none of these issues on their own explain why 90% of the pilots voted to strike.... What they say is that they're not being treated like professional adult people.' (*Financial Times*, 13 July 1996)
[52] *The Guardian*, 9 July 1997, p. 1,

Three days later, BA closed BASSA's Heathrow offices explaining that it was normal procedure for any company to withdraw support from a union that was threatening disruptive action. BA also disclosed contingency plans which would allow it to continue to fly in the event of industrial action. For several months, in secret locations, BA managers had been learning how to check passengers in, handle baggages and serve meals. This training was based on BA's 'snow plan' devised to keep services running in severe weather, but had been enhanced with the first threats of strike which had circulated the previous autumn.[53]

Notwithstanding these difficulties, by June 1997 some 30 000 of the airline's 45 000 UK based staff had agreed to pay freezes, pay restructuring and the introduction of practices promoting greater efficiency.

STAYING AHEAD OF THE PACK

Announced in May 1997, BA's results for the year ending March 1997 once again broke records with BA announcing £640 million in pretax profits. To thank employees for their efforts, they received 10 free shares (in celebration of 10 years of privatization) in addition to their bonus of 3.3 weeks salary (the share price was close to 600 pence compared to 125 pence when shares were offered for sale in February 1987). BA also announced its plans to step up its capital investment programme by about one-third.[54]

BA's stock surged by 22% in the first six months of 1997 (see Exhibit CS7.2.5 showing share price and events). The *Financial Times* voiced a few concerns about BA's performance, suggesting that given the slight fall in unit revenues (by 0.3%) efficiency gains were 'a necessity not a plus'.[55] Nevertheless, investor confidence seemed justified given the evident drive behind BA's efficiency programme, and investors' anticipation of the powerful financial rewards from the alliance with AA. The alliance was still awaiting clearance, having been delayed by the opposition of the European Commission to the deal and spring elections in the United States and United Kingdom. Nevertheless, the case for approval had been strengthened by the announcement in May of another powerful partnership, led by United Airlines and Lufthansa, together with Air Canada, SAS and Thai Airways, and known as the Star Alliance.

BA also looked well set to capitalize on the recent lifting of the last curbs on the operations of European Union airlines. From April 1997, it had become possible for EU airlines to operate domestic services in European countries other than their own. Few airlines looked tempted to emulate BA in starting domestic services in other countries—given that BA had not yet made any money from its French and German operations. But with two newly installed chief executives running these operations, BA expected to turn the situation around soon.

BA's global intentions were further emphasized by a dramatic change of identity, unveiled in June 1997 and beamed to 65 countries. BA discontinued its sober blue and red livery and crest—with its motto 'To fly, to serve' on the tail. The new livery to be introduced over three years consisted of 50 different designs commissioned from local artists across the world. The multicultural images, ranging from Kalihari Desert paintings to Japanese calligraphy or Canadian wood carvings,

---

[53] *The Daily Telegraph*, 9 June 1997, p. 8.
[54] *Financial Times*, 20 May 1997, p. 24.
[55] *Financial Times*, 20 May 1997, p. 24.

would progressively be transposed on to BA's tail fins, ticket jackets and cabin crew scarves. The new identity was intended to reposition the airline as 'a citizen of the world' in recognition of the fact that 60% of its passengers came from outside the United Kingdom.

The £60 million makeover drew tremendous controversy.[56] Some criticized it for betraying the airline's essential 'Britishness', while others considered it extravagant or confusing. There were also mutterings about spending so much money on redesign while employees were squeezed and this angered Ayling: 'Our new identity is about jobs; it's about training. People who talk about a "virtual airline" don't know what they're talking about.'[57]

Notwithstanding these criticisms, the redesign also generated a vast amount of free publicity for the airline with the striking new tail fins featuring on the front pages of most quality papers including both *Financial Times* and *The Times*. When the images of the new planes and ticket jackets flashed up on a big screen in front of BA employees in New York, 'there were audible oohs and ahs'.[58] Ayling stated that the new design was essential if the airline was to continue to compete in the next century. 'Some people abroad saw the airline as staid, conservative and a little cold. . . . To continue to be the world leader we have to do again what we did in the last decade—put clear blue sky between us and our rivals.'[59]

In fact, the new identity was only one element of BA's £6 billion investment programme over three years, suddenly announced in June 1997. The investment would include the acquisition of 43 new aircraft (already ordered from Boeing and awaiting delivery), as well as the building of BA's new head office, and enhanced facilities, in-flight services, products and training. These improvements were all part of the 'second revolution' launched by Ayling and reinforced the airline's new mission, first mentioned in the annual report: 'To be the undisputed leader in world travel'. The annual report also trumpeted BA's second place both in the list of Europe's most respected companies (an annual poll of 14000 senior executives run by *Financial Times*/Price Waterhouse) and in the British Quality of Management Awards (a Mori poll of 131 institutional investors, 86 business leaders and 35 business journalists).

FLYING INTO A STORM

The launch of the new corporate identity coincided with the start of two trade union ballots on possible strike action, which prompted union jibes that the airline should attend to its industrial relations rather than its image.

One dispute over the sale of the company's catering division was quickly resolved with management proposing a new deal. The other dispute, however, looked more intractable. It concerned the imposition of new pay and working conditions on BASSA-affiliated cabin crew, based on an agreement reached with a breakaway union, called Cabin Crew 89. This latter union, that Ayling himself had helped set up when in charge of human relations, comprised some 3500 members (mainly long-haul staff), compared to BASSA's 8500 members (mainly short-haul staff).

---

[56] *Financial Times*, 11 June 1997, p. 13.
[57] *The Independent*, 11 June 1997, p. 22.
[58] *Fortune*, 7 July 1997, p. 89.
[59] *The Times*, 11 June 1997, p. 5.

The pay restructuring deal meant that overtime and other allowances would be consolidated into basic pay (which would increase pensionable income by 14–24%). BASSA resented the way that management had introduced the new conditions without their consent, and also felt that some of their members would lose out from the changes. Management accepted that a minority of staff might lose money, but it offered a money-back guarantee to make up the shortfall and guaranteed that existing cabin crew would not suffer financially. Newly recruited cabin staff, on the other hand, would be paid about £2000 a year less than previously.

Not normally known for their militancy, the cabin staff voted for a 72-hour strike by a two to one majority in a secret ballot.[60] As the strike deadline approached, and talks broke down, BA warned those considering taking part that they would be heavily penalized: staff travel concessions would be withdrawn for three years; strikers would not be considered for promotion, early retirement or voluntary severance until the year 2000; and they might even be sued individually for loss of business—an unprecedented move in British industrial relations and a practical impossibility (see Exhibit CS7.2.6 for internal bulletin issued to employees).

On 9 July 1997, the first day of the strike, 1200 staff reported sick (over 10 times the seasonal average) and 70% of BA's flights from Heathrow were cancelled. BA claimed that the high absence rate was an indication that stewards and stewardesses did not want to take part in the strike, but added that staff claiming to be ill would have to produce a doctors' certificate or else would be assumed to have been on strike.

Contravening a BA ruling that staff should not speak to the press without authorization, one cabin services director (who stood to gain from the pay restructuring) told a journalist why she was taking action:

> This dispute isn't about money. It's the way we are being treated. British Airways was a disaster 15 years ago, and we've all worked really hard to make it the success it is. . . . We know there are cost savings that can be made, but they have to be negotiated. The management just want to walk all over us. The climate has really changed. When Lord King was in charge, he did unpopular things, but we felt he liked us. With Ayling, we feel he despises us and would rather get in cheap labour from overseas . . . No one wants a strike. But people feel that if we give in now, we're not going to have a job that's worth doing.[61]

The following day, BA's approach drew condemnation from 22 Labour MPs who published a parliamentary motion attacking 'the tactics of intimidation being pursued' by BA.[62] Union leaders also attacked the company: 'BA have tried intimidation. They are now trying litigation. We believe they should try negotiation.'[63]

Ayling also recognized the anger provoked by BA's approach and wrote in The Times:

> If we have appeared heavy handed or clumsy, I apologise. [He explained:] 'In order to continue to be the World's Favourite Airline and maintain our market leader position,

---

[60] Financial Times, 7 July 1997, p. 1.
[61] The Daily Telegraph, 9 July 1997, p. 6.
[62] This in spite of the fact that Robert Ayling had recently been invited by the Labour Prime Minister, Tony Blair, to head his policy unit.
[63] The Times, 10 July 1997, p. 1.

we have to be more competitive. To do that we have to persuade our workforce to give up old-fashioned working practices and pay structures. . . . What we want BASSA to do is to come back to the negotiating table and to accept the agreement we have reached with Cabin Crew '89.[64]

Ironically, the strike coincided with the publication of survey results showing that BA topped the list of companies for which UK graduates would like to work. Of the 2764 students polled, 13% ranked BA top, followed by Andersen Consulting and Marks & Spencer, the retailer.[65]

Unfortunately, this news was not of much comfort to Robert Ayling who, just three days after the end of the strike, was about to face the shareholders at the Annual General Meeting (AGM).

THE AGM

As expected, BA's management ran into considerable flak. Initially, private shareholders seemed much more agitated by the airline's new livery than by the previous week's cabin crew strike. Ayling denied that BA would ever drop the word 'British' but was heckled when he said there were elements of 'Britishness' that were standing in the way of the company's development plans: 'Sometimes when you are trying to do business abroad there are aspects which are not always helpful. We are seen to be slightly aloof.'[66]

The audience also comprised a core of past or present BA employees who were prepared to ask more hostile questions. One shareholder asked Ayling to justify his £64 000 pay rise the previous year when many of the airline's workers were having their wages frozen. Another accused Ayling of alienating staff, saying: 'There used to be a scheme called Putting People First (PPF). Now PPF has come to mean putting profits first—and to hell with the staff.'[67]

One BA employee told the directors that morale among workers was at all all-time low and asked what the company was doing to resolve the problem. Ayling replied that the radical changes were bound to upset people, but said that he hoped to push through his programme as quickly as possible before seeking to rebuild relations. He confirmed that the disputes involving the catering and cabin crew were 'now taking an altogether more rational, and cooperative turn' and 'an enhanced employee council' was among the measures being considered.[68]

One could not have blamed Ayling if he left the AGM feeling confused and disappointed by the attitude of the shareholders. BA remained the world's best-performing airline. It had not had a major crash for over 20 years. Under his guidance, it continued to lead competitors both on the ground—where it set the standard for arrival lounges with showers and valets—and in the air, with attractions such as seats that recline into beds. The airline's new look had been widely acclaimed outside Britain. He was pushing through an efficiency programme which would guarantee jobs and profits into the next century. He had paved the way for an alliance which, if cleared, would secure his place in aviation history—and even without the benefit of antitrust

---

[64] *The Times*, 10 July 1997, p. 22.
[65] *Financial Times*, 11 July 1997, p. 8.
[66] *Financial Times*, 16 July 1997, p. 27.
[67] *The Times*, 16 July 1997, p. 24.
[68] *The Times*, 16 July 1997, p. 24.

Case 7.2

immunity (which would allow collaboration on schedules and fares) could still yield substantial returns from code sharing and joint frequent-flyer programmes. What more did the shareholders want?

EPILOGUE

On 12 September 1997, BA issued a press release announcing the settlement of the cabin crew dispute which led to the strike in July. The airline and the union (BASSA) said the deal would ensure the £42 million worth of cost savings from BA's cabin crew budget as part of the airline's business efficiency programme.

As part of the agreement, BA lifted the sanctions imposed on the 300 cabin crew who went on strike in July—which had involved the removal of travel and pension entitlements and the blocking of promotional opportunities. In addition, BA agreed to return to the union the office facilities which management occupied during the summer.

The cost of the July dispute was far higher than expected. While analysts had anticipated losses of £40 million–£100 million, British Airways estimated the cost of the three-day stoppage at £125 million ($200 million).[69] The loss wiped out more than half the savings BA expected to achieve in 1997 from its '£1 billion business efficiency programme'. Combined with lower first quarter (April–June) earnings, mainly due to the strength of the pound sterling, the news sent BA shares down 5% to 610 pence, from a high of over 750 pence in early June.[70]

Later, commenting on the strike, the former chairman of the All-Party Parliamentary Aviation Committee, Sir Robert McCrindle, suggested that a few years ago all public and passenger sympathy would have been with the airline:

> Now . . . it would appear shareholders and travellers have lined up on the side of the crew and the trade unions are portrayed as responsible. It would seem that the management did not bargain for the widespread reporting of sickness by staff, nor that the effect of it would continue for weeks. BA is still suffering and many passengers, having been forced to use other carriers for the first time, may stay with them. And it is not certain that we have seen the last strike action. British Airways clearly has a lot to answer for.[71]

---

[69] *The Independent*, 5 August 1997, p. 14.
[70] *The Independent*, 5 August 1997, p. 14.
[71] *Executive Travel*, September 1997, p. 35.

Case 7.2

Exhibit CS7.2.1 The evolution of BA's financial and workforce figures

Year ending March	BA worfforce (000's)	Group turnover £bn	Operating result (airline only) £m	Pretax profit (Group) £m	Net profit (Group) £m
1980	56.1	1.92	16	20	11
1981	53.6	2.06	(104)	(140)	(145)
1982	47.8	2.24	6	(114)	(545)
1983	40.0	2.50	174	74	89
1984	36.0	2.51	274	185	216
1985	37.0	2.94	303	191	174
1986	38.9	3.15	205	195	181
1987	39.5	3.26	183	162	152
1988	42.7	3.76	241	228	151
1989	*49.7	4.26	340	268	175
1990	52.1	4.84	402	345	246
1991	54.4	4.94	166	130	95
1992	50.4	5.22	344	285	395
1993	49.0	5.66	310	185	178
1994	**49.6	6.60	496	301	274
1995	53.1	7.18	618	327	250
1996	55.3	7.76	728	585	473
1997	***58.2	8.36	546	640	550

*Includes employees from British Caledonian.

**Includes employees from Deutsche BA and TAT.

***Includes employees from Air Liberté.

Source: Company Annual Reports (*Note:* Some figures were subsequently restated, typically to reflect changes in accounting principles. The numbers presented in this exhibit are those that appeared in the given year's annual report.)

Exhibit CS7.2.2 Robert Ayling's career

Ayling started articles at a small, local law firm. By 1968, he had qualified and moved to a new firm in the City, specializing in transport and shipping law, a fast-growing sector. At 24, he was an equity partner in the firm, and by the age of 27 he was looking for fresh challenges. In 1973, the year Britain joined the European Community, he was shrewd enough to see the need for specialists in European law. Then, the only place he could do that was by working for the government, so he joined the Department of Trade.

By 1983, he was legal undersecretary at the department, and was responsible for drafting the bill to privatize state-owned BA. The task brought him into contact with King and Marshall, leading directly to his appointment as legal director of the airline in 1985, two years before it was privatized. An expert on international transport law and American antitrust legislation and a whole host of other things that impinge on BA's global business he was of interest to BA.

In 1988, he was named director of human resources, and by 1991 he had become director of marketing, a move made possible after Sears poached Liam Strong to run their operations. Recalling the day he offered Ayling responsibility for marketing, Marshall observed: 'I think he was more surprised by that than by anything else that has happened in his life.'[72]

As head of marketing and operations, he was thrust into the centre of the 'dirty tricks' row with Virgin which, at one stage, looked likely to cost him his job. That he survived and went on to run the airline is considered proof of his political skills.

Ayling's CV

1946 Born 3 August 1946.

1962 Articled to solicitor.

1968 Becomes equity partner solicitor in private practice.

1973 Legal assistant, Department of Trade.

1983 Legal undersecretary, Department of Trade.

1985 Legal director, British Airways.

1987 Company secretary, BA.

1988 Director of human resources, BA.

1991 Director of marketing and operations, BA.

1993 Group managing director, BA.

1996 Chief executive, BA.

---

[72] *Sunday Times*, 19 January 1997, p. 3.7.

Exhibit CS7.2.3 Comparison of executive team between 1995 and 1996 (*continues*)

Executive Team 1995	Position	Executive Team 1996	Position
Terry Butfield (54)	Head of network management		
Anthony Cocklin (52)	Head of communications		
Alistair Cumming (60)	Managing director, BA engineering	Alistair Cumming (61)	Chief operating officer
Dr Michael Davies (57)	Director health services		
Robert Falkner (47)	Group chief accontant		
Tony Galbraith (56)	Treasurer		
Valerie Gooding (45)	Director of business units		
Keith Hatton (50)	Managing director world cargo		
Bryan Haydon (49)	Director information management		
David Holmes (60)	Director of government and industry affairs	David Holmes (61)	Director of corporate resources
David Hyde (58)	Director of safety, security & environment		
Peter Jones (46)	Head of public relations		
Capt. Jock Lowe (51)	Director of flight operations		
Clive Mason (51)	Deputy MD, BA engineering		
Roger Maynard (52)	Director corporate strategy	Roger Maynard (53)	Director of investments and joint ventures

Exhibit CS7.2.3 *(continued)*

Executive Team 1995	Position	Executive Team 1996	Position
John Patterson (47)	Director operational performance	John Patterson (48)	Director of strategy
Valerie Scoular (39)	Director of human resources	Valerie Scoular (40)	Director of customer service
Mike Street (47)	Director of customer service	Mike Street (48)	Director of operations
Walter van West (52)	Group financial controller		
Ken Walder (52)	Secretary and legal director		
Mervyn Walker (36)	Director of purchasing and supply		
John Watson (51)	Director of regions and sales		
		Charles Gurassa (40)	Director of passenger business
		Capt. Mike Jeffery (51)	Director of flight crew
		Derek Stevens (57)	Chief financial officer
		Peter White (49)	Director of sales

Source: British Airways Annual Reports.

Case 7.2

Exhibit CS7.2.4 Evolution of key operating statistics

Year ending March	Passengers carried (millions)	Passenger load factor %	ATK: Available tonne kilometre (m)	ATK per employee (000s)	Yield: Revenue per passenger km (pence)	Punctuality (within 15 mins) %
1980	17.3	67.4	8 153	145	3.35	68
1981	15.9	62.6	8 243	154	3.74	81
1982	15.2	66.7	7 522	158	4.20	78
1983	14.6	66.5	7 208	182	4.89	84
1984	14.2	64.1	6 699	199	5.57	84
1985	16.0	68.5	7 275	213	5.87	85
1986	17.0	68.0	7 956	221	5.80	82
1987	17.3	67.0	8 141	222	6.00	81
1988	*20.2	70.2	9 427	236	5.82	80
1989	22.6	69.6	11 404	243	5.96	72
1990	23.7	71.5	12 035	247	6.37	72
1991	24.2	70.1	12 929	253	6.27	73
1992	23.8	70.2	13 379	274	6.50	79
1993	25.9	70.8	14 695	315	6.13	81
1994	28.7	70.0	16 240	334	6.32	85
1995	30.6	71.6	17 115	344	6.36	84
1996	32.3	73.6	18 508	343	6.41	82
1997	32.3	73.6	18 508	343	6.41	82

*Includes 600 000 passengers carried by British Caledonian.

Source: British Airways Annual Reports.

Glossary

Passenger load factor   RPKs expressed as a percentage of available seat kilometres (ASKs).

Revenue passenger kilometres (RPK)   The number of revenue passengers carried multiplied by the distance flown.

Available tonne kilometres (ATK)   The number of tonnes of capacity available for the carriage of revenue load (passenger and cargo) multiplied by the distance flown.

Exhibit CS7.2.5 BA's share price since Ayling took over

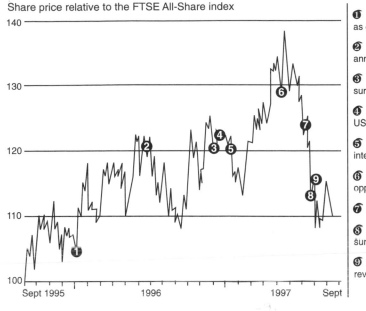

Share price relative to the FTSE All-Share index

❶ Robert Ayling takes over as chief executive

❷ BA and American Airlines announce alliance

❸ OFT recommends BA surrender 168 Heathrow slots

❹ BA announces sale of US Air stake

❺ European Commission intervenes between BA/AA deal

❻ Commission restates opposition to strike

❼ BA cabin crew strike

❽ Commission proposes BA surrender 350 Heathrow slots

❾ BA first quarter profits reveal strike cost £125 m

Source: *Financial Times*, 8 September 1997, p. 20.

Exhibit CS7.2.6 Internal bulletin issued to employees

IF YOU GO ON STRIKE

■ You will lose:

Staff travel for almost three years

Any prospects of promotion for almost three years

Any chance of early retirement or severance under BEP

■ You could be dismissed

■ You could be sued for damages

Full details inside

Source: *Le Canard Enchaîné*, 16 July 1997, p. 5.